When Judaism Meets Science

When Judaism Meets Science

ROGER L. PRICE

WIPF & STOCK · Eugene, Oregon

WHEN JUDAISM MEETS SCIENCE

Wipf & Stock
An Imprint of Wipf and Stock Publishers
199 W. 8th Ave., Suite 3
Eugene, OR 97401

www.wipfandstock.com

PAPERBACK ISBN: 978-1-5326-5355-1
HARDCOVER ISBN: 978-1-5326-5356-8
EBOOK ISBN: 978-1-5326-5357-5

Manufactured in the U.S.A. APRIL 18, 2019

Translations of the text of the Torah are, except as otherwise noted, taken from The First Five Books of Moses: Genesis, Exodus, Leviticus, Numbers, Deuteronomy, Schocken, New York, NY (1995), and used by permission of the translator, Everett Fox.

Translations from the Hebrew Bible, other than from the Torah, are taken from JPS Hebrew–English Tanakh, Copyright © 1999 The Jewish Publication Society, Philadelphia, PA.

The poem "A Rebbe's Proverb (From the Yiddish)" is taken from And God Braided Eve's Hair, The Town House Press, Spring Valley, NY (1976), and, as modified by the poet Danny Siegel, used with his permission.

To Marilyn,
Who continues to amaze, inspire, and enchant me.

Science without religion is lame,
Religion without science is blind.

ALBERT EINSTEIN

Contents

Preface

Hi there.

Thanks for opening this book. It's for people who have an interest in two of the great approaches to the big questions we ask, one based on a religious outlook and the other grounded in science, broadly understood. It is for people who want to learn about fact, fiction, and faith. And more specifically, it is for people who are curious about how, if at all, Judaism and science relate to each other; how, if at all, different perspectives inform one another. I'm excited to tell you how this book came to be and what you can expect to find in it.

In 2010, I retired from the practice of law. It was a good run, but I was ready to explore other subjects in greater depth than I had been able to do while working. I had been interested in both Judaism and science for as long as I could remember. Even an explosion in my organic chemistry class in college (caused by one of my classmates) did not damper my interest, though it did lead to a change in majors. That interest waxed and waned over the years as events unfolded. Retirement, however, provided the opportunity to drill down into these areas, especially at those points where they interface.

I began by seeking books that discussed some of the topics in which I was interested. I learned quite quickly that books on science and Christianity were being published on a weekly basis, or so it seemed. But books looking at science and Jewish texts or values were scarce. At the time, there were not more than half a dozen such books, and while each had its merits, none approached Judaism and science both systematically and broadly, and none addressed social issues (either the ones we face every day or the ones which loom ahead).

I found two books by Norbert Samuelson, a prolific professor of Jewish philosophy. Both *Judaism and the Doctrine of Creation* and *Jewish*

Faith and Modern Science offered serious, even bold, critiques of Jewish thought, but they said little about the world in which we live. Two other books, both by immigrants to Israel, attempted to reconcile Judaism with modern science, especially cosmogony and evolution, but as both *The Science of God* by Gerald Schroeder and *The Challenge of Creation* by Rabbi Natan Slifkin were written from Orthodox perspectives, neither applied science to Judaism's foundational texts, and neither discussed contemporary issues. *God and the Big Bang*, by Daniel Matt, and *Judaism, Physics and God*, by Rabbi David Nelson, were the best of the small bunch when addressing what science—primarily physics—teaches, but the journey in each instance, in one case sooner and one case later, led to the world of Kabbalah, a mystical place where science does not seem welcome. Finally, Rabbi Arthur Green's *Radical Judaism* stands out for his use of evolution, rather than creation, as the focus of his effort to rethink God and tradition. Here, too, though, the language of Kabbalah and mysticism set the tone.

Consequently, each of these books was instructive for what they said and how they said it, and for the lengths they would or would not go to make their arguments, even the quite problematic *The Science of God*, discussed in some detail in chapter 5. They provided perspectives worth considering. But, alas, none told me what I wanted to know. What, if anything, does modern science tells us about Judaism's foundational texts, the Torah and, indeed, the entire *Tanakh*, the Hebrew Bible? What, if anything, do Judaism and science have to say to each other about issues that matter to me today, issues ranging from abortion to fracking and from genetically modified foods to guns and on to vaccinations? And, finally, what, if any light, can Judaism and science together shine on the challenges that lie ahead, whether the near-term question of the proper role of robots, a more distant encounter with an alien species from an exoplanet, or, ultimately, the end of days, to name just a few?

Nobel-Prize-winning author Toni Morrison has said that "(i)f there's a book that you want to read, but it hasn't been written yet, then you must write it." I thought about doing that, but at the time, I lacked the materials necessary to write a book. My children suggested that instead of a book I should write a blog, drafting essays on topics of interest as they arose. After they patiently explained what a blog was, I created a website with the appropriate address of www.judaismandscience.com and started to write. More than seven years later, having written over fifty essays and 150,000 words, I had the materials. This book is based on some of those essays, which have been organized primarily according to their engagement with a particular time period. With a framework erected, repetition has been reduced and ideas have sharpened. Moreover, in the light of comments and constructive criticisms received over the years, and new information that has become available, some of the discussions in the original essays have been reconsidered and supplemented, revised, and otherwise modified as warranted.

All along this journey, I have tried to be guided by data, not dogma. Facts really do matter. This seems to have disturbed some readers of my essays because data sometimes interrupt and disrupt preconceived narratives. For me, though, when the data speak convincingly, I have tried to listen, regardless of what I may have thought previously or what others in my limited and limiting bubbles try to tell me. I have, as a result, argued with myself, among others, and changed my mind several times on certain topics. Should new discoveries suggest or require different conclusions than those presented here, I expect to do so again. It's called learning. That said, if you think that a discussion contained in this book is not well-founded, that it is based on misleading or incomplete references, or that there are other facts or thoughts that have not been considered and should be, please let me know. The learning should never stop, and I reserve my right to revise and modify my ideas as long as I am able.

I have found this journey to be exceptionally engaging and enriching. Yes, at times it has been frustrating, too. But Jews are nothing if not destined to wrestle with the mysteries of the universe. Traditional texts and scientific methodology are two ways to grapple with both the core questions of our existence and the more mundane challenges inherent in trying to live a decent life.

We will start, in chapters 1 through 3, by defining our task and setting forth our standard. Then, in chapters 4 through 12, we will look back in time, applying modern science, again understood broadly, to ancient texts to see how those texts stand up to the scrutiny. Next, in chapters 13 through 22, we will turn our scope around and look at issues we face in the world we occupy today to see if Jewish values have relevance and resonance. And, because we will live our lives in the future, in chapters 23 through 25 we will look ahead just a bit to see what, if anything, Judaism and science can teach each other about issues we will face within the coming generation and well beyond. We will conclude in chapter 26 by considering some hypotheticals, one involving a preeminent scientist and two learned rabbis, and a few others, each and all of which may help contribute to a positive and productive, a vibrant and vigorous Jewish future.

I hope that you will join me on this journey. If you do, I can promise you that you, like me, will learn something you did not know before, maybe find a source you wish to pursue further, maybe even consider a program or policy you would like to advance. And I hope that, like me, you will come to appreciate the many ways in which Judaism meets science, and in which our lives can be enriched as a result.

All the best,
Roger L. Price

Acknowledgments

I OWE A LIFETIME of debts to a great many people, many of whom I do not even know and a large number of whom I can no longer remember.

These debts extend back to my parents, who brought books into our home and encouraged me to read them. And they include all those who inspired my interests in Judaism and in science. Edward Oliver, at Hyde Park High School, introduced me to the wonders of chemistry. Eva Schull, also at HPHS, and Bernie Berlowitz at the University of Michigan showed me the integrity and elegance of mathematics. I thank them, albeit posthumously, and also whoever mishandled an experiment in the organic chemistry lab at Michigan, caused an explosion that sent smoke and glass flying by my head, and prompted me to ask some profound questions, like "What am I doing here?" I left the field of chemical engineering, but continued to be intrigued in science by the masterful essays of Stephen Jay Gould and the writing and television shows of Carl Sagan.

At my first congregation, Rodfei Zedek, I was fortunate to learn from Rabbis Benjamin Daskal and Ralph Simon, two quite different kinds of leaders, who expressed both the wit and wisdom of the Jewish tradition. They were supported by an educator, Dr. Irving Skolnick, whose modeling of personal integrity I have not forgotten.

While at Michigan, I learned about a unique program for college students at Brandeis Camp Institute in California. The scholar-in-residence for the session I attended was Rabbi Mordecai M. Kaplan. I can't say that Kaplan's answers satisfied me completely, but I could not get enough of his questions and his approach. Four years later, I met Kaplan's son-in-law, Rabbi Ira Eisenstein, as warm as Kaplan was tough. It turns out that the twenty-eight days I spent learning with Kaplan and the meeting with Eisenstein changed my life in ways I could not have and did not anticipate at the time.

Those who read the preface to this book know that when I retired from the practice of law, and wanted to read about Judaism and science, I could not find the book I really wanted to read. I thank my children, Matt Price and Sarah Price, for introducing me to the blogosphere, where I have operated for more than seven years. Thank you to Dan Caspi, who helped me create the blog I envisioned, and who has kept it up and running for longer than either of us probably expected. And thanks, also, to David Suissa and Rob Eshman at the *Jewish Journal* in Los Angeles for giving me the opportunity to republish some of my essays on their platform and reach a different and broader audience than I was able to do independently.

Anyone who does research today is in debt to all those who publish their studies and thoughts in an accessible way so that those of us who want to learn from them can do so. Without limitation, thank you to the Oriental Institute of the University of Chicago for making freely available the work of Daniel David Luckenbill on the Annals of Sennacherib. Thank you to Gary Rendsburg at Rutgers University and Lisbeth Fried at the University of Michigan for linking to your biographies numerous papers you have written. Thank you to the folks at TheTorah.com for collecting and publishing a growing library of contemporary Jewish analyses of our peoples' foundational texts. Thank you to the seminaries and rabbinic organizations for when they publish their resolutions and *responsa*. If we are to continue to be the People of the Book(s), institutions and individuals must open their hearts and minds and share with us as much as they can.

To a wide range of scholars who have offered their assistance on important points within their area of expertise (or more generally with their encouragement), thank you so much. Similarly, to those who were kind enough to use or cite my essays, even when you disagreed with what I said, thanks for your consideration.

To the readers who have offered their comments, sometimes complimentary, sometimes constructive and sometimes whacky and even mean, thank you. Each in your own way has been helpful either by sharpening my thought process or demonstrating why this discussion is important.

To those congregations and other organizations that have invited me to speak, allowed me to challenge them, and provided me with their thoughts, thank you. Two in particular stand out. First, *Kahal* at Beth Emet the Free Synagogue, Evanston, Illinois, is a community within a community that meets, among other times, for Shabbat morning services. There is nothing particularly unique about that, but during the Torah service on those occasions, there is normally a lengthy discussion led generally by a lay member of the group. Whether the leader is a teenager or college student making his or her first attempt at a *d'var* or a college professor for whom public speaking is the norm, the discussions are often serious, sometimes deep and even intense, and almost always worthy. That kind of discussion is relatively rare in the larger Jewish American community. I deeply appreciate the opportunity *Kahal* has granted me to test some of my ideas and to learn from its members. Second, for three summers, from 2015 through 2017, *Hevreh:* A Community of Adult Jewish Learners met at Capital Camps in Waynesboro, Pennsylvania. It was a unique experiment in the annals of American Judaism bringing together individuals from different denominations to study together with scholars from diverse movements. Here, too, I was afforded an opportunity to explore ideas and to engage and, on occasion, debate with well-educated and committed Jews about a variety of matters, some of which made their way into this book.

To Matthew Wimer, Assistant Managing Editor at Wipf and Stock Publishers, thank you for green lighting this project, and providing counsel along the way. Thank you also to the editing and production team at Wipf and Stock for their invaluable assistance, which helped turn a concept into a reality.

And above all these, more deserving of praise than the editors, and the communities, and the scholars, more even than the children who suggested the blog and the teachers from the distant past, there is one person, one special person, for whom written thanks is woefully insufficient. Not only has she read every word of this book and provided thoughtful advice, but she has offered inordinate comfort and support during this latest adventure, even as she has for over half a century. I am extraordinarily grateful to have Marilyn Price as my partner and friend. Indeed, I am blessed.

Roger L. Price

Part One

RECOGNIZING THE CHALLENGE AND THE OPPORTUNITY

Chapter 1

The Conflict Over Whether Judaism and Science Conflict

One week before the Jewish New Year in 2011, a group known as the Jewish Political Action Committee and self-identified as young Haredi activists, erected a display on the Upper West Side of Manhattan, home to a large Jewish community.[1] The display consisted of a large billboard which contained the following message, all in capital letters

THE JEWISH NEW YEAR
THE UNIVERSE TURNED
5772 YEARS OLD
AS JEWS WE PROCLAIM . . .
EVOLUTION IS SCIENCE FICTION
HUMANS GIVE BIRTH TO HUMANS
GIRAFFES GIVE BIRTH TO GIRAFFES
DOGS GIVE BIRTH TO DOGS
MONKEYS GIVE BIRTH TO MONKEYS
THERE IS NO LIFE ON OTHER PLANETS

1. Haredi (literally, those who tremble) refers to Orthodox Jews who have largely self-limited engagement with modern culture, often live in cloistered communities, and wear distinctive clothing. See a photograph of the sign at http://failedmessiah.typepad.com/failed_messiahcom/2011/09/haredi-anti-evolution-exhibit-opens-tomorrow-on-nycs-upper-west-side-345.html.

Of these statements, science teaches that the first two—regarding the age of the universe and evolution—are demonstrably false, that several are accurate, and that the last—concerning life on other planets—is as yet undetermined. We will discuss each of these issues, cosmology, evolution, and exobiology, on the pages that follow, often in detail. But before we do, let's be clear. For some people at both ends of the religious spectrum, and maybe even some in between, science presents a challenge to strongly held beliefs. While the example provided arises from what might be called the religious right, the reality is that there are those on the religious left who also take positions that are not supported by or consistent with modern science, for instance on issues like fracking, genetically modified crops, and vaccinations. Cue the chorus of militant atheists (as if it needs any encouragement): religion and science are incompatible. Moreover, religion is false and science is true.

So, from time to time, like a roiling cauldron that must overflow or an itch that just needs to be scratched, essays are written and debates ensue over the question of whether there is a conflict between Judaism and science. One such time occurred in the January–February 2014 issue of *Moment Magazine*, when nine rabbis were asked the following question: "In what ways, if any, do science and Judaism conflict?"[2] The rabbis were apparently selected as representative of, though not necessarily representatives for, various orientations and denominations. *Moment* even ordered their responses as if there were a linear line of Jewish thought from Independent to Humanist, Renewal, Reconstructionist, Reform, Conservative, Modern Orthodox, Orthodox, and, ultimately, Chabad. (Parenthetically, whether this means that *Moment* believed that Reform was at the center of Jewish opinion is unknown.) The rabbis' responses are illustrative of the problem inherent in these kinds of discussions.

Not surprisingly, some of the rabbis failed to respond directly to the question asked. One, for instance, focused on the challenges brought by the application of certain technologies, a related and interesting issue, but one distinct from the question posed. To the extent that they more or less addressed themselves to science, however, the responses were reasonably uniform. In general, the rabbis saw no conflict between Judaism

2. See generally Schwartz, "In What Ways." A related, but distinct, topic concerns the affinity for and the achievements of Jews in the sciences. For more on the importance of science to Jews and Jews to science, see generally Efron, *Chosen Calling*, and its predecessor, Efron, *Judaism and Science*.

and science, or no necessary conflict, or at least no apparent conflict that could not be resolved with greater study, understanding, and tinkering.

On the other hand, at first reading, that the opinions offered seemed compatible was a bit surprising. Was the result due to a patina of politeness, or was there a real consensus here? Is the notion of a conflict between Judaism and science imaginary, one asserted by troublemakers or, for instance, to sell books or magazines? Let's look more closely.

One reason for the apparent consensus on the Jewish side might be that the responses apparently needed to be constrained to between two and four hundred words, and that amount of space that does not allow for either nuance or development. In addition, the respondents were not in dialogue with each other, not asked to comment on, much less challenge, what their colleagues had said. Moreover, when the rabbis did reference a scientific topic, they tended to mention one or the other (or both) of just two topics: cosmology and evolution. Those are important subjects, to be sure, but they are also ones with long histories which have allowed for the emergence of some agreement.

Prior to the sixteenth century, Jews, like others, believed that the Earth was at the center of the then known universe. That is, there was a geocentric consensus. In 1543, however, Nicolaus Copernicus advanced a heliocentric model in his book *On the Revolutions of the Celestial Spheres*. His theory was rejected initially by the Catholic Church and, according to George Washington University professor Jeremy Brown, at first received a mixed response from Jews, as well.[3] Those who opposed it did so because the model was contrary to a literal reading of certain biblical verses, including one in Joshua about the Sun standing still (see Josh 10:13). By the eighteenth century, however, Jews were increasingly accepting the heliocentric model because they were increasingly rejecting a literal reading of the entire biblical text.

Reactions to Charles Darwin's publication in 1859 of *On the Origin of Species* evolved similarly. By positing generally that living organisms shared a common ancestry, and specifically that humankind descended from a line of apelike ancestors that diverged to give birth to both humans and apes, Darwin flatly contradicted a literal readings in Genesis 2:7 and 2:21–22, which talk about the formation of the first man from the dust of the earth and the first human female being fashioned from one of the man's ribs. Among Darwin's more vocal opponents was Reform leader

3. See generally Brown, "Can Judaism and Science Co-Exist?"

Rabbi Abraham Geiger.[4] Relatively soon, though, the main principles of evolution were accepted by most, if not all, Jews.

One could conclude, as does Professor Brown, that "Judaism and modern science are quite capable of co-existing. It just sometimes takes a little time."[5] But the various responses to *Moment*'s question reveal that coexistence is tenuous and uneven.

For several of the rabbis, there seems to be an easy acceptance of science, even seeing Judaism as "pro-science" and science as an "ally" of Judaism. Significantly, underlying those responses was a general sense that the Torah need not be read literally, that there were "mythic truths" and "scientific truths," and that one could and should "separate myth from fact."[6] But the language is inartful, as "myth" could mean a widely shared but mistaken belief, but also, and less pejoratively, simply an ancient story that tries to explain a circumstance or phenomenon.

In any event, neither of the rabbis assigned to the Modern Orthodox and Orthodox categories talked in terms of non-literal readings of Torah or myths. Both, however, did refer to the great twelfth century rabbi-philosopher Moses ben Maimon ("Maimonides" or "Rambam") in their responses, and both acknowledged that Maimonides was prepared to (re-)interpret Torah, as one said, "even drastically," to accommodate what science established.[7]

One of the two seemed hesitant, though. While initially rejecting the notion that Judaism and science conflict, and appearing to accept Maimonides' approach, Rabbi Shmuley Boteach set the scientific bar impossibly high. To accommodate a scientific theory, he reads Maimonides as requiring that the theory be "proven true by some infallible means."[8] Rabbi Boteach does not indicate what might constitute such "infallible means," however, and others in the Orthodox camp think that proof beyond a reasonable doubt—a tough, but achievable standard—would be sufficient.[9] That's good, because science is not and does not claim to be infallible. Indeed, a defining characteristic of the scientific method is

4. See Cantor and Swetlitz, *Jewish Tradition*, 13.
5. Brown, "Can Judaism and Science Co-Exist?"
6. See generally Schwartz, "In What Ways."
7. Schwartz, "In What Ways."
8. Schwartz, "In What Ways."
9. See, e.g., Angel, *Maimonides, Spinoza and Us*, 164.

continual testing and probing of a proposed hypothesis in order to confirm or disprove it.

What is the reason for Rabbi Boteach's hesitancy here? Nobody knows for sure the answer to "WWMD?" or "What Would Maimonides Do?" in response to current developments in science.[10] At least one Orthodox scholar, however, persuasively argues that given current science Rambam would have accepted the reality of evolution.[11]

Rabbi Boteach then, no doubt unintentionally, demonstrates why there are problems with these kinds of discussions. First, he states that there would be "no conflict" with the biblical creation story "[i]f evolution . . . (is) proven to be true" because the Bible clearly shows "a pattern of the inanimate being followed by the vegetable, animal and finally intellectual." In addition, he apparently thinks that he does not have to reach the conclusion that evolution is true because, with a nod to the late, great Harvard evolutionist Stephen Jay Gould, he thinks that there "remain holes in the evolutionary model" which preclude "full accommodation to the theory."[12] You don't have to be as irascible as University of Chicago's emeritus professor of evolutionary biology and militant atheist Jerry Coyne to *plotz* here (literally, burst, and also meaning to be outraged).[13]

First, the asserted biblical pattern is incomplete. The text in Genesis states, for example, that vegetation preceded the formation of the Sun (see Gen 1:12, 16–18). But the Sun would have been necessary for photosynthesis to occur and the vegetation to live. Rabbi Boteach could have taken the allegorical route, but he chose not to do so. Having made that choice, picking and choosing some passages and avoiding others is, to put it mildly, not helpful if you want an honest discussion. Second, while there may be debates about the mechanics of evolution in a particular setting, and there may be gaps in the fossil record of particular species, so what? No one who has read any of Stephen Jay Gould's writings would doubt for a nanosecond that he believed anything other than that the process of evolution of species was a reality.

Indeed, the process of adaptation and change over time has been established not only in fossils in the field, but also in deoxyribonucleic acid,

10. But see generally Price, "Science and Judaism."

11. See Slifkin, *Challenge of Creation*, 268, 271–72, 284–85, 289–90.

12. Schwartz, "In What Ways."

13. See, e.g., Coyne, "Rabbi Sacks" (criticizing Rabbi Jonathan Sacks).

or DNA, in the laboratory.[14] One of the most amazing demonstrations of the dynamic process of evolution is the long-term evolution experiment initiated in 1988 by Richard Lenski, a Michigan State University biologist who subsequently received a McArthur Genius Grant for his work and was later elected to the National Academy of Sciences. The project sought to track the development of twelve population strains of asexual E. coli bacteria. The populations reached the sixty-five-thousand generation in 2016. By comparison, modern humans have only evolved for about twelve thousand generations in the past three hundred thousand years. Over time, Lenski observed changes in the fitness of the E. coli populations, including with respect to the size and shape of cells, ecological specialization, possible speciation, the ability to consume a new food source, and other consequences.[15]

A brief look at the genomes of some species further illustrates the point. Humans share only about 7 percent of their DNA with bacteria, but about 21 percent with roundworms, 36 percent with fruit flies, and 79 percent with zebra fish.[16] By the time we reach African apes like gorillas and chimpanzees, the similarities in the genomes, by one count, reach 98.4 percent and 98.8 percent respectively.[17] In other words, the genetic data we now have supports the fossil record we have, all of which teaches the fact of evolution.[18] Any argument that cannot accept that fundamental truth cannot be considered as serious.

In the *Moment* survey, Chabad Rabbi Dov Wanger also evokes Stephen Jay Gould, though not by name, when he writes that "Science tells us *what*" and "Judaism tells us *why*."[19] This approach is akin to Gould's Non-Overlapping Magisteria or NOMA, which sees science as describing the natural world and religion as defining the moral world, with both coexisting in their separate, independent, and equal domains.[20] This approach has antecedents that arguably extend back as far as the psalmist who wrote that "The heavens belong to [YHWH], but the earth he gave

14. DNA is a molecule that contains the hereditary information for most organisms on our planet, including humans.

15. See Bucklin, "Conductor of Evolution's Subtle Symphony."

16. See the Tech Museum of Innovation, "Genes in Common."

17. See Smithsonian National Museum of Natural History, "Genetic Evidence."

18. See generally Coyne, *Why Evolution Is True*.

19. Schwartz, "In What Ways."

20. For more on NOMA, see Gould, *Rock of Ages*, 6, 49–67; see also Price, "Gould in the Fullness of Life."

over to man"[21] (Ps 115:16). While there are problems with this viewpoint, Rabbi Wanger advances the discussion by arguing that Judaism and science often have different roles.

Gould's vision of separate dominions is also discussed, and at greater length, by Orthodox Rabbi Avraham Edelstein in an essay published independently of the collection in *Moment*. In "Judaism and Science—Harmony or Conflict?," Rabbi Edelstein argues that Gould's formulation is limited. He agrees that Torah is not "a book of nature," but adds that it is a "book of what happens behind nature." He then acknowledges that Torah and science "can be in conflict," that "there are definite areas of incompatibility between modern science and Judaism." Unfortunately, he fails to specify what they are, other than to note "some tension" with evolution and "great tension" with science's secular worldview, its inability "to bring God into the picture." But he also sees modern science, including Big Bang cosmology and quantum mechanics, as "drawing closer to religion in general and Judaism in particular."[22]

Regrettably, none of the comments in the *Moment* collection addressed several major concepts and issues which need further exploration. After all, the direct answer to the question about whether Judaism and science conflict depends to a considerable degree on how one defines Judaism and, to a lesser degree, on how one defines science. Yet, no one bothered to define either of the topics under discussion. With respect to Judaism, we don't know whether the rabbis are talking about Judaism as a religion, or more inclusively as a civilization, to use Rabbi Mordecai Kaplan's classic formulation.[23] With respect to science, while there was some discussion about reasonably established scientific findings, there was none about scientific methodology. Was that intentional?

Moreover, when the rabbis did talk substantively about Judaism (however they defined it), none of them talked about the distinction between value propositions and truth statements in Judaism. This is odd for at least two reasons. First, words matter. They make a difference in how we define reality, how we understand it, and how we respond to it.

21. YHWH is a transliteration of the Hebrew Yud-Hay-Vav-Hay, which spells the name of God as first mentioned in Gen 4:26, and revealed to Moses by God according to Exod 6:2–3. For a historical and philological look at the adoption and development of YHWH as the God of Israel, see generally Römer, *Invention of God*.

22. See Edelstein, "Judaism and Science."

23. See generally Kaplan, *Judaism as a Civilization*.

Second, Judaism is built on the recognition of distinctions, and it is here that conflicts with science most likely may arise or be resolved.

To the extent that Judaism asserts value propositions such as "The world is good," "Love your neighbor," "Honor your father and mother," "Seek justice," "Welcome the stranger," "Repair the world," or even "Do not boil a kid in its mother's milk," science (meaning scientific methodology) has limited, if any, applicability. Yes, psychologists and social anthropologists can discuss the utility of those value propositions, but they are not phenomena that are observed and measured. Scientists will not test those propositions in repeatable experiments, with or without control groups, to determine their validity.

To the extent, however, that Judaism makes truth statements, such as "The first humans were created in adult form," "Over one million Israelites were slaves in Egypt, left that country, and encamped in the Sinai wilderness for forty years," and even "Priests wore pants in the tabernacle in the wilderness," then science, including but not limited to cosmology, physics, chemistry, evolutionary biology, paleontology, and archaeology most definitely can play a role, and the result of scientific analysis may well conflict with the text of the Torah and Jewish tradition.

In addition, when considering science substantively, the commentators tended to talk about areas in which science has reasonably well established its proofs. No one ventured into what might be more fertile grounds where science has yet to explain certain phenomena.

For instance, based on several independent measurements including those related to the age of white dwarf stars and globular clusters in our galaxy, the distance of Cepheid variable stars from us, and maps of microwave background radiation in the cosmos, Vanderbilt astronomy professor David Weintraub has placed the age of the known universe, however it began, at around 13.7 billion years.[24] Additionally, University of Michigan astrophysicist Fred Adams has described in detail a model of the evolution of the known universe from its birth in a process of sudden inflation and expansion that most of us think of as a Big Bang.[25]

24. See generally Weintraub, *How Old Is the Universe*. Subsequent to the publication of Weintraub's book, two sophisticated instruments, NASA's Wilkinson microwave anisotropy probe and the European Space Agency's Planck spacecraft, produced refined previous estimates of the age of the universe at 13.772 and 13.82 billions of years. See Redd, "How Old Is the Universe?"

25. See generally Adams, *Origins of Existence*.

At the same time, while mathematical models and recent observations have taken us on quite a journey, we have not yet reached the end of the inquiry. Scientists have not (yet) discovered what existed prior to the Big Bang, nor, importantly, what caused that event. Columbia University physics professor Brian Greene has acknowledged this gap directly. The standard Big Bang theory, he has said, tells us "nothing about what banged, why it banged, how it banged, or, frankly, whether it really ever banged at all."[26] A model with a preexisting inflation field provides an explanation for a repulsive push, a bang if you will, but raises other troublesome issues.[27] And these failures, while understandable, are, nevertheless, crucial. Without more knowledge, to claim, as does Professor Adams, that "[i]n the stark simplicity of the beginning, there was *only* physics" may not be quite accurate, nor is it sufficient.[28]

In addition, with mathematical theory now confirmed by experimental observation, we also know, among other things, the relative abundance of the lightest elements, the nature of the radiation footprint from the time of creation, and the rate of expansion of the universe. We can understand generally how galaxies coalesced and organized and how stars formed and died, spewing such heavy elements like carbon, nitrogen, and oxygen into space in the process, which formed the building blocks of life on our planet. And yet, while there is convincing evidence that Earth is close to 4.5 billion years old[29] and further evidence that primitive biological life arose around a billion years after Earth was formed,[30] how living cells emerged from the chemical stew remains a puzzlement. For over fifty years we have known how to synthesize amino acids, which are key to the formation of proteins, from basic inert chemicals. We have identified possible environments that might have been conducive to the emergence of biological life, but science has not yet been able to create autonomous, self-replicating organisms.

There are, then, many questions for which science currently has no answer, and each of them has implications for Judaism. Consider just six, ranging from the cosmic to the more personal:

26. See Greene, *Fabric of the Cosmos*, 272.
27. See generally Greene, *Fabric of the Cosmos*, 272–303.
28. Adams, *Origins of Existence*, 3 (emphasis mine).
29. See Weintraub, *How Old Is the Universe*, 6–39.
30. Coyne, *Why Evolution Is True*, 38.

- Prior to the origin of our universe in an event called the Big Bang, what, if anything, existed? Quantum chaos? Another universe? Something? Nothing?
- What, if anything, caused the Big Bang? A random event? A purposeful intervention?
- In what kind of universe do we live? The elements with which we are familiar, from hydrogen through carbon and on to lead and uranium, make up only 5 percent of the known universe. Stuff called dark matter and dark energy make up the rest. But what are they exactly? Where did they come from?
- How did life on Earth begin? How did inorganic chemicals combine into self-replicating molecules?
- What, if anything, really distinguishes humankind from all other animals? As we have seen, the human genome is exceptionally close to those of gorillas and chimpanzees, and University of Chicago anthropology professor Russell Tuttle teaches that gorillas and chimpanzees feel, fear, and think in a similar way to humans.[31] So, what makes us different? Is it, as he suggests, the ability to conceive ideas, hold beliefs, share information with symbolic language, or know the thoughts of others? If so, how did all of that happen?
- Are there other intelligent life forms in the universe at the present time? There may be untold billions of planets in the known universe, but some are only recently formed and others are associated with dying stars and, in any case, few are in the habitable zone of their host star. We know that intelligent life on Earth took over four billion years to emerge after our planet was formed. How likely is it that there is a planet out there right now, old enough, but not too old, and at the correct distance from its star, to have produced intelligent life?

Judaism and science may have much to share with each other on these and other questions, and the discussion needs to continue. But that discussion, on the Jewish side, is too important to be left just to the rabbis, many of whom are not well-versed in the sciences, some of whom may fear the consequences of such a dialogue, and a few of whom know just enough to say something foolish or dangerous. Whoever wants to engage

31. See Kelly, "Only Human."

and to be taken seriously needs to be careful to define the terms used, make distinctions between value and truth statements, and get the science as right as possible.

To do that, we need to expand our tool kit. What is important and useful about the classical sciences is that they depend on a considerable degree of skepticism and critical thinking. To advance, these sciences require a commitment to probing and learning, generally in an environment untainted by any political or personal preconceptions. They are not the only disciplines that do so, however. In this book, we will also draw on such areas as archaeology, anthropology, comparative art and literature, and linguistics.

And, because we seek to understand how science has affected Judaism, we necessarily must try to deal with Judaism's foundational texts as honestly as we try to understand the creation and evolution of our universe. To do so, we will—when appropriate—draw on an approach known as the historical-critical method, or source criticism, which seeks to learn about the history and development of the Torah we have received and what it may have meant to those who heard it thousands of years ago.[32]

Some may be concerned that such an approach is nothing less than a way to reduce the status of the Torah, but such a claim is misplaced and counterproductive. Irving "Yitz" Greenberg, an Orthodox rabbi, professor, and prolific author, has described the issue as follows: "The traditionalists absolutize their narrow, limited understanding of Scriptures in the name of defending God. In the process, they diminish God and the Torah's credibility in this culture. They turn the Torah into something antique or fossilized—dated instead of eternal."[33]

If we really seek truth, though, we cannot avoid dealing with reality, however inconvenient we may find facts along the way. The better approach, as Greenberg recognizes, is to use source criticism for the same reasons we use our other tools, like our telescopes and microscopes and our geological and molecular clocks, not to deconstruct our past or denigrate our traditions, but to learn and to derive meaning from what we learn. Says Greenberg, with characteristic insight and bluntness, "Studying the historical record and putting the Torah firmly in its comparative

32. See Brettler, *How to Read the Jewish Bible*, 3–5.

33. See Greenberg, "Meeting the Challenge," para. 29.

context gives it new power to speak to us today . . . It is a far better strategy to stop dumbing ourselves down when it comes to Torah study."

Happily, within the last generation, the use of source criticism has gained recognition across denominational boundaries for providing valuable insights. So, no dumbing down. Instead, as a prophet once urged, let us reason together (Isa 1:18).

Chapter 2

Faith in Religion, Confidence in Science

At the outset, we have seen how important language is, how we have to be careful about defining words and concepts. Now we can drill down a bit deeper into some of the issues we have identified earlier.

In response to a theoretical physicist's article regarding developments in cosmology and the then current debate about whether the universe had a finite age or was in a steady state without beginning or end, the late Lubavitcher Rebbe, Rabbi Menachem Mendel Schneerson, initiated a brief but revealing correspondence.[1] The correspondence was prompted by Schneerson's deep concern over what he considered to be widespread misconceptions about science and his perceived urgent need to correct those misunderstandings. In this correspondence, Schneerson demonstrated an expected devotion to the text of the Torah and traditions relating to it, but also a certain and perhaps unexpected awareness of technical issues, for instance whether light was an electromagnetic wave or "corpuscular" or both. More importantly, in the course of the correspondence, he articulated his approach to faith and science and what some asserted was a conflict between them.

Schneerson thought the purported conflict was the result of a misconception of the nature of science. The "sciences," he said, "are at bottom nothing more than assumptions, work hypotheses and theories which are only 'probable.'" By contrast, he viewed "religious truths" as "definitive

1. See "On Science and Its Truths."

and categorical." Consequently, science could not challenge religion because "science can never speak in terms of absolute truth."

Speaking in 1961, Schneerson stated that "our world came into being 5721 years ago" (in mid-2019, that would correspond to the age of Earth being 5779, according to the traditional Jewish calendar count). He recognized that it would be "*impossible* to cram within a period of 5722 years a process of evolution . . . which . . . would require . . . billions of years" (emphasis original). But he disagreed with evolutionary cosmological theory and considered the traditional annual dating to be "historic" based on the language of the Torah, which was reaffirmed by *Halakhah*, that is, traditional Jewish legal principles expressed in accepted writings like the Talmud.[2] For Schneerson, the test of "the matter is *Halachah*. Where *Halachah* is concerned there can be no alternatives, for the rule of *Halachah* is the rule of reality."[3]

Such language tends to light the already short fuse of a group known as the New Atheists. One of the more prominent members of this group is Richard Dawkins, a biologist and professor of science at Oxford University in England. Dawkins considers the kind of faith displayed by Rabbi Schneerson to be a great evil. "Faith is an evil," he contends, "precisely because it requires no justification and brooks no argument."[4] Worse, teaching children that "unquestioned faith is a virtue primes children . . . to grow up unto potentially lethal weapons for future jihad or crusades."[5] Another New Atheist leader, neuroscientist, and philosopher Sam Harris concedes that humankind cannot live by reason alone, and acknowledges with favor "spiritual" and "mystical" experiences.[6] But he, like Dawkins, criticizes "faith," defined as the kind of unreasoned life orientation toward "certain historical and metaphysical propositions" that has motivated many for millennia. He compares this kind of faith not just to ignorance, but to mental illness and violent fanaticism.[7]

Is there a third way, one less rigid and one which disparages neither science nor faith? Another approach, often articulated by "faithful"

2. The Talmud is a compilation of Jewish teaching assembled in the third through sixth centuries of the Common Era.

3. "On Science and Its Truths."

4. Dawkins, *God Delusion*, 347.

5. Dawkins, *God Delusion*, 347–48.

6. See Harris, *End of Faith*, 43.

7. See Harris, *End of Faith*, 64–65, 80–107, 131.

scientists, attempts to bridge the divide by arguing that science is, at its core, no different than faith.

The late physicist and astronomer Charles Townes won a Nobel Prize in 1964 for his part in the development of lasers, and subsequently was one of the discovers of the black hole at the center of our Milky Way Galaxy. He was also a devout Christian. Townes thought that religion and science were two methods which could be used to understand the universe and, moreover, were complimentary. More specifically, he claimed that faith is a part of science, because science has "postulates and we believe in them but can't prove them."[8]

Paul Davies, another physicist and director of the Beyond Center at Arizona State University, has made similar statements. In an essay first published in 2007 by *The New York Times*, Davies took aim at the conventional argument that science is seen as based on testable hypotheses, and asserted that science is based on faith. That is, according to Davies, "science has its own faith-based belief system." "All science," he continued, "proceeds on the assumption that nature is ordered in a rational and intelligible way." Indeed, in Davies' view, to be a scientist, "you had to have faith that the universe is governed by dependable, immutable, absolute, universal, mathematical laws of an unspecified origin. You've got to believe that these laws won't fail, that we won't wake up tomorrow to find . . . the speed of light changing by the hour." In sum, in Davies's view, both religion science and religion rest "on belief in the existence of something outside the universe, like an unexplained God or an unexplained set of physical laws, maybe even a huge ensemble of unseen universes, too." Thus, "[c]learly, . . . both religion and science are founded on faith."[9]

This third approach is far from clear, however, and hardly satisfying. In fact, the argument contains a least two fundamental flaws, one an overstated premise and the other a conclusion of false equivalence. First, the premise that "all" science proceeds on an assumption of "ordered" nature, and that the ordering is "rational," is not at all obvious. The formulation smacks of intentional design and the presence of some controlling and sensible Orderer. It does not account for a universe that is replete with seemingly random activity, some of which is not so kind. Supernovae explode. Galaxies collide. Comets hit planets. Volcanoes explode. Plagues spread. Genes mutate. Quantum events occur, by

8. See "Religion and Science."

9. Davies, "Taking Science on Faith."

definition, unpredictably. In short, stuff happens far and near and there are real life consequences. Science may be able to explain specific events to a degree, even a significant degree, by reference to certain laws of physics, chemistry, and biology, but explaining any such activity is not the same as determining it to be "rational" and "ordered." Second, the conclusion that both religion and science rest on "faith" is based on a distortion of both the word "faith" and the essence of science, or, more precisely, the scientific method.

In response to an essay by professor of science and society Daniel Sarewitz, one of Davies's colleagues at Arizona State University, Jerry Coyne dissected the argument that science and religion both rest on faith. Quoting Princeton philosopher Walter Kaufmann, Coyne defined religious faith as an "'intense, usually confident belief that is not based on evidence sufficient to command assent from every reasonable person.'" Worse, argued Coyne, the belief rests on "revelation, authority, and scripture." By contrast, "scientists don't have a quasi-religious faith in authorities, books, or propositions without empirical support." They neither proceed based on a personal revelation nor swear allegiance to a creed.

Does science *assume* that nature is ordered, as Davies has suggested? No, says Coyne, but scientists have *observed* that there is regularity in the universe. Does science *assume* that reason will lead to truth? No, but scientists use reason because it's a "tool that's been shown to work . . . it produces results and understanding. Even *discussing* why we should use reason employs reason!" Does science have faith "that it's good to know the truth?" No, Coyne continues, but scientists *prefer* to "know what's right because what's wrong usually doesn't work."[10]

Consider two examples. The first deals with the relationship of the Sun and the Earth. Some may believe that the Sun will "come up" tomorrow because God wills it, and Little Orphan Annie may be willing to bet on that occurrence because—well, she's a cockeyed optimist. But science will tell us that until the Sun stops burning its remaining multibillion year supply of fuel or Earth's orbit or axis is altered, the Sun will certainly appear to rise tomorrow in the east as it has each and every day for billions of years.

Now consider the difference between thinking that the first human was fully formed out of a lump of soil less than six thousand years ago and

10. See Coyne, "No Faith in Science."

thinking that *Homo sapiens* emerged after billions of years of evolution. We weren't around for either event, so how can we know which, if any, is factually true? The sole support for the first proposition is a religious text thousands of years old. You can believe it or not, but you cannot interrogate the author of the text or read reports of any witnesses. There is, in short, no evidentiary support for the proposition.

Support for the second proposition rests not on an unverifiable text, but on a reasonably well-established, if not fully complete, sequence of evolution evidenced both by bones and samples of DNA. The stories told by the analyses of both corroborate each other and lead to confidence in the shared conclusion: modern humans, *Homo sapiens*, did not emerge fully formed within the last six thousand years. Rather, our order of mammals, characterized by placentas, opposable thumbs, and relatively large brains, begat a smaller family of apelike creatures, *Hominidae*, which have such distinguishing features as thirty-two teeth and extended parenting. About seven million years ago, give or take, that family generated two branches.[11] One led ultimately to chimpanzees and bonobos, and the other to a group collectively called *homonims*. Perhaps five million more years passed until the emergence of the genus *Homo*. Our species, *Homo sapiens*, appeared about two to three hundred thousand years ago.[12]

The process by which science attempts to determine truth is called the scientific method. It consists of a series of discrete, though interrelated, steps that loop back at one or more points so that the idea at issue is constantly refined and, if possible, falsified or verified. The process can be summarized as follows:

1. Observe phenomenon
2. Ask questions
3. Develop a hypothesis
4. Predict an outcome
5. Test the hypothesis
6. Gather data
7. Evaluate results
8. Falsify, modify, or confirm hypothesis
9. Share conclusions

11. See Coyne, *Why Evolution Is True*, 195.

12. See Coyne, *Why Evolution Is True*, 4, 8, 190–212.

Once we understand the nature of the scientific method, it is clear how different religion's approach is to the resolution of perceived puzzles. Religion may begin with observations, but then its methodology departs from the scientific framework. Religion may, for instance, tell a story about how one person's walking staff miraculously turned into a snake or generated sprouts, blossoms, and fruit (see Exod 7:10–12; Num 17:16–23), and you can choose to accept those stories as historical facts, unique and sacred, or as literary devices, but certainly the text contains no prediction that the outcomes would ever be the same if the incidents were repeated and no attempted replication is ever attempted.

Astrophysicist Neil deGrasse Tyson, director of the Hayden Planetarium and popular science communicator, likes to say, essentially, that "[t]he good thing about science is that it's true whether or not you believe in it."[13] It's a great applause line. But it is not quite right, as he well knows. Science, even at its best, is not truth, but a process for attempting to find truth.

And, as Tyson also surely knows, the scientific method has its limits. Sometimes our tools and techniques are not sufficient or are not used correctly and rigorously enough to measure natural phenomena accurately or completely. The geocentric model of the universe advanced by the second-century Greco-Roman mathematician Ptolemy appeared to work reasonably well for centuries to explain the movement of planets and stars, and even successfully to predict events like eclipses. But it was a flawed model, and ultimately replaced by the Copernican view, which itself was refined by (among others) Galileo Galilei, who had a telescope, then Sir Isaac Newton, who utilized calculus, and Albert Einstein, who applied the mathematics of relativity. Further, some scientific studies are not well-designed or executed for financial or political reasons. The result is that many research findings are less reliable than they suggest.[14]

Today, our view of the universe is neither geocentric nor heliocentric. Science now tells us that our planet is just part of one of many solar systems in one of many galaxies generally moving away from each other as space expands.[15] What this teaches us, however, is not that the scientific method is not to be trusted so much as that, over time, science tends to self-correct.

13. See https://www.youtube.com/watch?v=yRxx8pen6JY.

14. See, e.g., Goodman and Greenland, "Assessing the Unreliability."

15. See Tyson and Goldsmith, *Origins*, 120; Weintraub, *How Old Is the Universe*, 227–33, 257–59.

And to be fair, though Dawkins, Harris, and Coyne might not agree, religion can and sometimes does too. Judaism today is surely not the Judaism of the temple periods (c. 900 BCE—70 CE), when the biblical stories were written, collected, redacted, and canonized.[16] Nor is it the Judaism of the talmudic period (c. 170–500 CE), when oral conversations about a myriad of topics were reduced to writing and became precedential and even binding. Similarly, Judaism transitioned through its medieval and modern periods.

Today, some may still follow Rabbi Schneerson in his belief in the literal truth of the biblical creation story, but not all or even most. Today, most understand that the story was not meant to assert a scientific truth as much as to teach a social or political lesson, that it was not meant to describe the origins of the cosmos as much as set the stage for a historical drama which, in turn, helped forge a nation.

In short, Jewish thought has evolved from the biblical perspective on everything from the grand question of the origin of the universe to the less cosmic but very serious issues of abortion and genetics and violence with firearms, among others. And it has not done so by hierarchical decree, because for two thousand years Jews have not had a high priest or an accepted religious governing structure. Rather, Jews have developed their Judaism organically, and for the last several centuries they have done so largely in the wake of a European Enlightenment in which science has been a dominant factor.

Like Rabbi Jonathan Sacks, former chief rabbi of the United Hebrew Congregations of Great Britain and the Commonwealth, Jews today (with some notable exceptions) accept that science and religion both "seek to decode mysteries," but do so with different techniques and for different purposes, that they are "different intellectual enterprises," one about "explanation" and the other about "interpretation." For them, the "Bible is not proto-science, pseudo-science or myth masquerading as science."[17] Consequently, for the overwhelming majority of Jews there is no need to rationalize the nonrational or to engage in contortions to conflate ancient stories and modern science.

Nor is it necessary to engage in word tricks that define faith to encompass that which it does not. Rather, modern Jewish thought can

16. The notations BCE and CE refer respectively to Before the Common Era and Common Era. The dating is the same as systems that utilize BC and AD, but avoids explicit Christian references.

17. See Sacks, *Great Partnership*, 284–85.

respect both faith and science, even as it denies both that "religion holds a monopoly on virtue" and that science is the sole source of truth.[18] Indeed, for Jews who respect and affirm science, Judaism today can be reality-based. And reality-based Judaism can be as vibrant and even more compelling than science-rejecting, myth-based Judaism.

18. Sacks, *Great Partnership*, 287, 289.

Chapter 3

The Greenberg Hurdle

As we go forward, exploring what happens when Judaism meets science, what shall be our metric, the standard by which we evaluate the interface of fact and fiction and faith? Here again, Yitz Greenberg provides some guidance, though unintentionally. Writing over forty years ago, and writing not about science but about the Holocaust, Greenberg argued that in the future, "no statement, theological or otherwise, should be made that would not be credible in the presence of burning children." A few years later, Greenberg repeated that proposition in an extended essay entitled "The Third Great Cycle in Jewish History."[1] He did not, and did not have to, discuss in depth that in the enormous and unforgiveable horror of the Shoah, there was not much, if any, theology that was credible. The notion that God was present, but impotent, obviously raises profound questions. But so does the idea that God was present, but passive, or alternatively, that God was absent, in hiding. Without engaging in an extensive theological discussion, Greenberg suggested that "a period of silence in theology" was in order. He concluded, somewhat paradoxically, that in the future "the primary scene of religious activity must be the secular."[2]

Obviously, the nonideological engagement of science is dramatically different on many levels than the evil enterprise of Nazism. So how

1. Greenberg, *Third Great Cycle*, 16.
2. Greenberg, *Third Great Cycle*, 17.

does Greenberg's idea relate to our exploration of Judaism and science, aside from underscoring that serious issues ought to be treated seriously?

The issue both yesterday and today is one less of philosophy and more of credibility. A child dying in the Holocaust deserved compassion which, in turn, required respect. Respect calls for honesty, or at least the avoidance of dishonesty. Nothing would have been gained by shattering the illusions of one with little time to live. Indeed, it might even be cruel to do so. Hence, silence.

Yet if Greenberg makes a good point with regard to a person losing his or her life, why is it not a better point for one who might reasonably be expected to live a life to its fullest? For such a person, respect calls for honesty, as well. And for that person, a strong basis for future well-being requires not the soft comfort of gauzy illusions, but the firm foundation of reality. What we say to the living, even more than to the dying, must be credible because it can affect their present and their future.

For over a hundred generations, Jews have self-identified as *b'nai Yisrael*, the children of Israel. One result is that Jews often approach Judaism, including but not limited to its primary texts, in a childlike way. This condition is what some call pediatric Judaism. But where the community was once bound in ways that could survive that status, the situation really is different now, partially because of the new and dramatic teachings of science, especially within the last generation or so.

To be sure, Jews (like others) have been confronting modern science since the Enlightenment. We have already mentioned the revolutions in cosmology and evolution sparked by Copernicus and Darwin, respectively. We could add the revolution in physics prompted a century ago by Albert Einstein.[3] What is different today is that fantastic scientific discoveries are being generated in more disciplines and more frequently than ever before. And because of advances in personal communication, we are surrounded by them, immersed in them. The bombardment is so continuous, comprehensive, and cumulative, and its force so intense, that it threatens to overwhelm the ancient pillars of Judaism—God, Torah, and Israel (the people, not the state)—at a time when mid- and late-twentieth-century supporting elements, such as resolve in remembrance of the Holocaust and enthusiasm regarding the establishment of the state of Israel, are also challenged.

3. See generally Isaacson, *Einstein*, 90–139.

Two bits of information illustrate the impact of science on one of the historic supports. The first data point relates to the rate of higher education for young Jewish adults. Estimates vary, but perhaps 85 to 90 percent or more of young adult Jews go to college.[4] What do they see and what do they hear on campus? For most, the experience is mind-expanding, as it should be. They encounter new subjects, from archaeology to zoology. They drill deeper into formerly familiar concepts. No doubt old ideas are challenged, old assumptions questioned. Their heads are stuffed with dates and facts. Their brains are asked to engage in critical thinking. In the majority of cases, our young adults will succeed in college. Not only will they graduate, but many will continue their education. In the Chicago area, for instance, about forty percent (40 percent) will earn graduate degrees.[5]

The second data point concerns the orientation of graduates when they emerge from their collegiate chrysalis. What do they know? Many things, of course, but here's one thought they think is true. According to a 2012 survey at a Reform congregation in Springfield, Massachusetts, a majority (51 percent) of Jews aged twenty-something who responded agreed with the proposition: "Science explains everything, making God an unnecessary hypothesis."[6]

Let's pause for a moment. From 2005–8, Rice University sociologist Elaine Howard Ecklund surveyed nearly 1,700 scientists at twenty-one elite research universities in the United States (e.g., Chicago, Duke, Michigan, Stanford, and Yale) in order to determine their religious beliefs. In general, she found that fewer scientists (36 percent) expressed a belief in God than did the population as a whole (over 90 percent).[7] But it is not at all clear that a majority of scientists would think that "science explains everything." It is not even clear that they would agree on what "everything" is.

4. See, e.g., Redden, "Why More Colleges Want"; Falk, "American Jews."

5. See Sheskin and Dashefsky, "Jewish Population in the United States," 28.

6. See Shapiro, "God Survey," 36.

7. See Ecklund, *Science vs. Religion*, 32–36. Among her findings is that there are a statistically significant and disproportionate number of Jewish elite scientists. Specifically, while Jews are only 2 percent of the US population, 16 percent of the elite scientists were Jewish. By contrast, 28 percent of the population is evangelical Protestant and 27 percent is Catholic, but only 2 percent of the elite scientists were evangelicals and only 9 percent were Catholic.

While Ecklund's survey sample was sizeable and her methodology well-constructed, we need to be careful about reading too much into the results of the Massachusetts survey. It appears to be flawed on several levels. The sample was small, and here size matters. The questions did not define key terms, "God" and "everything," and usually definitions matter. The questions did not allow for subtle responses. There was no control or comparison group.

And yet, the Massachusetts results should not be surprising and remain instructive. Let us remember that religion in general and God in particular once functioned, among other things, to explain the origin and evolution of the universe and our place in the scheme of things. Today much of what once was totally mysterious and inexplicable, while still wondrous, can be described by science to a reasonable degree of certitude, without primary or, for some, any reference to a supernatural force. And, as we shall discuss more than once, facts really matter. So the traditional view of God—understood as the creator of light and life, the author of freedom and salvation, and the source of law and ethics—has been shattered. And, for many, God is not just constrained, withdrawn, or hidden, but nonexistent, or at least superfluous. In Ecklund's survey, almost 75 percent of Jewish scientists reported that they were atheists. That a younger generation, brought up in an era of discovery, should agree that "science explains everything" may be imprecise, and may be exaggerated, but it is also understandable.

Moreover, we don't need a PhD in sociology or psychology to understand that these these new BAs and BSs and MBAs and MDs and JDs have read too much literature to be moved by mere lore, studied too much science to be captivated by fable, learned too many facts to be swayed by fiction, and, not incidentally, met too many different kinds of good and decent people to be committed without question to their tribe. We need to engage with these young Jews on the level they have now reached, help them move beyond pediatric Judaism by showing them that there can indeed be a vibrant Judaism that their minds can affirm, a positive and worthwhile Judaism for the adults of Israel.

There is, then, an important message to be derived from the Massachusetts survey. Whatever they learned in Hebrew school or Sunday school, in whatever congregation they belonged just a few years earlier—another version of pediatric Judaism—that they seem to have rejected. What they have replaced it with, if anything, is uncertain. According to a 2018 survey by the Pew Research Center, like other highly educated

Americans, Jews seem to be less likely than their ancestors to believe in the God of the Bible, but perhaps are open to a higher or spiritual power of some sort.[8] What does seem clear is that the pediatric Judaism that might once have been sufficient to sustain the children of Israel is no longer suitable for many of the new adults of Israel. If Jewish college graduates in North America are the future of the Jewish people in North America (and they are), we need to do better by them, to speak more clearly and much more credibly about how science and Judaism interact. This is true not just with respect to young Jewish adults. The same Pew survey also found that only 33 percent of all Jews believed in the God described in the Bible, while 56 percent believed in a higher or spiritual power and 10 percent believed in neither.[9]

Let's be clear: science is not the only challenge Judaism faces today. Volumes have been written, and will not at any length be rehashed here, regarding the ways in which we are considerably distanced in time and tone from the circumstances and culture, often idealized, of the Old Country inhabited by our great-grandparents, our grandparents, perhaps even our parents. Similarly, shelves creak with books arguing about the benefits and detriments of assimilation in America, even the ironic possibility that we are being loved to death.[10] Plainly, the recent increase in anti-Semitic incidents, tragically including the murders in a Pittsburgh synagogue complex, offer proof that this love is not shared universally. Still, the post-war commitment to never forget the Shoah has waned with the passing of time and the survivors. The post-war enthusiasm of joining with the pioneers to drain swamps, plant in deserts, and build and defend a national homeland has encountered resistance from many who fear that the behavior of the Jewish state is not the light to the nations that they had envisioned.

But Jews have gone through cycles of destruction and exile, and of migration and assimilation, before. The new challenge for Judaism comes not so much from physical migration away from familiar settings or old confinements and persecutions, but from mental migration in brains filled with new discoveries in a variety of disciplines and conveyed through new modes and systems. It is in the *Yiddische kup*, the Jewish

8. See Pew Research Center, "When Americans Say They Believe," para. 18.

9. See Pew Research Center, "When Americans Say They Believe," fig. 5.

10. See, e.g., Tobin, "Loving Us to Death."

mind (literally, "head"), as much as in the Jewish heart, that the future of the Jewish people will be decided.

As a result, Judaism, which has its roots in a conception of powerful deity who engages historically with a particular people, is now challenged in a way in which it was not before the Enlightenment or even before this past generation, by science as well as the Shoah. In Greenberg's view, theology needs to be credible. We can adopt and adapt the credibility standard as our own. Call it the Greenberg Hurdle. It is, and perhaps should be, an obstacle that is hard to overcome.

Because science is, at least ideally, about unbiased inquiry and enhancing human learning and living, in stark contrast to the Nazi ideological commitment and its goals of suppression, dehumanization, and death, we depart from Rabbi Greenberg in one respect. We respectfully decline the call for silence. Conversation should not cease just because it is difficult. If, for instance, traditional talk about an omnipresent, omnipotent, and loving Supreme Being is not credible in the ashes of innocents at the base of a crematorium, then we need to reexamine that language and perhaps attempt to find other, more credible words. The Greenberg Hurdle presents a challenge to that understanding and to any proposed alternative. What kind of deity has what kind of role in the universe described by professors Adams, Weintraub, and Coyne, among other scholars? And how do we address or relate to it? What are the consequences for Judaism in such a universe? For Jewish values and for the Jewish people?

Of course, faith, by definition, is not dependent on a fully confirmed factual foundation. One does not need faith to hold to that which is proven. Rather, faith concerns the unknown. And commitment to Judaism and the Jewish people are not necessarily dependent on faith in any particular deity and any particular set of rituals. That is, one can be Jewish and believe in a traditional God, but also be Jewish and an atheist.[11] One can be Jewish and follow biblical dietary rules, but also Jewish and depart from those rules.

If, however, Judaism is to be meaningful and valuable today, worth living for and worth dying for, that is, both attractive and compelling, it should at least account for and be consistent with what we know about our universe. In short, it must be credible. Consequently, and beyond theology, we must study our foundational texts with rigor, as well as

11. See Price, "Jewish Atheism and Jewish Theism."

respect, and the same is true with regard to the application of our values to the world in which we live. If science teaches us something about our texts that is contrary to what we have heard before, if our values are based on premises that turn out to be flawed, then we must deal with the consequences forthrightly. A vibrant Jewish future demands no less.

Conversely, because we are concerned with science as well as Judaism, to the extent that science seeks to explore and explain, it, too, must confront the Greenberg Hurdle. It, too, must be credible. In recent years, for instance, scientists have attempted to resolve some of the questions listed earlier, but they have not yet reached the required level of credibility. And that list was non-exhaustive. Are there more than three spatial dimensions in our universe? Why are there gaps in the evolutionary tree? Do we have free will?

This much we do know: from their different perspectives, science and Judaism can react with amazement at the universe we know and our place in it. And whether the universe burst forth by some quantum fluctuation or by the word of God, humility, as well as awe and wonder, is in order. Now we are ready to see what happens when Judaism meets science. First, we'll look back, then around, and, finally, into the future. It should be fun.

Part Two

ILLUMINATING THE PAST

Chapter 4

The God Assumption and the Anthropic Principle

When the Torah starts telling its history of the Israelite nation, God already exists (see Gen 1:1–3). There is no discussion of God's origins or what, if anything, God may have been doing previously. God simply just is, and proceeds with the work of creation. This all may have seemed quite plausible to our ancestors. God was present at creation of the universe as we know it; and, moreover, God is confirmed by nature. We are here, are we not, they must have thought.

Given a number of relatively recent discoveries, however, this unquestioned assumption creates a bit of a dilemma for traditional believers and atheists alike. Both may hear that the initial conditions of the universe, certain laws of nature, or the location and chemistry of our planet are set within a limited range that allows for human existence, a proposition sometimes called the anthropic principle. If you are a traditionalist, you may understand those conditions and characteristics as demonstrating a "fine-tuned" universe, which you may further take as proving or at least strongly implying the existence of a Fine-Tuner, a Divine Designer, a supernatural God. But if you are also used to being skeptical about certain claims of science, such as the origin of the universe in a Big Bang, the age of the universe, the nature of the evolution of life on Earth, or some other proposition, how do you now accept this science, this proof? And if you do, will you now accept what you have rejected previously as contrary to what is revealed in some sacred text? Conversely, if you are an atheist who readily accepts what science teaches, how do you deal with the numerous, seemingly improbable, features of our existence and not at least admit the possibility that there is some Prime Cause, some God, behind it all?

The appeal of the Fine-Tuning argument comes in large part from the apparently unique circumstances in which we humans find ourselves. In his popular book, *Just Six Numbers,* British cosmologist Martin Rees describes six constant numbers that he claims "constitute a 'recipe' for a universe."[1] One is the ratio of the "strength of the electrical forces which hold atoms together" to the "force of gravity between them." It is huge, 10^{36} to 1. Were it smaller, however, only a miniature universe could exist, and then not for enough time to allow evolution to take hold. Another number is small, 0.007. While not meant to invoke James Bond, ironically it "defines how firmly atomic nuclei bind together, . . . (and) how stars transmute hydrogen into all of the atoms of the periodic table." If this number were a thousandth larger or smaller, you would not be reading this chapter. You would not exist.[2]

Two other numbers relate to the size and texture of the universe. One "measures the amount of material in our universe" and the relationship of gravity and "the expansion energy in the universe." The fourth number "controls the expansion of our universe." Were there even slight deviations in either number, cosmic evolution would not have occurred.[3]

The final two numbers "fix the property of space itself." One "represents the ratio of two fundamental energies and is about 1/100,000 in value." A smaller number would have meant an inert universe while a larger one would evidence a violent one. The final number is three, the number of spatial dimensions in the universe. With either less than or more than three dimensions (exclusive of time), life as we know it would not exist.[4]

Others have produced different and longer lists. Some years ago, astronomer and devout Christian Dr. Hugh Ross generated a list of over two dozen "parameters" that he contended set the boundaries for life on Earth.[5] Some of these factors relate to the universe as we know it. They

1. See Rees, *Just Six Numbers*, 4.
2. Rees, *Just Six Numbers*, 2.
3. Rees, *Just Six Numbers*, 2–3.
4. Rees, *Just Six Numbers*, 2–3.
5. See Ross, "Design and the Anthropic Principle."

include, in addition to the factors discussed by Rees, the age and entropy level of the universe, the energy levels of beryllium, carbon, and oxygen, and the distance between and the luminosity of stars. Other parameters apply to the Earth's capacity to be a fit habitat. Some of these concern the Earth's parent star, including that the Sun is the only star companion for the Earth, the birth date and age of the Sun, the Sun's distance from the center of the galaxy, as well as the Sun's mass, color, and surface gravity. Still others are more Earth-specific, including the Earth's period of rotation, gravitational interaction with the Moon, magnetic field, axial tilt, and seismic activity, plus the levels of carbon dioxide, water vapor, and ozone, as well as the oxygen-to-nitrogen ratio, in the atmosphere. According to Ross, each and every condition identified operates within a relatively narrow range beyond which human life would not be possible. Ross updated his list in 2006 to include over ninety values.[6]

Drawing conclusions about the existence and attributes of God from certain natural phenomena is not, of course, a new idea, especially to Judaism. Well over two thousand years ago, the author of Psalm 19 anticipated an early, nonscientific, and unrefined version of the idea when he wrote that the "heavens declare the glory of God" and "the sky proclaims His handiwork" (Ps 19:1).

Centuries later, but still almost two millennia ago, we find a Jewish version of the Watchmaker argument. Responding to a heretic who asked him who made the universe, Rabbi Akiba (c. 50–c. 135 CE) reportedly said, "Just as a house attests to its builder, a garment to its weaver or a door to its carpenter, so too does the world attest to the Holy One who created it."[7]

Even twentieth-century theological rationalists such as Mordecai Kaplan have been attracted to the argument from nature. Kaplan found evidence for God in "the oneness that spans the fathomless deeps of space" and "the elemental substance of stars and planets, of this our earthly abode and of all it holds."[8]

The anthropic principle has been adopted formally on occasion by Orthodox Jewish scientists such as Nathan Aviezer, who seeks to promote the notion of a close and harmonic relationship between the Torah

6. See Ross, "Fine-Tuning For Life In The Universe."

7. Midrash Temurah 5, http://www.jtsa.edu/stuff/contentmgr/files/0/48344b3d8eb66a3d747d9ecd4d5f0962/pdf/torah_20from_20jts_20bereishit_205772.pdf; see *also*, Slifkin, *The Challenge of Creation*, 28.

8. Scult, *Radical American Judaism*, 150–51.

and modern science. Aviezer, a physics professor at Bar-Ilan University in Israel, offers a few examples of the anthropic principle. He refers to the strength of "the nuclear force" in thermonuclear reactions in the Sun, the presence of water on Earth, and the delicate balance of Earth's atmosphere. In addition to these often-discussed factors, Aviezer adds that human life would not have been possible except for the highly improbable event of the sudden destruction of all dinosaurs due to a meteor impacting the planet with a force strong enough to cause the extinction of the then dominant life form, and yet not so strong as to kill all life on the planet.[9]

Aviezer concludes his argument by discussing the probability that human life would ever arise. Rather than revealing any particular number, however, he merely concludes that the events leading to human life on Earth were "extremely unlikely." He adds that the "extreme rarity of the events . . . is well established." Consequently, he claims that "the anthropic principle has become a scientifically established fact."[10] That is just not true, though. There are problems with the anthropic principle—some definitional, some philosophical, some empirical.

The first problem with the anthropic principle is that there is no agreement on what it is. The term itself was invented relatively recently, apparently by British cosmologist Brendan Carter around 1974. In contrast to the Copernican revolution, which denied that humankind was at the center of the universe, the purpose of the anthropic principle, according to Timothy Ferris, emeritus professor at the University of California, Berkley, is to "portray the cosmos less as an impersonal machine and more as . . . a 'home to Man.'"[11]

Unfortunately, most writers do not seem overly concerned with the definitional dilemma. Even well regarded scientists we have mentioned before, like Neil DeGrasse Tyson and Fred Adams, tend to define the anthropic principle in a cursory and general manner. According to Tyson and co-author Donald Goldsmith, the argument is that "[b]ecause we exist . . . the parameters that describe the cosmos, and in particular the value of the cosmological constant, must have the values that allow us to exist."[12] Professor Adams describes the basic idea as restricting the laws of nature in order to "allow for the appearance of living beings capable of

9. See Aviezer, "Anthropic Principle," 4–5.

10. Aviezer, "Anthropic Principle," 7.

11. Ferris, *Whole Shebang*, 292.

12. Tyson and Goldsmith, *Origins*, 102.

studying the laws of nature." That is, "our universe had to take its observed form for us to be here to argue about these issues."[13]

In the last forty years, others have proposed variations on the theme. For instance, some claim that there is a weak anthropic principle ("WAP") and a strong anthropic principle ("SAP"), among other anthropic principles. WAP asserts that the laws of nature must be such as to "permit the emergence of life," while SAP asserts that the universe must also allow for "'the creation of observers within it.'"[14] In addition, some writers have discussed PAP, the participatory anthropic principle, and FAP, the final anthropic principle. The late, great science writer Martin Gardner called the last of these the completely ridiculous anthropic principle, or CRAP.[15]

One need not go as far as Martin Gardner to see a problem. When, in 1905, Einstein formulated the relationship of energy and matter, a fundamental feature of the entire universe, he did so with a simple and elegant formula, $E = mc^2$. The precision of his formulation has allowed his theory of special relativity to be tested thoroughly over the course of a century, leading to greater understanding and new technologies.[16] By contrast, articulation of the anthropic principle is neither simple, nor elegant, nor even agreed upon. Far from being "a scientifically established fact," as Nathan Aviezer has argued, the anthropic principle is confusing and flawed.

For starters, the anthropic principle is misnamed. It is not even a principle as that word is commonly understood—that is, as a rule or a basic truth, an explanatory fact or law of nature. Tyson and Goldsmith call it an "approach," rather than a principle,[17] but even that may be generous. The anthropic principle does not really prove anything, nor does it predict anything. Indeed, to assert that the universe must be fine-tuned for life because we are here is simply to make a circular argument. The World War I ditty "We're here because we're here because we're here because we're here" makes as much or as little sense.

Nor is it clear that the underlying theme of the anthropic principle is, or ought to be, anthropic—that is, of or related to humanity. The argument that the universe was made for humans because we are here could

13. Adams, *Origins of Existence*, 209, 218.

14. See Ferris, *Whole Shebang*, at 299.

15. See generally "WAP, SAP, PAP, and FAP" in Gardner, *Night is Large*, 40–49.

16. See, e.g., Ost, "Einstein was Right."

17. Tyson and Goldsmith, *Origins*, 104.

be made with regard to other organisms that have found their niche, like Darwin's finches, or which are long-time survivors like turtles, or ubiquitous like cockroaches. Under this approach, one might even argue that a hole in a street was designed for the puddle that fills it perfectly.

Professor Ferris asserts that the anthropic principle "is less scientific than philosophical."[18] Even so, the anthropic principle fails. Martin Gardner considered himself a philosophical theist, but he had a "dim view" of the anthropic principle, and thought that the fine-tuning argument for God was "logically fragile."[19] As Gardner recognized, if the anthropic principle is read as proclaiming that "because we exist the universe must be constructed as to allow us to have evolved," then it is nothing more than a "trivial tautology."[20]

More formally, the anthropic principle contends that, because B (humankind) chronologically follows A (the original conditions for the universe), then A caused B, or in the phrasing of the anthropic principle, that A was fine-tuned for B. This backward reasoning is a form of "retrograde analysis" in which observations about the present are the basis for speculation about the past.[21] The fallacy is known as the *post-hoc* fallacy.[22] What is more accurate is that, through a natural process of cosmological and then organic evolution, including natural selection and adaptation, over a long period of time human life emerged. Rather than the universe being suited for humankind, humankind and its predecessors adapted to the universe that was available.[23]

The anthropic principle also fails to conform to reality. To assert as a principle that our universe is fine-tuned for life is contrary to some rather obvious facts. Only about 4 percent of the universe is made up of conventional (baryonic) matter, with almost all of the balance being composed of dark energy (~70 percent) and dark matter (~25 percent), about both of which we know precious little.[24] Almost the entire universe is, in fact, empty and frigid, and, because of accelerating expansion, get-

18. Ferris, *Whole Shebang*, 300.

19. See "Proofs of God" in Gardner, *Night is Large*, 539–40, 546.

20. See "WAP, SAP, PAP, and FAP" in Gardner, *Night is Large*, 41.

21. See "Proofs of God" in Gardner, *Night is Large*, 539–41.

22. See Ferris, *Whole Shebang*, 300.

23. The logic of the arguments from design and fine-tuning are dissected briefly, but effectively, in Goldstein, *36 Arguments*, 350–55, 357–58.

24. There are also very small amounts, relatively, of other components like neutrinos and radiation. See Adams, *Origins of Existence*, 55.

ting visually emptier and colder.[25] So, it is not surprising that to date we have no evidence of any life, much less intelligent or humanlike life, on other planets.

Even on our planet, life flourishes in a limited number of locations. And where it does is often a scene of harsh and bitter struggle. If the universe really were fine-tuned for life, especially human life, then why isn't it flourishing and visible? And why would a Fine-Tuner create such vast amounts of wasted space and a system dependent on violent conflict? Conversely, if our search for extraterrestrial life is successful, what does that mean for the anthropic principle? Does that confirm that the universe is fine-tuned for life, or does it suggest that life is not as unique as we once thought? And what if that life is not human?

The anthropic principle is also premature. The universe is just shy of fourteen billion years old. Our genus emerged about two million years ago, with our species, *Homo sapiens*, arriving about two to three hundred thousand years ago.[26] So, we have been around for just two ten thousandths of one percent of the life of the universe. For the other 99.9+ percent of the time, it would not have been at all obvious or perhaps even plausible that humankind would arise.

Similarly, the complete story of human existence on the home planet has not yet been written. If, in the next one thousand or one hundred thousand years, another massive meteor struck Earth triggering extinctions, including all humankind, would that mean the anthropic principle was false? Or just a temporary principle?

But what about the long odds on us being here, of each cosmological constant and ratio being just right to allow for the universe to form and evolve sufficiently for heavy elements to be formed in stellar nuclear reactions, spewed into space, and collected in a solar system which includes a planet at just the right distance from a correctly sized and aged single star and with just the right chemistry to allow for life to emerge?

Let's start by considering the probability that any human now alive would be alive today. Each of us is the product of some twelve thousand generations of human evolution over the last three hundred thousand years or so. Each of those generations is the product of the fertilization of a reasonably random egg by an even more random sperm. Change the egg or the sperm in any one of those encounters and the current beneficiary

25. See Adams, *Origins of Existence*, 60, 62–63, 219–20.

26. See Coyne, *Why Evolution Is True*, 203, 206.

of the process would not be here. The retrograde analysis of the anthropic principle would argue that we should not be here, but we all are—despite the odds. Apparently, improbable things are happening every day.

We can learn from British mathematician Roger Penrose. Having once calculated an extreme unlikelihood that we would find ourselves in an environment suitable for life, Penrose also acknowledged that the improbabilities of the anthropic principle do not fully explain the improbabilities of our universe. He subsequently characterized the anthropic principle as a "highly contentious set of ideas," upon which "too much reliance is frequently placed" to support "implausible-sounding themes."[27]

Or, perhaps, as the late physicist and astronomer Professor Victor Stenger has argued, the probability question is misplaced. He contended that there is, in fact, nothing unusual about the processes that preceded our existence or the situation in which we find ourselves. Rather, the values identified by Martin Rees or Hugh Ross are within the range one would expect from "established physics."[28]

Now, if you are having problems with the Aviezer's claim that the universe is as God designed it, and you don't like Stenger's view that the universe is what it naturally is, there is another option. Rees, among others, finds "compellingly attractive" the admittedly speculative theory that ours is but one of many universes.[29] Postulating a theoretical multiverse naturally improves the odds that a universe like ours would be among other universes, and Rees is, therefore, not surprised our number came up.[30]

But is it necessary to imagine a multiverse in order to explain our existence? Doesn't that just substitute one mystery for another? Moreover, doesn't a multiverse violate a principle of problem-solving known as Occam's Razor which favors, among competing theories, the one with the fewest assumptions? Rabbi Jonathan Sacks, thinks so, which is one reason why he favors "a single unprovable God over an infinity of unprovable universes."[31]

There are, for now at least, limits to cosmology, as there are to theology. In a rare objective and balanced view, Professor Ferris writes that

27. See Penrose, *Emperor's New Mind*, 354; see *also*, Penrose, *Cycles of Time*, 171.

28. Stenger, *Fallacy of Fine Tuning*, 288.

29. See Rees, *Just Six Numbers*, 166.

30. See also, Carroll, "Welcome to the Multiverse."

31. Sacks, *Great Partnership*, 269; but see also, Stenger, *Fallacy of Fine Tuning*, 249–52, 292.

cosmology tells us nothing about God. "Cosmology presents us neither the face of God, nor the handwriting of God, nor such thoughts as may occupy the mind of God. This does not mean that God does not exist, or that he did not create the universe, or universes. It means that cosmology offers no resolution to such questions."[32]

In sum, the anthropic principle is an intriguing and seductive but ultimately unsatisfying concept for believers and non-believers alike. Its adoption by believers in support of a supernatural God is understandable, but it does not really advance their claim, does not provide proof. Nor is it even necessary for a believer to rely on, or even refer to, the anthropic principle. As Rabbi Sacks teaches, "faith is the defeat of probability by the power of possibility."[33]

Nonbelievers, too, are attracted to the anthropic principle for its purported nontheistic explanation of the seeming miracle of our existence. Some, conscious of the apparent overwhelming improbability of that existence, seek to improve the odds by changing the number of universes under discussion. But, based, on the available data, there is no reason to think that what we observe is anything other than what one could have expected from the normal operation of physics and chemistry and biology, over time, and adding new universes, for which there is no current observable evidence, does not make the argument stronger or resolve the logical fallacies inherent in the anthropic principle.

At its core, the anthropic principle represents a return to pre-Copernican thinking, placing humanity, if not at the physical center of the universe, certainly as the reason for the origin and evolution of the universe. But we do not need to regress. The fact of our existence, even if not at the center of our universe but just on a speck of rock at the outer spiral of a conventional galaxy in an obscure region of space, is in and of itself reason enough for wonder and joy. We are, as Carl Sagan wrote, "the local embodiment of a Cosmos grown to self-awareness." We are, he said, "starstuff pondering stars; organized assemblages of ten billion billion billion atoms considering the evolution of atoms; tracing the long journey by which, here, at least, consciousness arose."[34] And the truths of our emergence from stardust and our historic relationship to each other, to all of life, to all matter and all energy, are truths to be cherished and nurtured with humility and gratitude.

32. Ferris, *Whole Shebang*, 304.

33. Sacks, *Great Partnership*, 283.

34. Sagan, *Cosmos*, 345.

Chapter 5

Let's Start Near the Very Beginning

HAVING ASSUMED GOD, THE Torah begins with a description of beginning of our known universe, our solar system, our planet, and us. The saga starts in literally murky circumstances. The first two verses of the well-known King James translation of the biblical Hebrew say: "In the beginning, God created the heaven and the earth. And the earth was without form and void; and darkness was on the face of the deep. And the Spirit of God moved upon the face of the waters." For centuries, this translation dominated.

More contemporary Jewish translations render the Hebrew somewhat differently, though. Here are just two examples:

- Everett Fox: "At the beginning of God's creating of the heavens and the earth, when the earth was wild and waste, darkness over the face of Ocean, rushing-spirit of God hovering over the face of the waters—"[1]
- The Jewish Publication Society: "When God began to create heaven and earth—the earth being unformed and void, with darkness over the surface of the deep and a wind from God sweeping over the water—"[2]

1. Fox, *Five Books of Moses*, 11, 13. Except as noted, all quotations from the Torah, also known as the five books of Moses, are from Everett Fox's translation. All quotations from the remainder of the Hebrew Bible are from the Jewish Publication Society's *JPS Hebrew-English Tanakh*.

2. Stein, *Tanakh*, 1.

There are interesting philosophical questions that arise from the text and the various translations, questions about whether the universe is eternal or started at a fixed point in time, but fundamentally the story in the Torah starts, if not at the very beginning, close enough to it for all practical purposes, including ours. It starts with a dark, watery, and chaotic situation, out of which God, in a systematic separation of aspects of a universe, creates order. Former University of Chicago professor Leon Kass (crediting Umberto Cassuto and Leo Strauss) has used a graph and a cladigram in his commentary on Genesis to parse and depict the parallel double creation of the first three days and the second three days. One distinction between the first three days is that of place, and one distinction between days 1 and 4, days 2 and 5, and days 3 and 6 is that of motion. Further separations are effected for life, first nonterrestrial and then terrestrial, and for animals not in God's image and humans in God's image.[3]

The current consensus of scientists is that the biblical rendition of creation is not an accurate portrayal of the origin of the universe or of our species, nor anything in between. Here is another view, based on present scientific understanding:

When the cosmos as we know it was about to be created—the fundamental forces of nature being unified in an exceedingly dense, uniform, and hot volume that was very small, but of unknown size, with even stable matter itself yet unformed—there was no recognizable space, no measurable time.[4] There was no darkness over the surface of the deep because there was no deep, no surface, no over, and no under. No wind hovered over any water, as there was not yet any hydrogen or oxygen, much less any combination of them in the form of water. And there was no wind, either. What there was—all that there was—was chaotic, pulsating *potential.*

At some moment, for reasons yet unclear, what was began to change into what is. Gravity separated first from the combined strong nuclear and electroweak forces. Then the strong force emerged and the electroweak force devolved into the electromagnetic force and weak nuclear force.[5] The nascent universe, still small and unbelievably hot and turbu-

3. See Kass, *Beginning of Wisdom*, 31–36.

4. Tyson and Goldsmith suggest that everything that was could have fit "within a pinhead" (Tyson and Goldsmith, *Origins*, 25), but Adams has speculated that the entire cosmos "could have been a billion times smaller than a single proton" (Adams, *Origins of Existence*, 40). See also, Siegel, "No Big Bang Singularity."

5. See Tyson and Goldsmith, *Origins*, 25–26, 39.

lent, was an ever-changing soup of energy and subatomic particles.[6] It was all good, and about to become better.

Within one second from the mystery of beginning, our mini-universe inflated and started to expand. In that exceedingly brief time, by one estimate, it grew amazingly to span across a distance of about eighteen trillion miles, give or take.[7] That's almost the distance from our Sun to our nearest stellar neighbor, Proxima Centauri. There was no one to hear any explosive sound, but some call it the Big Bang, anyway. The temperature dropped from an unfathomably hot state of one hundred nonillion degrees Kelvin to only one trillion degrees, but that relative cooling was sufficient for subatomic particles to become protons and neutrons and other heavier particles.[8] Around the two-minute mark, with the temperature now down to a cool billion degrees, particles fused into atomic nuclei, mostly hydrogen nuclei, some helium nuclei, and other kinds as well, even as electrons remained unbound and they and photons were free to move around.[9] Chemistry now joined its sibling, physics. This, too, was good.

For 380–400,000 years after expansion began, the universe was an opaque "glowing fog."[10] But, as the universe aged, the temperature in the considerably expanded and expanding universe dropped to less than three thousand degrees Kelvin. Existing nuclei could "capture" electrons, which began to settle into orbits around existing nuclei, forming neutral atoms.[11] Still-roaming photons lost energy and morphed into microwave radiation.[12] The universe was now transparent to visible light.[13] Although there were no stars yet to shine, this was very good.

6. See Tyson and Goldsmith, *Origins*, 40–41; see also "Timeline of the Big Bang."

7. See Tyson and Goldsmith, *Origins,* 25–45, 84; see also, Tyson, *Astrophysics*, 26, and Chow, "Universe," step 2.

8. See Choi, "Our Expanding Universe."

9. See Tyson and Goldsmith, *Origins*, 42–43; see also "Timeline of the Big Bang."

10. See Tyson and Goldsmith, *Origins*, 43, 53–54; see also Harman, *Evolutions*, 197.

11. See Tyson and Goldsmith, *Origins*, 54; see also "Cosmos: The Big Bang."

12. The presence of this radiation in the cosmic microwave background ("CMB") was discovered in 1965 by Arno Penzias and Robert Wilson. See National Aeronautics and Space Administration, "Tests of Big Bang: The CMB."

13. See Tyson and Goldsmith, *Origins,* 54, 60; Carroll, *From Eternity to Here*, 54, 60; see also Chow, "Universe," steps 3 and 4.

A second stage in the life of the cosmos now began. In this interstellar stage, light elements, primarily hydrogen and helium gasses, begin to coalesce to form stars and galaxies.[14] Based on a new discovery published in May of 2018, the first stars may have formed as early as around 250 million years after the Big Bang.[15] Astrophysicist Ethan Siegel theorizes that stars could have formed when the universe was only fifty to one hundred million years old, and that improved observatories and data collection processes will ultimately disclose the truth.[16] We shall see, perhaps not long after the James Webb Space Telescope is launched in 2021.[17]

As stars burned hydrogen, and then helium, they created heavier and heavier elements. Over long periods of time, some stars died, and in the process sent forth the heavier elements created in their solar furnaces. As that process repeated and repeated, the ever-expanding universe became seeded with heavier elements.[18] This was very good.

One of the earliest galaxies to form is one now known as the Milky Way.[19] Billions of years later, this galaxy assumed a spiral shape. On one of the outer arms of this spiral galaxy, light gasses gravitated together and then ignited to become a conventional yellow star we know as the Sun.[20] Heavier stellar dust surrounding the Sun accreted into various objects, some large enough to be called planets. The third planet now in orbit around the Sun is known as Earth. Due largely to its distance from the Sun, the temperature of Earth relatively soon came to allow for liquid water and a protective atmosphere. This was very, very good.

Shortly after conditions permitted, simple life emerged on Earth. Just as we do not know what initiated the explosive growth in the original small, hot, and dense universe, we do not know exactly what forces changed inorganic chemical compounds into self-replicating lifeforms.[21]

14. See Adams, *Origins of Existence*, 67–123.

15. See Parks, "First Stars Formed."

16. See Siegel, "When Did The First Stars Appear."

17. For more information on the James Webb Space Telescope, see https://jwst.nasa.gov/launch.html.

18. See Adams, *Origins of Existence*, 108–23; see also NASA Jet Propulsion Laboratory, "Dying Stars." The discovery of the presence of oxygen in a very old star is what led to the conclusion that the star was formed 250 million years ago.

19. See Weintraub, *How Old Is the Universe*, 4, 361; see also Britt, "Milky Way's Age."

20. See Tyson and Goldsmith, *Origins*, 27; see also Choi, "Earth's Sun."

21. See Harman, *Evolutions*, 202–5; see also "Emergence of Complex Life."

All that we know (now) is that the forces of biology then joined those of physics and chemistry, and, for us, that the change was very, very good, as well.

Life evolved over time, all kinds of life, slowly at first and then profusely and profoundly. Subsequent natural disasters caused mass extinctions of many lifeforms, but those situations also allowed others to flourish. Following the impact of an asteroid or comet on Earth about 65 million years ago, non-avian dinosaurs died, but small mammals now had an opportunity to multiply in number and evolve in form.[22] And they did, ultimately generating a species of primates who could stand erect and wonder and think and speak and proclaim that all that had come before was very, very, very good.

How does Judaism respond to this science-based view of our universe and our place in it? Let's start by stepping back. The literal accuracy of the biblical description of the origin of the cosmos and of life itself has been the subject of controversy and reinterpretation for millennia. Even before recent scientific discoveries made the story of a six-day creation simply untenable as fact, many Jewish scholars, among others, readily acknowledged that the opening chapters of the Bible do not reflect a true portrayal of historic creation.

The tradition of reading parts of the Torah as allegory or metaphor is certainly both rich and lengthy. Maimonides surely viewed the Bible as God's word, and therefore true and of binding authority, but he held that "The account given in Scripture of creation is not, as is generally believed, intended to be in all its parts literal."[23] Consequently, he took biblical anthropomorphisms and anthropopathisms as symbolic. Similarly, Rabbi Avraham Yitzchak Kook (1865–1935), the first Ashkenazi Chief Rabbi in Palestine, said that the "Torah certainly obscures the [meaning of] the act of creation and speaks in allegories and parables."[24]

Today, the view that the biblical creation story ought not be taken literally has achieved a sort of consensus. With respect to *B'reishit* (Genesis), the Reform movement's latest commentary asserts that the Bible "has a great deal to tell about God's relationship to the world and about human beings and their destiny," but concedes that the opening chapters are "unscientific, antiquated myths" that may be approached "in much

22. See "What Comes After Mass Extinctions?"; see also "Mass Extinctions."

23. See Maimonides, *Guide for the Perplexed*, 2:29.

24. As quoted in Slifkin, *Challenge of Creation*, 206–7.

the same manner as one approaches poetry."[25] *Etz Hayim*, the Torah commentary published by the Conservative movement, holds similarly: "The opening chapters of Genesis are *not* a scientific account of the origins of the universe. The Torah is a book of morality, not cosmology."[26] The more traditional *Chumash* accepts the understanding of the great French commentator Rabbi Shlomo Itzhaki, known as Rashi (1040–1105), that the Torah starts with creation in order to establish God's supremacy, but acknowledges that "the Torah is not a history book."[27]

Of course, for some people, Jewish and non-Jewish, the notion that the biblical creation story is not literally true is unacceptable. They still insist on the accuracy of the description of some or all of creation as written in first verses of *B'reishit*. We have previously noted the existence of a group of Haredi Jews in New York who insist that evolution is science fiction. And in the *Chumash*, when considering the story of the serpent in the garden of Eden (Gen 3:1–7), the editors state, without citation or elaboration, that the "consensus of the commentators is that the *serpent* of the narrative was literally a serpent."[28]

AN ATTEMPT TO RECONCILE TORAH STATEMENTS AND SCIENTIFIC FACTS

Can these approaches be reconciled? Can the Torah be a book of history or cosmology and yet conform to what modern science teaches? Dr. Gerald L. Schroeder, a Massachusetts Institute of Technology trained physicist and longtime resident of Israel, seems to think so. His view, presented in his book *The Science of God* ("*TSOG*"), is that modern science and biblical creation stories fit comfortably together. More specifically, Schroeder states that physics and contemporary cosmology confirm the account of creation in the first two chapters of Genesis. One of his more audacious arguments is that the universe was created *both* in six days, as stated in *B'reishit*, *and* billions of years ago, as shown by modern science. In other words, according to Schroeder, both versions of creation are entirely correct.

25. Plaut and Stein, *Torah*, 6.
26. Lieber et al., *Etz Hayim*, 3 (emphasis mine).
27. Scherman and Zlotowitz, *Chumash*, 2.
28. Scherman and Zlotowitz, *Chumash*, 15n1–14 (emphasis original).

While taking this approach, Schroeder is fully aware that one of the tensions between religion and science is biblical literalism.[29] He suggests that one resolution of this tension is to read the Bible as a poem.[30] Indeed, one could devote an entire book to the nature of biblical poetry.[31] But that is not what Schroeder does. The uniqueness of his effort is not a review of meter or metaphor, but a purported application of mathematics and science. He wants the Torah to be true, and he wants to prove that truth with established scientific principles.

Schroeder is, unapologetically, a believer in the God of the Torah, the creator of the universe and one who intervenes in history. "God runs this world," Schroeder says, and when appropriate "God steps in and redirects the way."[32] Based on a medieval rabbi's interpretation of a psalm, Schroeder asserts that "[t]he first Divine creation was wisdom" and he seeks to use "that wisdom" to "explore the workings of God in our magnificent universe."[33] This he calls "the science of God."[34] While he sometimes speaks about objectivity and sometimes even speaks objectively, Schroeder has a clear goal. He seeks "to fit fifteen billion years into six twenty-four hour days."[35] The arguments he musters are designed to help him do just that.

The result, in *TSOG*, is a generally easy read, almost too easy. Schroeder so thoroughly mixes textual commentary and scientific commentary that a reader might have difficulty separating out the elements of his stew. And, if a reader is not careful, s/he can get swept away before realizing that some statements are at best misleading. For instance, after Schroeder makes his argument that the Torah treats the time before Adam differently than the time after Adam, he states the proposition as one of three "facts," known "*with complete certainty*."[36] You can accept Schroeder's construction, if you like, but exegesis is just that, an explanation or interpretation of a text. It is not fact.

29. See Schroeder, *Science of God*, 10.

30. Schroeder, *Science of God*, 11.

31. See generally, e.g., Alter, *Art of Biblical Poetry*.

32. Schroeder, *Science of God*, xiii.

33. Schroeder, *Science of God*, xiii–xiv.

34. Schroeder, *Science of God*, xiv.

35. Schroeder, *Science of God*, 47. Subsequently in *Science of God* (at 69–73), Schroeder uses a different and larger number for the age of the universe. He later revised both numbers downward, as discussed later in this chapter.

36. See Schroeder, *Science of God*, 52 (emphasis original).

Leaving Schroeder's style alone, and addressing the merits of his case, we find that there are two obstacles to Schroeder's approach to the Torah and physics. One is the Torah. The other is physics. First, we will discuss three examples of his use of the text, looking at source selection and reliance, translation, and textual accuracy and completeness. Then we will discuss Schroeder's physics, again discussing three examples of his methodology. Finally, we will see where his analysis takes us and whether he has successfully accomplished his goal, i.e., whether he has fit billions of years of cosmic life into a week of biblical creation.

THE SELECTIVE USE OF TEXT, THREE WAYS

Example 1—The trouble with source selectivity

Schroeder correctly recognizes at the outset that the creation text contained in the opening thirty-one verses of *B'reishit* is quite sparse, and, in and of itself, cannot be reconciled easily, if at all, with modern science. Taking the stance of the dispassionate analyst, he says that he wants "to avoid the subjective tendency of bending Bible to match science or science to match Bible."[37] He proudly announces that, with respect to theological sources, he intends to restrict himself primarily "to works that predate by centuries the discoveries of modern science."[38]

Two consequences follow, neither good. First, and most obviously, Schroeder has deprived his readers of the wisdom of hundreds of years of biblical scholarship. Wouldn't it have been more intellectually honest to have heard their voices and accepted their teaching, or not, on the merits of their arguments rather than on some preconceived notion of scientific taint?

Second, for his scientifically untainted biblical scholar, Schroeder looks primarily to a thirteenth century kabbalist, Moshe ben Nachman, known as Nachmanides or Ramban (1194–1270). He likes Nachmanides because, in his view, Nachmanides seems to have anticipated Big Bang cosmology when he "described" the process over 700 years ago "with uncanny accuracy."[39] Nachmanides did not describe anything, of course, but he did speculate that the universe began as a speck, not bigger than

37. Schroeder, *Science of God*, 19.
38. Schroeder, *Science of God*, 19.
39. Schroeder, *Science of God*, 58.

a mustard seed, and grew into all the matter we are and see.[40] Then, too, Nachmanides, like many learned scholars of his time, also thought that the world was made of just four elements or properties—water, fire, dirt, and air.[41] In this respect, his viewpoint was conventional. That is not a crime, but it is not the mark of an insightful mind either. In short, Nachmanides was not exactly a science maven. By selecting a person "untainted" by science, Schroeder has skewed his argument by tilting toward, if not fully embracing, ignorance.

Let's be clear. There is nothing inherently wrong with looking to Nachmanides as a source of biblical knowledge. And relying on Nachmanides solely (or even primarily) is certainly a permissible way to make an argument, to offer a polemic. But Schroeder's selectivity violates his own dictum against bending the Bible. It may be clever. It is just not objective.[42]

Example 2—The trouble with translation

Schroeder's premise is that the Torah treats six days of creation differently than all other recorded time.[43] He tries to buttress his argument with an unconventional translation of the repeated biblical reference at the start of Genesis to an evening and a morning. The Hebrew words at issue are *erev* for evening and *boker* for morning. Schroeder asks, fairly enough, how there can be an evening and a morning before the biblical day four, when the Sun was made.[44]

His answer, in turn, depends on the root of each word, i.e., the two or three Hebrew consonants that form the word. Schroeder tells us that the meaning of the "root of *erev* is disorder, mixture, chaos," and that the meaning of the root for *boker* is "orderly."[45] Therefore, he argues, the real meaning of the phrase "and there was evening and there was morning"

40. See Schroeder, *Science of God*, 55–58, 62, 184.

41. See Ramban on Genesis 1 at https://www.sefaria.org/Ramban_on_Genesis,_Introduction?lang=bi.

42. For those interested in a fuller, more textured view on the issue of creation, see generally Samuelson, *Judaism and the Doctrine of Creation*, and Seeskin, *Maimonides on the Origin of the World.*

43. See Schroeder, *Science of God*, 47.

44. See Gen 1:14–16; see also Schroeder, *Science of God*, 10, 102.

45. Schroeder, *Science of God*, 102.

is not that twenty-four hours passed, but that there was chaos and then there was order.[46]

Here, Schroeder makes a serious methodological error, one that was unnecessary, and one that ultimately undercuts his main argument. As University of California Bible professor Richard Elliot Friedman notes in his commentary on the Torah, while biblical Hebrew is built around three letter roots, and studying them can be helpful, "we must be cautious not to commit the etymological fallacy."[47] Consequently, "*we should not automatically derive the meaning of a word from its root*."[48] Language scholar Dr. Joel Hoffman agrees: "internal word structure is interesting, and it has a solid foundation as a cool way to look at words. But it doesn't tell us what words mean."[49]

In this instance, changing the meaning of *erev* and *boker*, of evening and morning, distorts the obvious definitions, runs counter to the poetic tempo of the text, and creates considerably more problems than it solves. All other verses in the Bible that refer to the six days of creation (see, e.g., Exod 20:8–11) and texts that rely on six creation days, like the traditional blessing over wine (the Kiddush) would, if Schroeder were right, require drastic reinterpretation. Indeed, taking Schroeder's approach would upend the biblical creation story itself, causing it to become a saga of cycles of chaos and order instead of a statement of linear and orderly progression.

Example 3—The trouble with incomplete citations

Part of Schroeder's argument regarding time post-Adam relates to the generations that issued through Adam's son, Cain, and ending with Tubal-cain (Tuval-Kayin). Schroeder asserts that Genesis, at 4:22, "attributes the start of sophisticated forging of copper and brass" to Tubal-cain.[50] By Schroeder's calculations, Tubal-cain lived "some seven hundred years after Adam, or about five thousand years ago."[51] For Schroeder, this places

46. See Schroeder, *Science of God*, 102.
47. Friedman, *Commentary on the Torah*, 9.
48. Friedman, *Commentary on the Torah*, 9 (emphasis original).
49. Hoffman, *And God Said*, 30.
50. Schroeder, *Science of God*, 136.
51. Schroeder, *Science of God*, 136.

him at roughly the same time as the start of the Bronze Age, and provides proof that the Torah and science are compatible.[52]

Around the turn of the present century, the late Mark Perakh, former professor of physics at California State University and commentator on science and religion, dissected Schroeder's claim as contained in Schroeder's first book and as modified in the original version of *TSOG*.[53] As Perakh noted, Schroeder's description of Tubal-cain is a truncated version of verse 4:22 in Genesis. The complete verse regarding Tubal-cain states that he sharpened "every blade of bronze and *iron*."[54] This omission is highly significant, Perakh contends, because the use of iron started about 1,500 years after the use of bronze. If Tubal-cain made iron tools, as well as copper or bronze ones, then, according to Perakh, he lived much later than Schroeder would have us believe, and Schroeder's post-Adam chronology collapses.[55]

Of course, dating historic eras is not a precise science. In the Near East, the Bronze Age ran from about 3300 to 1200 BCE, and the Iron Age from about 1300 to 600 BCE. The use of iron instruments may well have occurred intermittently before iron assumed dominance. But Perakh has a point. Schroeder could have and should have addressed that possibility directly. He did not, undercutting his credibility. His 2009 edition of *TSOG* makes no attempt to complete the reference to Tuval-Cain or explain the problem raised by the timing of the Bronze and Iron Ages. So here we have a deceptive biblical argument (one based on a significant text omission) and bad science (a failure to deal with archaeological times).

52. See Schroeder, *Science of God*, 137.

53. Schroeder's first book was *Genesis and the Big Bang: The Discovery of the Harmony Between Modern Science and the Bible*, published in 1992. See generally Perakh, "Not a Very Big Bang."

54. Fox, *Five Books of Moses*, 28 (emphasis added). The *Chumash* translation renders the key phrase somewhat differently: "all cutting implements of copper and *iron*." See Scherman and Zlotowitz, *Chumash*, 23 (emphasis added). But both Chumash and Fox include iron as one of the metals used by Tubal-cain.

55. See Perakh, "Not a Very Big Bang," paras. 34–46.

BENDING PHYSICS AND MATHEMATICS TO MATCH THE TORAH

Setting the standard

Having looked at Schroeder's approach to text, we can now turn to his application of science. To make his argument that the six biblical days of creation and the billions of years of the evolution of the universe as measured by contemporary scientists actually occurred over the same time period, Schroeder looks to physics and mathematics. Again, he asserts a standard for himself. He says that he wants to avoid bending "science to match the Bible."[56] That is an appropriate approach and a fair, if somewhat ambiguous, standard.[57] Unfortunately, Schroeder fails to pass his own test, as he fails to apply science in a manner generally accepted by the scientific community. Of a variety of possibilities, three examples will suffice to illustrate his tactics.

The misuse of relativity

To make his point that billions of years of cosmic evolution and six days of biblical creation occurred simultaneously, Schroeder must show that time is tolled differently in the two situations. Apparently to lay the groundwork for that conclusion, Schroeder writes about "understanding time" and refers to Albert Einstein's "*law* of relativity."[58] The effort is somewhat confusing as Schroeder, who undoubtedly knows better, does not distinguish between Einstein's Special Theory of Relativity, published in 1905, and his General Theory of Relativity, published in 1915.

56. Schroeder, *Science of God*, 19.

57. Another guideline that may be useful would be the current standard for the admission of scientific testimony and evidence in litigation in a federal court in the United States. Federal Rule of Evidence 702 currently permits testimony of a proffered expert if, among other things, the proposed testimony is based upon "sufficient facts or data" and is the product of "reliable principles and methods," which the witness has "reliably applied to the facts of the case." See "Federal Rules of Evidence," 15. Of course, Dr. Schroeder is not presenting a case to a trier of fact in a courtroom, and he did not formally commit himself to prove anything sufficiently to satisfy Rule 702. But he has sought to present his case in the court of public opinion through his books and other media. Consequently, using Rule 702 as a supplemental check on the sufficiency of Schroeder's science does not seem at all unwarranted.

58. Schroeder, *Science of God*, 49 (emphasis original).

Nevertheless, pursuant to what he calls the "*law* of relativity," Schroeder asserts that time passes differently in two different systems relative to each other due to differences in the gravity and velocity of those systems.[59] The difference in the passage of time, Schroeder says, is known as "time dilation."[60]

Schroeder then contends that "[e]ach planet, star, each location within our universe has its own unique gravitational potential, its own relative velocity and, therefore, its own unique rate as which local proper time passes, its own age."[61] How then do you find the appropriate points of reference among the billions of locations in the universe? For Schroeder, the answer cannot be Earth, because—according to Schroeder's reading of the Torah—there was no Earth during the first two of the six biblical days of creation.[62] Instead, he asserts, "The *only* perspective for the *entire* Six Day period is that of the total universe, one that encompasses the entire creation."[63]

This approach raises at least two problems. First, even given his understanding of the absence of Earth, Schroeder's exclusion of Earth as a reference point should apply only to the first two days of the week of biblical creation, but Schroeder does not seem to have considered that. Secondly, as Schroeder has recognized, relativity pursuant to Einstein requires two perspectives, time being measured in relationship of one to the other. But Schroeder has rejected Earth and all other discrete points of reference as the location of one of those perspectives. Even assuming that there is a universal perspective, to what will it be compared, if not Earth?

Schroeder attempts to resolve that last problem by claiming that, for the purpose of "exploring the brief biblical age of the universe relative to the discoveries of cosmology," it is erroneous "to view the universe from a specific location."[64] Because the "clock of Genesis starts with the creation of the universe and continues till the creation of humankind," he argues, without citation to any authority (much less a peer review), that the measure of "the relative passage of time" is "not between particular

59. Schroeder, *Science of God*, 49.

60. Schroeder, *Science of God*, 49.

61. Schroeder, *Science of God*, 51.

62. Schroeder, *Science of God*, 53.

63. Schroeder, *Science of God*, 53 (emphasis original).

64. Schroeder, *Science of God*, 54.

places in the universe but between *moments* in the universe" as it evolves subsequent to the Big Bang.[65] That is, "we must maintain the undifferentiated frame of reference that pervaded the universe at its beginning."[66] That reference, for Schroeder, is the cosmic microwave background radiation emanating from the Big Bang.[67]

Notice what Schroeder has done. He starts a discussion about understanding time by discussing Einstein and his concept of the relativity of time between two systems, an approach which has been thoroughly tested and proven in the last century.[68] Then he completely abandons Einstein and relativity based on gravity or velocity for a cosmic clock with a single and undifferentiated frame of reference. It is nothing less than a bait and switch, with no apparent reason other than to invoke Einstein and his theories initially in order to provide some gloss of respectability for what would follow. At best, this is a disingenuous exercise and should caution the reader as to Schroeder's credibility.

The misuse of exponential decay.

The key to Schroeder's argument that billions of years of cosmic evolution fit into six biblical days is his allocation of all of those years to particular days. Indeed, the match of the events reported for each biblical day to scientific discoveries is the test he asserts will show the validity of his approach.[69]

Fundamental to his calculation is his claim that time slowed in an exponential fashion. He does not use the phrase "exponential decay," but that is what he is asserting. Before disclosing his crucial allocation formula, however, Schroeder refers to Joseph Bernoulli and Cartesian coordinates and instantaneous ratios and natural logs.[70] He then presents an equation complete with superscripts and parentheses.[71] It is all mathematical mumbo-jumbo, apparently designed either to impress the reader or numb his mind.

65. Schroeder, *Science of God*, 54 (emphasis original).
66. Schroeder, *Science of God*, 55.
67. Schroeder, *Science of God*, 55.
68. See, e.g., Witze, "Einstein's 'Time Dilation' Prediction Verified."
69. See Schroeder, *Science of God*, 61.
70. See Schroeder, *Science of God*, 66–68.
71. See Schroeder, *Science of God*, 68–69.

For instance, at one point in this discussion, Schroeder argues that "Genesis has chosen a base that occurs throughout the universe, a base known in mathematics as the natural log, e."[72] He asserts that the value of the natural log, *e*, is 2.178, but then makes no serious use of that number. Instead, he invokes a new concept, the existence of a "half period" which he then equates to "one Genesis day."[73] Again, no authority is cited for either the concept or the asserted equivalence. When Schroeder finally gets to the heart of his case, we find that all he is really doing to determine the age of the first biblical day is simply dividing his selected age of the universe by two, and then dividing that number by two to reach the age of the next biblical day and so on.[74]

This is both odd and, yet, familiar. As we will discuss later, exponential decay commonly exists in nature in continuing processes,[75] and base *e* is a rate of decay (or growth) that commonly governs such processes. But here Schroeder is arguing that the process abruptly halted on the biblical day six, after which there was no further slowing of time. And, in any event, his calculation of the rate of "time decay" depends not on base *e*, but on the number two. We have, it seems, another bait and switch.

Why did Schroeder decide that the first Genesis day was a part of a process of exponential decay, and why did it have a life half as long as that of the entire universe? How did Schroeder decide to divide each subsequent day by two? He does not say in his book or on his website. In other words, for the critical concept and the crucial mathematical allocation absolutely necessary to support his entire theory, Schroeder does not cite any established scientific principle or practice, any peer-reviewed work, or indeed anything at all. He seems to have just pulled the number out of the blue. With equal or maybe better logic, he could have divided thirteen billion eight hundred million years by six and claimed that each biblical day approximated 2.3 billion conventional years.

To try to secure the rationale for his half-period theory, and his selection of the number two for his divisor, I sent an inquiry to Dr. Schroeder. In his reply, he conceded that he was talking about exponential decay.[76] As to why the first Genesis day should necessarily be one-half of

72. Schroeder, *Science of God*, 66.

73. See Schroeder, *Science of God*, 69.

74. See Schroeder, *Science of God*, 69.

75. See chapter 8 of this book.

76. Personal correspondence between the author and Dr. Schroeder on October fifth, sixth, and eleventh in 2011.

the entire age of the universe, and each successive day, half of the prior day, he replied, "In the exponential decay, which matches much of nature, the first half life encompasses half, as the term implies. And I chose that since it is so vastly common in nature."[77]

Apart from its circularity, the answer makes no sense. It is true that decay is common in nature and it is also true that decay can sometimes, as a convention, be measured in terms of half-lives. For example, the decay in the radioactivity of an isotope can be measured by noting that half of it depletes over a given period. This is known as the isotope's half-life, and different isotopes will have different half-lives. In such measurements, though, the calculation seeks to determine how long it will take for a designated reduction of a specified property. Schroeder has turned this approach upside down. He is not measuring the time required to achieve a particular reduction of a property against time. Rather, he is claiming that time itself decayed, and a decay of time is not something that is common in nature at all.

The misuse of thermodynamics

Later, Schroeder claims that the "flow from the big bang to the abundance and diversity of life we find today . . . has been a passage moving upstream against the relentless tendency of nature to proceed from order to disorder."[78] He then invokes the second law of thermodynamics, which, he tells us, states that "all nonmanaged, or random, systems always pass to a state of greater disorder."[79] How does one account for what he perceives as a trend toward order? Says Schroeder, "[t]his move toward order from chaos" in the universe, from hot energy to simple life and then humanity, "is not impossible *provided the system had direction*."[80]

The fatal science flaw in this kind of argument, as physicist Sean Carroll has pointed out, is that the second law of thermodynamics does not say what Schroeder says it does. Rather, it says that entropy (a trend toward decay and disorder) "always increases (or stays constant) in a *closed* system, one that doesn't interact noticeably with the external

77. Personal correspondence between the author and Dr. Schroeder on October fifth, sixth, and eleventh in 2011.

78. Schroeder, *Science of God*, 101.

79. Schroeder, *Science of God*, 101.

80. Schroeder, *Science of God*, 101–2 (emphasis original).

world."[81] But living organisms on our planet, indeed the biosphere itself, are functioning in an *open* system.[82] There may or may not be a Director, as Schroeder believes, but once again he has misread and misapplied basic science.

DESPITE METHODOLOGICAL AND OTHER FLAWS, DID THE UNIVERSE NEVERTHELESS UNFOLD IN ACCORDANCE WITH SCHROEDER'S CALCULATIONS?

In *TSOG*, Gerald Schroeder establishes the ultimate test of his unusual claim. He invites us to "compare day by day the fidelity by which the events of Genesis map onto the corresponding discoveries of science."[83] So, let's do that. Let's ask whether, despite all of his methodological flaws, his overstatements and misleading references, his curious selectivity of data, and his omissions, Schroeder's results nevertheless demonstrate a convergence of science and Torah. We will look at Schroeder's comparison as contained in *TSOG* adjusted for the revised timeline discuss on his website.[84]

Day One: The text is clear that, during the first biblical day, God created light and separated light from darkness (Gen 1:1–5).

According to Schroeder's latest calculations, the first biblical day lasted from about 14 billion years ago to 6.97 billion years ago. Current science teaches that, from the moment of the Big Bang, and for much of the next 380–400,000 years or so, there was a "glowing fog," but discrete points of light could not be seen. Only after that time period had passed, and the universe had cooled to about three thousand Kelvin, did the universe become transparent, with the result that light could be observed.

The roughly seven-billion-year time span Schroeder assigns to day one is long enough to encompass the coalescing of stars and galaxies. The first of these stars, as noted earlier, may have formed as early as 250 million years after the Big Bang. According to astronomer David Weintraub, 13.4 billion years ago, or within four hundred million years after the Big

81. Carroll, *From Eternity to Here*, 191 (emphasis added).

82. Carroll, *From Eternity to Here*, 191; see also Dawkins, *Greatest Show on Earth*, 413–16.

83. Schroeder, *Science of God*, 61.

84. See Schroeder, *Science of God*, 70–74; see also Schroeder, "Age of the Universe," para. 28.

Bang, there were globular clusters in what was about to become our own galaxy, the Milky Way.[85] So galaxies and stars generally, although not our Sun, were already forming. Stars are not mentioned in the Bible's description of the events of the first day, but are first discussed on the fourth day (see Gen 1:14–19). The biblical description of events for day one is, therefore, at least incomplete and, conversely, the Schroeder length of day one is at least partially inconsistent with what the Torah describes to have occurred on day four.

As we have also noted, the Torah also states that, during this initial phase, the Earth was shapeless and formless. More specifically, it says that the spirit or presence of God hovered over the surface of "*hamayim*," generally translated as "the waters" (Gen 1:2).[86] But according to cosmologists, our Sun and the Earth were not formed in the first seven billion years of the life of the universe, so there was no earthly surface in existence, and there could not have been a shapeless, formless mass of water present on or around it.

Apparently to avoid this dilemma, Schroeder translates *hamayim* not as "the water," but as "the universe."[87] Yet he refers to no basis in or outside of the text for doing so, nor why the word should be translated differently in this verse than it is in several verses that follow shortly later when the Torah speaks of a divine separation of waters (see Gen 1:6–10).

In sum, the physical presence of stars and the absence of a watery Earth are inconsistent with the text for the first biblical day.

Day Two: Schroeder claims that the second biblical day ran from 6.97 billion years ago to 3.87 billion years ago, a period of over three billion years. He summarizes the biblical event of the day as the forming of the "heavenly firmament."[88] But it was more complex than that.

On the second day, according to *B'reishit* (Gen 1:6–8), God erected or fashioned a "*rakia*," a division like a shell or dome, in the midst of the watery chaos mentioned in Genesis 1:2 in order to separate the waters above that space from the waters below it (Gen 1:6–7). The resulting space was called "sky" or "heaven" (Gen 1:8). This description reflects an

85. Weintraub, *How Old Is the Universe*, 4.

86. Or, singularly, as "the water."

87. See Schroeder, *Science of God*, 71.

88. See Schroeder, *Science of God*, 70–71.

understanding of a multistoried universe, filled with water that needed to be separated.[89]

With a good deal of evidence, however, science teaches that the Earth was formed about 4.5 billion years ago, as relatively small objects, planetesimals, coalesced into a fuller-scale planet.[90] The science of the origin of water on Earth, however, is less precise. A variety of processes, including but not limited to outgassing in a cooling environment and asteroid bombardment, appear to have contributed to the creation and retention of liquid water.

However water came to be, within one hundred million years—a relatively short time after the planet itself was formed—oceans may have existed. According to members of NASA's Astrobiology Institute, mineral grains or crystals called zircons (zirconium silicate), found in granite rock formations in Western Australia, have been dated to 4.4 to 4.3 billion years old. The existence of granite implies that continents existed at that time, and the presence of zircons implies that sufficient water was available to allow for the incorporation of the crystals into rocks then being formed.[91]

While there is evidence, then, of water on Earth during Schroeder's assigned time period for day two, there is no scientific evidence for the multistoried universe described in the relevant passages, no instance of a watery chaotic mass into which a space was inserted, protected by a dome, to separate levels or areas of water. Certainly, Schroeder does not cite any such evidence. And neither oceans of water which may well have existed, nor land masses which were elevated above sea level, are mentioned in the Torah until day three.

In short, we had sky and probably sea, but we also had the Sun and land.

Day Three: In Schroeder's scheme, the third biblical day extended for about 2.3 billion years between 3.87 and 1.57 billion years ago. On the third day, according to *B'reishit* (Gen 1:9–13), the waters under the sky were gathered together to form seas, and the land that then appeared was called earth. Schroeder notes simply that "plant life" followed,[92] but the

89. For examples of pictorial representations of this structure, see, e.g., Friedman, *Commentary on the Torah*, 8; see also Samuelson, *Judaism and the Doctrine of Creation*, 161–62.

90. See Weintraub, *How Old Is the Universe*, 26.

91. See National Aeronautics and Space Administration, "Precocious Earth."

92. See Schroeder, *Science of God*, 71.

Bible is way more precise, stating that seed-bearing plants and fruit trees emerged (Gen 1:11–12).

Of course, there would have been no seed-bearing plants and fruit-bearing trees before the photosynthesis powered by the Sun, an event the Torah places on day four. Putting that major problem aside, anaerobic cells, resembling bacteria and photosynthetic blue green algae, developed early in the Precambrian era, 3.8–3.5 billion years ago (3,800–3,500 million years ago).[93] Three billion years would pass before vascular plants evolved. When they did, 410–360 million years ago, they released spores, not seeds. Terrestrial seed-bearing plants did not emerge until the Devonian period, about 410–365 million years ago. Flowering plants did not develop until from 140 to sixty-five million years ago during the Cretaceous period.[94] The latter event falls within Schroeder day six.

In sum, Schroeder day three is too late for the emergence of oceans and land, but too early for land-based seed- and fruit-producing plants.

Day Four: The fourth biblical day spanned a period between 1,570 to 680 million years ago, according to Schroeder. On the fourth day, according to *B'reishit* (Gen 1:14–18), in order to separate the day and night and serve as signs for festivals and days and years, God created lights to be placed in the sky under the dome created on day two. These lights included the Sun, Moon and stars. As previously mentioned, science currently teaches that stars and galaxies had already started to appear more than ten billion years before this time period, and our Sun and Moon were already about 3 billion years old.[95] Science and the biblical day four are not in harmony.

Day Five: Schroeder asserts that the fifth biblical day spanned 450 million years, from 680 to 230 million years ago. On the fifth day, according to *B'reishit*, God created swarming creatures in the water and birds that could fly across the sky (Gen 1:20). These marine and avian forms included giant sea creatures, creeping things, and all winged fowl (Gen 1:21).

Jellyfish, sponges, worms, and other sea creatures arose in the Cambrian explosion around 600 million years ago. Jawed fish appeared about 440–410 million years ago. Crawling insects arose early in the Devonian

93. See Whitfield, *From So Simple a Beginning*, 45–46; Coyne, *Why Evolution Is True*, 26–27.

94. See Whitfield, *From So Simple a Beginning*, 19, 34, 38, see also, Coyne, *Why Evolution Is True*, 28.

95. See Weintraub, *How Old Is the Universe*, 39.

period, and flying insects later in the Carboniferous period, between 365 and 290 million years ago. Amphibians that could walk on land first appeared about 375–350 million years ago, and reptiles about fifty million years later.[96] All of these facts are consistent with Schroeder day five. The first birds with feathers, however, did not show up until the Jurassic period about 210–145 million years ago, too late to be included in Schroeder's fifth biblical day.[97]

Day Six: On the sixth and final creative day of that first biblical week, according to *B'reishit*, God caused the land to bring forth animals, presumably other than fish and fowl, and creeping things (Gen 1:24; see also Gen 2:19). Then God created fully formed humans, male and female, in God's image (Gen 1:27; but see Gen 2:7, 21–22). Schroeder places the sixth biblical day at a period between 230 million to almost six thousand years ago, when he says that God also implanted a soul in Adam.[98]

As noted earlier, wingless insects were present as far back as the Devonian period, and amphibians and reptiles were around for seventy to 120 million years before Schroeder's assigned times for biblical Day Six. Dinosaurs originated in the early Triassic period 245–210 million years ago. Mammals are almost as old as the dinosaurs, having begun to evolve about 190 million years ago, but they really blossomed with the death of the dinosaurs around sixty million years ago.[99]

The human branch on the evolutionary tree split from other primates about seven million years ago.[100] Fossil records of *Homo habilis*, a true tool-using human, have been dated back 2.5 million years ago. Migrations of *Homo erectus* from Africa to the Middle East, China, and Europe occurred between 1.8 million and three hundred thousand years ago. Modern *Homo sapiens*, the species to which we belong, arose in Africa around three hundred thousand years ago.[101]

For the purpose of this analysis, the creations identified in *B'reishit* emerged prior to the Schroeder timeline for day six. That is, land-based animals, including dinosaurs, roamed the planet prior to the time

96. See Whitfield, *From So Simple a Beginning*, 18, 30–36; Coyne, *Why Evolution Is True*, 27–28.

97. See Whitfield, *From So Simple a Beginning*, 40–41; see also Coyne, *Why Evolution Is True*, 27.

98. See Schroeder, *Science of God*, 17, 72, 131, 143–46.

99. See Whitfield, *From So Simple a Beginning*, 40–41.

100. See Coyne, *Why Evolution Is True*, 28.

101. See Coyne, *Why Evolution Is True*, 205–6.

Schroeder set day six, and various species of the genus *Homo* were both fully formed and using tools long before the time Schroeder assigns for biblical Adam.

Schroeder certainly knows the long history of human evolution, including that of *Homo sapiens*, so to make his argument work, he needs to distinguish a particular human being six thousand years ago from his ancestors. His effort to do so underscores the non-scientific approach he brings to his argument. First, Schroeder accepts as historic fact that there was an Adam, even though he refers to no independent scientific proof for the existence of that person. Then he insists that God placed a *neshama*, which he defines as a soul, in Adam, even while acknowledging that "[a]rchaeologists can never discover the fossil remains of *neshama*."[102] The problem is not Schroeder's faith, to which he is obviously entitled, but his pretense.

What of all of the humans who preceded Adam over the course of hundreds of thousands of years? Schroeder calls them "human-looking creatures," that is, "animals with human shapes but lacking the *neshama*."[103] Not surprisingly, he never provides a clear definition of what this *neshama* is. He alludes, for instance, to a connection to free will, to a spiritual link, to the "spirit of the Eternal,"[104] but he never supplies the nature or elements of the *neshama* in a way that we could use to determine whether any, if not all, of the pre-Adam humans had one. Even to say, as he does, that this intangible *neshama* makes "us moral beings rather amoral animals" does not help.[105] It certainly does not explain adequately why pre-Adam humans acted ethically despite not having a *neshama*, nor does it explain why contemporary moderns act unethically with one.

Schroeder does suggest that the invention of writing and the transition from village to large city is evidence of the effects of the *neshama*.[106] But what actually supports that conclusion? We congregate in larger urban masses today and have more forms of social communication, but does anyone seriously contend that we are more moral today because of those circumstances? If there were anything to that theory, then surely ethical behavior should increase as forms of communications become

102. Schroeder, *Science of God*, 151.

103. Schroeder, *Science of God*, 123.

104. Schroeder, *Science of God*, 17, 143, 179.

105. Schroeder, *Science of God*, 143.

106. Schroeder, *Science of God*, 149–50.

more sophisticated and commonly used. Yet, more widespread activity on forums such as Facebook and Twitter seem to contradict rather than support such an idea. And, more importantly, where is the evidence that pre-Adam humans lacked a spiritual component or were not moral? We know that they were sensitive enough to bury their dead, artistic enough to draw pictures and form pottery, smart enough to invent agriculture and weave baskets, and social enough to live in those villages that Schroeder disdains. On what basis, other than a need to impose a religious value, can one disparage such individuals?

STANDARDS SET, STANDARDS UNMET: THE FAILED ATTEMPT TO CONFLATE TEXT AND SCIENCE

Today, many people who consider the implications of science on Genesis either simply reject science as contrary to revealed truth or accept, willingly or grudgingly, that the biblical story of the creation of our world was not intended to be a scientific treatise; or, in any event, was written at a time long before what we have come to know. Schroeder, however, has taken an unconventional approach. He has claimed that the Bible and modern science converge, specifically that the billions of years of cosmic evolution and the six days of biblical creation occurred simultaneously. We have gone into considerable detail, in part because Schroeder's approach is so unique, in part to allow his views to be illustrated by his own words, and in part because his claim has, apparently, attracted a certain following who may benefit from having available an objective analysis.

As we have seen, Schroeder's uses of both the Bible and science are more than questionable. He promises "pure, peer-reviewed physics and traditional Genesis."[107] Instead of traditional Genesis, though, what we get are novel and unwarranted translations and textual omissions, cobbled together in service of a theme that is largely dependent on the mystical sensibilities of a medieval rabbi who is neither mainstream nor persuasive, except to those whose preexisting narrative he may confirm. Instead of science rigorously applied, we get a verbal stew of scientific terms which are distorted or misapplied and statements about science for which there are no references. While Schroeder assures us that his approach "has been given the stamp of approval" by "prestigious, peer reviewed journal(s)," he undercuts his own claim by criticizing the articles

107. Schroeder, *Science of God*, 54.

he cites as based on a "nontruth."[108] How this helps Schroeder is not clear. What is clear is that Schroeder has not identified any recognized journal of physics or cosmology to which his unique conflation of cosmic time and biblical time has been subjected to peer review. Similarly, he has not identified any reputable physicist who supports the critical components of his approach: use of universal cosmic clock, exponential decay of time, or division of conventional years into biblical years using a factor of two.

To the contrary, the physics of Schroeder's work has been criticized by professionals such as the late Victor Stenger and the late Mark Perakh. Stenger was a professor of physics and astronomy at the University of Hawaii and adjunct professor of philosophy at the University of Colorado. Stenger has attacked Schroeder's physics in *TSOG* and other works from several perspectives, including his selection of quark confinement and his red shift factor, matters not discussed here. At first, he apparently thought that *TSOG* was actually "a clever spoof on religious apologetics."[109] Perakh was a former professor of physics and materials science at California State University, Fullerton. Perakh's criticisms of Schroeder are more extensive and more detailed. They address matters we have discussed here, plus Schroeder's explanation of the diffraction of light waves, his review of the "photoelectric effect" and his comments on the effect of light on metal, and his claim that a "maser" can "fire one atom at a time," among other problems.[110]

Walter Isaacson tells us that at one point Einstein thought that he had resolved remaining issues on his special theory of relativity, only to review his calculations and realize that his resolution was seriously flawed. He admitted his error, and went back to the drawing boards.[111] Making a mistake is not a sin in science, or elsewhere. Failing to acknowledge and correct the mistake, however, is much more serious. Yet, where has Schroeder responded publically and substantively to the criticisms of Stenger and Perakh?

With his numerous disturbing misdirections, glaring omissions, and statements that are unsupported, when not incomplete and misleading, Schroeder's approach, which originally seemed incredible, now seems simply not credible. Violating his own promise to avoid matching

108. Schroeder, *Science of God*, 61–62.

109. See Stenger, "Flew's Flawed Science," para. 12.

110. See generally Perakh, "Not a Very Big Bang."

111. See Issacson, *Einstein*, 189–224.

science to the words of the Bible, he has engaged in pseudoscience in order to squeeze a square peg into a round hole.

The attempted manipulation of time can only go so far, so when we reach the conclusion of and presumed proof for his thesis, the actual comparison of scientific fact to biblical text, we should not be surprised that the results do not meet Schroeder's final self-imposed standard: the purported "phenomenal" match of "the discoveries of modern science" and "the biblical account of our genesis."[112] But there is no such match. There are, to be sure, some consistencies between events science reports and the periods Schroeder has designated for biblical days, but there is no "match too good to be relegated solely to fortuity."[113] There are, instead, just too many inconsistencies and anachronisms. Consequently, Schroeder has failed his final exam as well, and his approach certainly does not pass the Greenberg Hurdle we discussed in chapter 3.

The more serious problem is not that Schroeder tried too hard and failed. Rather, when he takes this epic story and tries to make more of it than there is, to imbue it with a modern scientific foundation that does not exist, he does a true disservice to the story itself. Centuries past, Maimonides cautioned against both biblical literalism and unquestioned acceptance of the teachings of the sages in all circumstances. To do so not only violated reason, but was dangerously counter-productive because it would, in the end, make the text or the sages look foolish and lead to a broad rejection of what was being taught.[114]

Efforts like Schroeder's have, in fact, provoked the very response about which Rambam was concerned. For instance, in one of his books, Professor Stenger wrote that he saw "little resemblance in Genesis to the picture drawn by contemporary science. All these facts can lead to only one conclusion: the biblical version of creation is dead wrong."[115] Now, Stenger did not need to be provoked by Schroeder. Rather, he wrote often and vigorously about what he called the failed hypothesis of God.[116] What this highlights, though, is the wisdom of Rabbi Greenberg's understanding of the importance of the words we use if we want to be credible. Schroeder would have been wise to heed Rambam's warning.

112. Schroeder, *Science of God*, 61.

113. Schroeder, *Science of God*, 61.

114. See Angel, *Maimonides, Spinoza and Us*, 8–10.

115. Stenger, *God: The Failed Hypothesis*, 175; see also Stenger, *Fallacy of Fine-Tuning*.

116. See generally Stenger, *God: The Failed Hypothesis*.

Of course, Stenger, like Schroeder, overreached, too. It is not quite accurate to say that Genesis is "dead wrong," bearing "little resemblance" to what science teaches. For all its flaws, seen in retrospect, the story presented is one of developing order and differentiation, if not detailed Darwinian evolution. And that was sufficient for the purpose for which the story is offered, as the setting of the stage for the greater story to come.

Stenger may not have cared for that greater story either, but in his literalness, he forgot the nature of literature. Moreover, he avoided the real uniqueness, even genius, of the biblical creation story. Instead of sun gods and sea gods and serpent gods, as we find in other Near Eastern creation stories, here we have essentially demythologized nature, a system without magic or rituals. We find nature dependent on a single powerful force. Some might even call it a unified field theory of creation. As cosmology, as evolution, as modern science, it fails. But as a perspective, a point of view, an orientation regarding that which surrounds us, it seems quite valuable, even if not fully prescient about contemporary discoveries, because it reflects an intuitive understanding of the development of existence from universal and grand to earthy and particular, from water to life, stationary objects to those in motion, simpler forms to those more complex, from plants to lower animals to humanity itself. Not bad at all for about 2,500 years ago.

Chapter 6

An Ark Is a Terrible Thing To Waste

WITHIN TEN GENERATIONS AFTER creation, the Torah tells us (without irony) that God's initial plan was in serious trouble due to the perversity of the humans made in God's image. To remedy this situation, God must reverse certain prior action and create anew. The waters above, which were separated from the waters below, must be brought down and merged with the oceans of the world until the Earth is totally covered in water. Supposedly, the sins of humankind can then be washed away.

Instead of creating a new set of humans, though, God will rely on a righteous man, Noah, who with his family will construct an enormous vessel called an ark, gather representative samples of all land-based animals, and survive the extended flood to repopulate an otherwise desolate world (see generally Gen 6:9—9:29).

As with creation itself, there are Jews who take the story literally. But, to date, none of them has tried to replicate Noah's endeavor. None has built an ark.[1] One non-Jewish group has, however, and what it has done (and not done) is another example of what happens when faith encounters modernity. The results are important for those who presumed to replicate the biblical ark and those who visit the site, of course, but also

1. The Steinhardt Museum of Natural History, which opened in Tel Aviv in July, 2018, has a façade part of which simulates the exterior of a biblical ark, but the purpose of the museum is hardly to prove the historicity of the biblical text. Rather, the museum, consistent with modern science, seeks to promote "biodiversity research, education and conservation." See https://english.tau.ac.il/campus/steinhardt_building.

for the story itself and for the Torah of the Jewish people in which that particular story was first written.

Ark Encounter is a theme park in Williamstown, Kentucky, premised on the belief in the literal truth of the biblical story of a global flood and of an ark constructed by Noah that floated on flood waters that covered the Earth. That is, Ark Encounter holds the Noah story to be "historically authentic" and aims to equip visitors "to understand the reality of events that are recorded in the book of Genesis."[2] Consequently, it claims that it has built a timber frame ark "according to the dimensions given in the Bible," more specifically, "[s]panning 510 feet long, 85 feet wide, and 51 feet high" (see Gen 6:15). Moreover, Ark Encounter promises that that its animal and other exhibits, which include sculpted dinosaurs, "will amaze and inspire you." It invites you to "be prepared to be blown away" all for the single-day price of forty-eight dollars per adult, twenty-eight dollars for youth over twelve years of age, and fifteen dollars for a child over five years old.[3] Seniors get a discount. A parking pass and tickets for the zipline ride are not included. Combination rates are available if you also want to go to Ark Encounter's "sister attraction," the Creation Museum, just north in Petersburg, Kentucky.

For Young Earth creationists, like the proponents of Ark Encounter, history only dates back to about six thousand years ago, when, they believe, God created heaven and earth. Following the reckoning of seventeenth century Irish archbishop James Ussher as to the date of creation, this equates to 4004 BCE.[4] The traditional Jewish calculation of the date of creation is somewhat different, occurring 3,761/60 years before the start of the Common Era.[5] The flood then commenced 1,656 years later (After Creation, sometimes Anno Mundi ["the year of the world"]), when Noah was six hundred years old (see Gen 7:6).[6] This calculation places the flood at about 2105 BCE.

An analysis undertaken in 1994 evaluated major safety parameters, and concluded that a floating ship of biblical dimensions would have been seaworthy.[7] Maybe. Without question, though, building a wooden

2. See generally https://arkencounter.com/.
3. See https://arkencounter.com/.
4. See Weintraub, *How Old Is the Universe*, 9–13.
5. See "Jewish Calendar Year."
6. See "Lunar Flood, Solar Year."
7. See Hong et al., "Safety Investigation of Noah's Ark."

ship the length of one and a half football fields and seven stories high is quite a feat.

Yet if Ark Encounter's engineering and construction skills are impressive, its understanding of the biblical story of Noah is not. In fact, to sustain its undertaking, Ark Encounter purposely has had to (1) ignore the Mesopotamian antecedents of the biblical flood story, (2) reject a consensus among modern academics that the Noah story as contained in the Torah is really a combined and edited version of two quite separate and, in key locations, contradictory stories, and (3) avoid what current science teaches. Ark Encounter, in short, can only sail on a sea of denial.

THE CULTURAL ANTECEDENTS FOR NOAH

Ark Encounter recognizes that many cultures, over two hundred by its count, have flood stories. That there are myriad stories from every corner of the globe is essentially correct, and flood stories from around the world can now be easily accessed. On its website, however, Ark Encounter discusses only a few of these, and then in a cursory fashion, taking the position that the only "true account" is the one found in the Bible. At the same time, it contends that the existence of so many flood stories "point(s) to a universal truth—there was a worldwide flood in the ancient past."[8] The argument doesn't hold water, though. The variety of detail in these stories is so extensive, and the dating, such as it is, so inconsistent, that collectively the vast number of stories actually refutes the notion that there was one global deluge.

Conversely, there is at least one nation that apparently has no encompassing flood story. Surprisingly, given its history and geography as an archipelago, that nation is Japan. Rather than invent or adopt such a story, some Japanese argue that the absence of a flood tradition demonstrates Japan's uniqueness and superiority to other nations, as Japan alone seems to have existed on a higher plain than other nations, spiritually and physically, and, so, has been spared from an otherwise worldwide destructive event.

Ark Encounter's failure to address the flood stories of one particular region is especially disingenuous. In his lectures on the book of Genesis, Rutgers professor of Jewish history Gary Rendsburg points out that the "only geographical location mentioned in the biblical account is the

8. See "World Flood Myths."

mountains of Ararat, which are located in . . . modern-day eastern Turkey . . . near the headwaters of both the Tigris and Euphrates rivers."[9] This is in the northern region of the ancient territory of Mesopotamia, literally the land between the rivers. Similarly, the biblical stories that follow the flood account—the Tower of Babel saga and the journey of Abram's family—are also rooted along the Euphrates and throughout Mesopotamia (see Gen 10:10; 11:1–9, 31; see also Josh 24:2).

Mesopotamia, in stark contrast to Canaan, the land ancient Israelites later claimed to be promised to them, is prone to flooding.[10] The Mesopotamian plain receives ample rainfall, and both the Tigris and Euphrates overflow their banks with regularity. Canaan benefits from no similar experience. So, it would not be surprising if the land of frequent flooding also was the source of the biblical flood story, and, indeed, current evidence suggests that is the case.

Perhaps the most famous of these Mesopotamian stories was discovered less than a century and a half ago when the remarkable George Smith, a young printer by trade and largely self-trained archaeologist, was working on a collection of cuneiform tablets at The British Museum in London. In 1872, Smith succeeded in translating a portion of the eleventh of twelve tablets that make up the Babylonian Epic of Gilgamesh, King of the city of Uruk in southern Mesopotamia (modern day Iraq). Smith subsequently deciphered the entire flood story within the Gilgamesh Epic.

The flood story Smith uncovered was written about 3,200 years ago.[11] Tablet Eleven tells of Gilgamesh's search for immortality when he learned of Utnapishtim, who survived a massive flood and was awarded immortality by the gods.[12] Ark Encounter's website acknowledges the Epic of Gilgamesh, but avoids any serious analysis of it, characterizing the saga as "clearly fiction" and a "myth."[13] In fact, the Noah story appears to incorporate not only the general theme of Utnapishtim's tale, but also its structure, its details, its phrasing, and, especially tellingly, a particular

9. Rendsburg, "Book of Genesis," Lecture 7, sec. 4.C; see also, Gen 8:4.

10. See "Ancient Mesopotamian Geography and Location."

11. See Wiener, "Animals Went In Two by Two."

12. See Kovacs, "Epic of Gilgamesh."

13. See "World Flood Myths."

word. To the extent that they both are traditional stories that "attempt to explain how and why something is as it is," both are myths.[14]

In both stories, there is a massive divinely ordained flood, a ship serves as a sanctuary for a selected group of humans and certain animals, the ship is built with wood, and pitch is used to provide waterproofing. Eventually the ship comes to rest on a mountaintop, birds are then sent forth sequentially to determine whether the flood has ended and dry land is again available for occupation, the inhabitants of the ship are released, a sacrifice is offered, and the smell of the burnt offering is found pleasing by the divine power(s).

As Professor Rendsburg teaches, while one could tell essentially the same story with considerable variation, the version found in Genesis maintains many of the key elements in the same order as they are found in the Gilgamesh epic. Even the order of the references to materials, dimensions, and decks on the ship is the same. There are differences, of course, large and small. For instance, the theological assumptions expressed and the rewards granted the hero are quite dissimilar. Moreover, the boat in the Noah story is shaped differently than Utnapishtim's vessel, and it is smaller with fewer decks. Still, the similarities are striking.

Two unique non-Israelite pieces of evidence seem to confirm the story in Gilgamesh Tablet Eleven as a prior source of the biblical account. Yeshiva associate professor Shalom E. Holtz teaches that the word for waterproofing "pitch" in the Noah story is *kofer*, the "cognate to Akkadian *kupru*, which is what Utnapishtim uses."[15] By comparison, Holtz observes, the word used for the pitch that sealed the container in which the baby Moses was placed was a native Hebrew word, *zefet* (see Exod 2:3). Holtz concludes that *kofer* was "borrowed directly from Akkadian, and provides the strongest evidence for the Mesopotamian origin of the (Noah story)."[16]

Further, after Noah sacrifices to God, God reportedly "smelled the soothing savor" (see Gen 8:21). This, according to Professor Rendsburg, is the only time among the many biblical references to Israelite sacrifices where it is said that God smelled a sacrifice. Rendsburg then calls attention

14. See Hayes, *Introduction to the Bible*, 34, 52–55.
15. See Holtz, "Flood Story," para. 7.
16. Holtz, "Flood Story," para. 7.

to line 161 in Tablet Eleven of the Gilgamesh Epic, where it is written that "the gods smelled the sweet savor" of Utnapishtim's sacrifice.[17]

The Gilgamesh Epic, however, is not the oldest of the Mesopotamian flood stories. It was preceded four to five centuries earlier by another flood story, the Epic of Atrahasis.[18] Citing the work of the late Tikva Frymer-Kensky, Holtz argues that a "theme of creation, deconstruction and re-creation drives the plot" of the Atrahasis Epic. The story arises out of the circumstances of ancient gods who had grown weary of doing physical labor, including digging ditches for the Tigris and Euphrates rivers. They created humans to bear their burden, but the human population grew substantially and they were noisy, so the gods could not rest comfortably. The solution was first a flood and, then, new rules to regulate marriage, fertility, and infant mortality.

The primeval historical background in the Atrahasis legend is much closer to that in Genesis than is the travelogue of King Gilgamesh. As noted above, the story of Noah is in many respects a story of recreation following the destruction of God's original world, one in which the people created in God's own image had become wicked and lawlessness corrupted the land (see Gen 6:5, 11–13). In the biblical version, the deity that separated the waters above from the waters below in Genesis (1:6–7) now reverses course, allows the waters that had been separated for the benefit of the land to recombine and envelope Earth. God will start over with a newly selected group of humans.

Even older is the Sumerian flood story featuring Ziusudra of Shuruppak.[19] Sumer was in the southernmost part of Mesopotamia, roughly from contemporary Baghdad to the Persian Gulf. Sumerian civilization was well-established by early in the fourth millennium BCE, and Ziusudra is listed in the Sumerian king list. Here, too, the antediluvian gods sought to destroy mankind, but one god, Enki, urged Ziusudra to construct a large boat in order to survive, which he did and, having left his boat, offered a sacrifice.

The archaeological and anthropological records clearly show, then, that several classic flood tales predated the biblical story of Noah. But would the authors of the biblical tale have known of them? And, if so, how and when? Minimally, as the book of Kings shows in particular, the

17. See Rendsburg, "Book of Genesis," Lecture 7, sec. 4.G.3; see also Kovacs, "Epic of Gilgamesh."

18. See "Epic of Atraḫasis."

19. See "Great Flood: Sumerian Version."

kingdoms of Israel and Judah were in constant contact with the greater powers of Egypt, Assyria, and Babylon, either through commerce or war. No doubt cultural tales, even phrases and words, were shared.

Professor Rendsburg provides some more concrete clues. He notes that a fragment of the Gilgamesh epic, dating to late in the second millennium BCE, has been found in Megiddo in Northern Israel, and another tablet about Gilgamesh's life, apparently not part of the twelve-tablet epic, has been found in the coastal city of Ugarit, formerly Northern Canaan and presently in Syria. Neither refer to a flood, but both provide evidence that Mesopotamian stories were transported physically to core biblical lands.[20]

Needless to say, if the exchange of ideas in the normal course was not a sufficient opportunity to learn of ancient Mesopotamian flood stories, the exile of the ruling and literate Judahites to Babylon following the destruction of Jerusalem around 586 BCE surely provided it. Those who were taken or, later, grew up in Babylon, could hardly have failed to hear of the classic tales that had already traveled beyond their home of origin.

THE REFUSAL TO ACKNOWLEDGE THE CULTURAL HERITAGE

The failure of Ark Encounter to acknowledge the intellectual history preceding the development of the biblical flood story is more than troublesome. It also precludes a careful reader of the biblical story from fully appreciating the insights and inventiveness of the biblical authors and the dramatic changes they brought to an ancient Mesopotamian tradition.

For instance, the gods in the Atrahasis story resort to flooding the land because they are irritated by the noisiness of humankind. Their clamor prevented the gods from sleeping well. By contrast, the text in Genesis states that humans were destroyed because of "*hamas*," which Yale Bible professor Christine Hayes translates as "'violence and bloodshed' but including all kinds of injustice and oppression."[21] In Gilgamesh, the gods are seen fighting with each other, and once the flood is unleashed, they become terrified of what they have wrought, in part because of the power of the flood and also because they now do not have any food and are starving. That is, as Professor Hayes teaches, the Mesopotamian gods

20. See Rendsburg, Lecture 7, sec. 4.D.2.

21. Hayes, *Introduction to the Bible*, 53.

are in disarray rather than control. The biblical God, however, is neither at the mercy of the elements of nature nor subject to anthropomorphic needs. Similarly, the old gods acted capriciously, but God in the Noah story has ethical standards, and punishes immorality while rewarding righteousness.

The reader may learn no grand lesson from the survival stories of Ziusudra, Atrahasis, and Utnapishtim. But the reader of the Noah tale will easily grasp the basic lesson being taught, which Professor Hayes summarizes as follows: "inhumanity and violence undermine the very foundations of society."[22] The "cosmic catastrophe" of the flood is not due to religious sins, offensive as they may be, but for a more fundamental breach of basic moral law. With this modified, transformed story, then, the biblical authors changed the discussion of the nature of the universe and provided guidance for future generations.

Obviously, the Noah fable can stand on its own. It has successfully done so for well over two thousand years, effectively replacing the tales of Ziusudra, Atrahasis, and Utnapishtim as the paradigmatic world flood story. This result is not unprecedented in great literature. Hamlet, too, can be appreciated easily without realizing that William Shakespeare relied on earlier sources such as one written by Thomas Kyd in the late sixteenth century and the much older, twelfth century *Historia Danica* by Saxo Grammaticus.[23] Understanding that there were such stories, though, does not diminish Shakespeare's contribution. Rather, such knowledge highlights his skill and the marvelous nature of his text. Similarly, Ark Encounter customers could glean certain moral lessons of the Noah tale from the biblical text alone. But by restricting the customer to the biblical text, and avoiding the back stories, Ark Encounter flattens a textured work and disrespects its author(s).

THE REFUSAL TO RECOGNIZE THE DUAL NOAH STORIES

Ark Encounter also fails to address the current consensus among contemporary biblical academics—Professor Rendsburg being an exception—that the story of Noah told in the Torah is the product of a conflation of parallel texts representing two Noah traditions. By utilizing principles of

22. Hayes, *Introduction to the Bible*, 54.

23. See "Dates and Sources."

the documentary hypothesis and source criticism, mentioned earlier as part of our toolkit, a reader might be able to discern the original texts.

The documentary hypothesis is the idea that the Torah is a compilation of works from several discrete sources. The intellectual roots of the documentary hypothesis extend back to the seventeenth-century philosopher Baruch Spinoza, who many see as the father of the Enlightenment.[24] For our more limited purposes, Spinoza was the first notable scholar, or at least among the first, to suggest that the Bible should be reviewed critically and without religious or other preconceptions. That is, it should be treated like any other book.[25] Subsequently, others argued that the text contained anachronisms and other errors, that many passages seemed unnecessarily duplicated, and that human hands rather than divine ones must have written the Torah.

The documentary hypothesis reached its classical exposition in 1878 with the publication of Julius Wellhausen's *History of Israel*. There, Wellhausen identified four sources that were combined into the Torah we know today. Those sources are characterized by their writing style, the language used, and content. Thus, for example, the J source uses the personal name of the Israelite God, YHWH, while the E source does not do so, but refers instead to Elohim, a more generic reference. Similarly, J treats God in an anthropomorphic manner, while E views God more remotely. J writes of a covenant at Mount Sinai, while E places similar activity at Mount Horeb. J seems to reflect the interests of the northern kingdom of Israel, while E's orientation is more that of the southern kingdom of Judah.[26]

The other two sources in Wellhausen's analysis are the P and D documents. The P document focuses on the role and responsibility of the priests (especially the Aaronid priests), rituals in the Tabernacle, and themes of order and purity, often expressed as rules regarding the Sabbath, the holidays, and matters of personal behavior, including diet.[27] While passages attributed to P are concentrated in the book of Leviticus, they can also be found elsewhere, as in the opening chapter to Genesis and throughout the book of Numbers. By contrast, in the Torah, D is

24. For more on Spinoza, see generally, Nadler, *Spinoza*, and Goldstein, *Betraying Spinoza*.

25. See Hayes, *Introduction to the Bible*, 58; see also, Brettler, *How to Read the Jewish Bible*, 3.

26. See Hayes, *Introduction to the Bible*, 58–60.

27. See Hayes, *Introduction to the Bible*, 60–61.

found only in, and occupies almost all of, the book of Deuteronomy.[28] Thematically, D includes a revised and extended legal code, a broad view of the constituency of the priesthood, and advocacy for the centralization of worship of a single god in one location.

Wellhausen's approach was criticized both substantively and for his perceived intention to destroy the Jewish past. And, over time, others have argued about the contents of J, E, D, and P and when and where they were formed.[29] But today there is a general (though not uniform) acceptance by Bible scholars that the Torah was based on different sources. This approach, this historical critical method, this source criticism is another tool in our kit. We accept it, as do most (though not all) contemporary Jewish scholars, because the approach can be constructive, can enrich our understanding of the text, and can help us deal with Judaism's encounter with science.[30]

Applying the technique to the Noah story, one can indeed find two separate stories.[31] In one, attributed to J, YHWH tells Noah to bring one pair of unclean animals, but also seven pairs of clean animals; the flood is caused by rain, which lasts for forty days; Noah ultimately sends out a dove three times; and YHWH enjoys the sweet savor of Noah's post-journey sacrifice of certain clean animals. In the other, longer version, attributed to P, Elohim directs the activity, including specifying the dimensions of the ark; only one pair of each kind of animal boards the ship; the source of the flood waters is the waters above and below land, which flow through gates Elohim opens; the flood lasts for 150 days until Elohim closes the gates; and, after more than a year has passed, Elohim tells Noah to leave the ark and repopulate the world.[32]

28. In the *Tanakh*, D is seen as one of the elements of a broader history, known as the Deuteronomistic History. This work extends through the story of conquest in Joshua, past the period of Judges, and into the creation and dissolution of the kingdom of Israel, concluding with the destruction of the kingdom of Judah and the exile of the royal family to Babylon. Deuteronomistic History is discussed more in chapter 12.

29. A key question, for instance, is whether P was written before or after D. See generally, e.g., Friedman, *Who Wrote the Bible?*, and Ska, *Introduction to Reading the Pentateuch*; see also Schniedewind, *How the Bible Became a Book.*

30. See Brettler, *How to Read the Jewish Bible*, 5.

31. See "Textual Study of Noah's Flood," http://thetorah.com/textual-study-of-noahs-flood/.

32. Similarly, using the documentary hypothesis, one can find two related stories in the book of Exodus about the Egyptians chasing the Israelites into the sea. Separating the stories helps eliminate both repetitions and inconsistencies in the text. See

Ultimately, the reader must decide whether to accept or not the utility of documentary hypothesis and source criticism. According to Orthodox Rabbi Hacham Isaac S. D. Sassoon, source criticism can enhance our appreciation of the Torah because it allows us to accept the text without "running away or going into denial." It paves a path "for a celebration of this Torah of many visions whose combined and consolidated whole is greater and nobler than the sum of its parts."[33] You may disagree, but surely failing to provide the opportunity to consider this approach by claiming that the received text is not to be questioned reflects a determination that a closed mind is to be preferred to an open one.

THE REFUSAL TO ACKNOWLEDGE PHYSICAL FACTS

Ark Encounter's failure to address the literary antecedents of the Noah story, a story it embraces, is accompanied by its silence in the face of challenges raised by a variety of scientific disciplines, apparently none of which Ark Encounter embraces. The issues are extensive, if not endless. Here are some of those issues:

- Where is the geological evidence, in the nature of silt formations or otherwise, of a flood that Ark Encounter describes as a "worldwide catastrophic event" about forty-four centuries ago?[34] There doesn't seem to be any.
- If a flood covered the land surfaces of the world, fresh water lakes and rivers would have been swamped by saltwater oceans. Most fish are stenohaline, able to live only within a narrow range of salinity.[35] How did saline sensitive fish from either environment survive?
- To have covered the world, over the top of Mount Everest, would have required an enormous amount of water. For the sake of discussion, given the improbability of such a quantity appearing by rain or the opening of gates to mythical upper and lower bodies of water, when the time came for the flood to cease, where did all the water

"What Really Happened at the Sea." Those interested in seeing how the documentary hypothesis may be applied to the entire Torah may want to review *The Bible with Sources Revealed* by Richard Elliot Friedman. Using different fonts and colors, Friedman claims to identify the sources for almost all of the Torah.

33. See Sassoon, "Source Criticism."
34. See https://arkencounter.com/about/.
35. See Melina, "Can Saltwater Fish Live in Fresh Water?"

go? The ocean basins were already filled, the land was saturated, and the atmosphere could not contain all that water vapor. Where did the water go?

- Where is the archaeological evidence, in the form of ruins of dwellings and other structures and of bone layers, of such a global flood? There doesn't seem to be any here, either. To the contrary, the antediluvian Great Pyramid of Giza and the circle of large sarsens at Stonehenge testify to the absence of a worldwide deluge.
- Why is there no record of civilizational disruption in the annals of the Old Kingdom of Egypt, or even in the records of the great Mesopotamian cultures just over four millennia ago?
- Did Noah collect, say, pandas from China and koalas from Australia? If so, how did he feed them the massive amounts of unique food (bamboo and eucalyptus leaves, respectively) they would have required to survive? How did he get them down the steep slope of Mount Ararat and back to their native homes? And, if pandas and koalas were not on the ark, how can they be here today?
- Given that science tells us that dinosaurs died about sixty-five million years ago, where did the dinosaurs that Ark Encounter depicts on the ark live 4,100 years ago, before Noah brought them on the ark? After they left the ark, where did they go, and when and how did those kinds of dinosaurs then die?

Honest answers to these questions should confirm the conclusion of Orthodox Rabbi Norman Solomon that the story of a massive, universal flood rising fifteen cubits over Mount Ararat some forty-one centuries ago "never happened; [rather] there is overwhelming evidence that most life around the planet continued in its normal course."[36] Conversely, we have good evidence, from thick deposits near Shuruppak, the Sumerian home of Ziusudra, that a substantial flood occurred there around 2900 BCE.[37] This event is of the right kind at the right time to have prompted the saga of Ziusudra and, in turn, the other stories upon which the biblical tale is based.

36. See Solomon, "Torah's Version of the Flood Story," para. 2.

37. See Collins, "Yes, Noah's Flood;" see also Collins, "Twenty-One Reasons."

ARK ENCOUNTER IS REALLY ARK AVOIDANCE

Although it extends an invitation to witness history, Ark Encounter in practice simultaneously rejects both the flood literature that predates the writing of Noah and also modern science, which conclusively demonstrates in a variety of ways why the biblical story did not and could not have happened as it is written. Instead, it focuses narrowly on a story which the author, not claiming to be either an historian or a scientist, carefully constructed for a particular purpose at a particular time. We have not yet sufficiently identified that time, much less that author, but at least we should be able to understand the inspirational precedents with which the author was working, marvel at the vision that guided his work, and appreciate the talent necessary to forge the resulting work to advance his needs and goals.

Moreover, in this case, the teachings of disparate disciplines like geology, hydrology, archaeology, cultural anthropology, and evolutionary biology converge harmoniously. Accepting the results of their lessons does not require that person to abandon belief in a Higher Power, just that such a person render to science that which belongs to science. And how can one not? When faith fears facts, and opts for fiction, it risks looking foolish. Rather than demonstrating character, it invites caricature.

We have seen the results in our own community, and it is tragic. When any person of faith succumbs to such fear, whether promoted by Ark Encounter or the leaders of a Jewish sect, he or she actually displays a lack of true faith both in the god s/he professes to worship and the reality that god presumably created, including the creature the Bible uniquely claims was made in the image of the God of Israel. Conversely, if the phrase "image of God" means anything, it must mean, as the first chapter of Genesis makes abundantly clear, a being capable of making meaningful distinctions among various possibilities, a being that exercises the brain with which it has been blessed.

In the end, and in addition to its other failures, Ark Encounter does a disservice to its customers. Though it claims to have spent over one hundred millions dollars, the Ark Encounter project has missed a marvelous opportunity. It has chosen to comfort the choir, instead of informing, challenging, and elevating them. Worse, it encourages them, rewards them for keeping their eyes and ears shut and their minds closed. There may be short-term financial gain in that, but not much future.

Moreover, Ark Encounter does a disservice to a story in the Torah, the sacred text of the Jewish people. To invoke Yitz Greenberg's phrase, it dummies down a multifaceted story with a complex history. In an effort to simplify, it becomes simplistic, and so fails the Greenberg Hurdle—it is not credible.

What a pity. An ark is a terrible thing to waste.

Chapter 7

Eyes And Evolution

Why Do Some with Eyes Not See?

IN THE GREAT SATIRICAL movie *Duck Soup*, first released in 1933, Mrs. Gloria Teasdale (Margaret Dumont), the financial underwriter of the nation of Freedonia, recruits Rufus T. Firefly (Groucho Marx) to be the insolvent country's new president. The opposition then retains two spies, Chicolini (Chico Marx) and Pinky (Harpo Marx), to work for them. Toward the end of the film, in a bedroom scene with Teasdale, the spies both dress like Firefly in order to secure the combination to a safe. After she gives the combination to one Firefly (the disguised Pinky), Mrs. Teasdale watches him leave the room, but suddenly another Firefly (the disguised Chicolini) appears. Teasdale confronts Chicolini, who denies leaving and blusters, "Well, who you gonna believe, me or your own eyes?" He knew that Teasdale had convincing evidence against the spies. She was, after all, an eyewitness. And yet, she did not understand what was happening.

Perhaps more than any other of our senses, humans rely on sight. Our camera eyes allow massive amounts of information to enter our brain, first through the lens at the exterior of the eye, and then from the inverted image on the retina at the back of the eye by way of the optic nerve. In the plains of Africa, our ancestors once stood to gather information about their surroundings. Today we focus on ironically named smartphones, not so much to hear audio transmissions, but to stare at screens with text or other visual data.

So important is sight to us that over two-thirds of the sensory cells in our bodies are the light-sensing cells in our eyes. So energy-consuming is the human retina that it uses more oxygen per gram than does the brain.[1] Our eyes are not only the mechanism by or through which we focus on our surroundings—they are also at the focal point of a broader picture and a serious controversy. One hundred and thirty years after the death of Charles Darwin, some who do not accept the validity of the theory of biological evolution pioneered by him argue that the complexity of the eye disproves evolutionary theory. In Jewish circles today, the argument is advanced in several quarters, including Rabbi Joseph Ginsburg and Professor Herman Branover writing for Chabad, and media executive Jonathan Rosenblum, among others.

The arguments tend to overlap. United by a fundamentalist belief in a supernatural Author of creation and centered on a fundamental disbelief that a complex organ such as the human eye could have evolved by small, adaptive mutations over time, Jewish opponents of evolution contend that the statistical probabilities of life emerging spontaneously are somewhere between nonexistent or abysmally low, that macroevolution has not been proven, that there is no fossil or other record that demonstrates how a human eye could have developed, and that an intelligent designer in the form of a supernatural God is a much better explanation for the existence of the human eye.

For the general proposition that the statistical probability of life forming from an inorganic soup is extraordinary small, Ginsburg/Branover and Rosenblum each refers to British cosmologist Sir Fred Hoyle (1915–2001). It's an argument from authority, without the authority. Hoyle was an astronomer, not a biochemist. Moreover, Hoyle rejected the notion that life could have emerged on Earth because he believed it more probable that life on this planet was seeded by extraterrestrial sources.[2] Do Hoyle's acolytes really want to go down that road? None of them offers any comment on Hoyle's extraterrestrial origin theory, let alone its validity.

In the last twenty years, there have been serious scientific studies conducted which suggest possible mechanisms for the creation of life from prebiotic conditions on Earth. The science involves the geology of deep-sea hydrothermal vents and the impact of protons on molecules

1. See Shubin, *Your Inner Fish*, 150, and Lane, *Life Ascending*, 175.

2. See Lane, *Life Ascending*, 9, 12.

through a process called chemiosmosis.[3] Of course, no one can be sure about the conditions on Earth some four billion years ago. There may, consequently, be more than one explanation for how raw materials developed into the DNA/protein world in which we find ourselves. This is not a deficiency of science, merely an example of how science probes for the truth. In the end, though, "the existence of different abiotic mechanisms by which biochemical monomers can be synthesized under plausible prebiotic conditions is well established."[4] The mystery about the specifics remains, and the specifics are obviously important, but there is no reason to believe that the problem is unsolvable.[5]

Evolution deniers do not stop at questioning the origin of life, they attack biological evolution generally and specifically. Ginsburg/Branover claim that macroevolution, the descent of life from a common ancestor through a process of mutation and selection, is neither "empirical fact nor a scientific theory."[6] Rosenblum complains that "Darwinists" have not observed the "mechanism" by which a common ancestor to, say, whales and humans have been "fashioned," nor "can Darwinists explain how complex systems, such as human sight, none of whose component parts would alone provide any advantage, could have come into being by a long series of micro-mutations."[7]

What, if any, scientific literature, the evolution deniers have read is unclear, but it does not seem to be much, and certainly nothing current, nothing that would be critically accepted today. A good starting place for those seriously interested in exploring the development of species is *Your Inner Fish* by University of Chicago paleontologist Neil Shubin. Wonderfully written, Professor Shubin looks at fossils in rocks and genes in cells which mutually trace and reinforce a story of development and speciation from ancient fish to modern vertebrates, including primates, including us. Where Shubin looks at shapes and functions, biochemist Nick Lane tends to discuss processes in his book, *Life Ascending*.

Both Shubin and Lane address vision and the history of the eye. As a paleontologist, Shubin is troubled by the absence of a substantial fossil record for eyes, but notes that "the really important work in the

3. Lane, *Life Ascending*, 8–33.
4. See Ruse and Travis, *Evolution*, 73.
5. See Ruse and Travis, *Evolution*, 72, 74.
6. Ginsburg and Branover, "Appendix 4," para. 36.
7. Rosenblum, "Charlie Darwin's Angels," para. 11.

light-gathering cells happens inside the molecule that actually collects light."[8] Moreover, despite the incredible variety of eyes, all animals use "the same kind of light-capturing molecule," one which involves a protein known as an opsin.[9] Looking at opsins, tissues, and genes, Shubin sees the bridge back in time to our primitive ancestors.[10]

Lane's discussion of sight drills down further.[11] He acknowledges the challenge to natural selection by the existence of the complex human eye and proceeds to deal with it. He demolishes the notion of eye absolutism by showing that nature balances the competing interests of resolution and sensitivity in different ways over time. He shows how, in some circumstances, half an eye is better than no eye at all. He shows how the lens could have been assembled in the first instance, using available proteins. His conclusion: "there are no particularly difficult steps in the sequence to make an eye."[12]

Lane also describes a 1994 study led by Dan-Eric Nilsson and Susanne Pelger of Lund University in Sweden which considered the specific adaptive steps necessary to evolve from a flatworm eyespot to a complex eye of a vertebrate. Professor Coyne refers to the same study in his book *Why Evolution Is True*. Given the approximate number of generations that would be required to reach the goal, the investigators determined that the entire process of evolution could be accomplished in about four to five hundred thousand years. In other words, there was ample time for independent evolution of eyes in dozens of groups of animals.[13]

For those who want even more, in 2011, ophthalmologist Dr. Ivan Schwab published what appears to be the definitive work to that date on the evolution of the eye.[14] Over the course of more than 250 pages, Schwab discusses and illustrates in detail the development of the incredible variety of eyes we know today from initial photoreceptors that existed over 3.7 billion years ago through various geologic periods and eras as manifested across various animal classes. Bird eyes, snake eyes, mollusk eyes, octopus eyes, and human eyes—they and more are all here.

8. Shubin, *Your Inner Fish*, 152.
9. Shubin, *Your Inner Fish*, 152.
10. Shubin, *Your Inner Fish*, 148–57.
11. See Lane, *Life Ascending*, 172–204.
12. Lane, *Life Ascending*, 199.
13. Lane, *Life Ascending*, 182–86; Coyne, *Why Evolution Is True*, 142–43.
14. See generally Schwab, *Evolution's Witness*.

And the discoveries continue. The following year, scientists led by Dr. Davide Pisani of the University of Bristol presented evidence to the National Academy of Scientists that LOCA, the last opsin common ancestor, was a placozoa that lived over seven hundred million years ago, earlier than previously believed. Within eleven million years, through genetic changes, the opsin became able to sense light.[15] From there, vision in all its glorious forms evolved.[16]

All this data, all these studies, all of the articles, chapters, and books place a heavy burden on the deniers and rejectionists. Their objections have been met and persuasively countered. No longer should the argument be about macroevolution or speciation. Rather, the onus is clearly on those who continue to deny evolution generally and that of the eye specifically to tell us why Shubin is wrong, why Lane is wrong, why Nilsson and Pegler are wrong, why Schwab and Pisani are wrong, why the history, chemistry, molecular biology, anatomy, genetics, physiology, geology, and paleontology to which the scientists refer, and which importantly are mutually reinforcing, are wrong. While questioning and testing are appropriate, rote recitation of unsupported rhetoric is no substitute for reason, and the deniers need to elevate their arguments or give them up.

The criticism from the Jewish sources identified here (and those not) fails in large part because it is outdated, shallow, and made casually, without serious regard for definition of terms or reference to serious studies. The question is why the critics persist and in such strong terms? There are at least two answers: one found in history and the other in social psychology.

The Jewish reaction to Darwin originally was quite mixed. Some reformers, like German-born Reform Rabbi Kaufmann Kohler (1843–1926), favored progressivism and transformation, embracing evolution, while others did not share such sentiments. Similarly, some traditionalists opposed evolution, while others found its gradualism appealing, as well as its seeming to confirm the durability and adaptability of the Jewish people. What appeared to be at stake at that time was not so much science as the nature of contemporary Judaism itself.[17]

15. See "New Study Sheds Light."

16. See also Harman, *Evolutions*, 109–19, 216–19, for a lyrical approach to the development of the eye, from the perspective of a tribolite, and additional commentary and sources on the evolution of the eye.

17. See Cantor and Swetlitz, *Jewish Tradition and the Challenge*, 47–70.

Today one is hard-pressed to find opponents of evolution across a wide spectrum from the Reform tradition, through the Reconstructionist and Conservative movements, to and including Modern Orthodoxy. While individuals might attribute a greater, lesser, or no role to a Supreme Being in the evolutionary process, Jews as a group overwhelmingly tend to accept the reality of the process. Some, however, do not, and they are often, if not exclusively, the traditional and Haredi-Orthodox.[18] The question, again, is why? Why are these Jews different from all other Jews on this issue?

The answer cannot be that the rejectionists are necessarily smarter (or dumber) than the others, nor are they necessarily more devout or more internally consistent in their religious orientation than the others. There is no evidence to support any of those conclusions. The answer may be much more basic. Rabbi Natan Slifkin has suggested that there is a social aspect to this phenomenon of some Jews rejecting generally accepted science, especially evolution: their personal identity is at stake.[19] The notion is that if "They" believe in evolution, "We" must not.

There is some academic support for Rabbi Slifkin's insight, evidence that human nature, in part, is "groupish."[20] In this view, a decision to reject science is less a decision on the merits than an unconscious tactic to foster (or at least maintain) cohesiveness within a group. It is less a fear of some slippery scientific slope, than an ethos of conformity which has been stretched beyond religious belief and behavior and has become a committed unwillingness to engage, to question. It is, in a Jewish context, an effort to maintain a smaller tribe within the larger one, and it operates by emphasizing a more particular outlook over a more universal one.

A corollary principle of tribalism is that its morality both "binds and blinds."[21] If this is so, then the denial of reality, in the form of evolution or otherwise, does not seem destined to be a successful model for a healthy long-term future. What may have worked well in confined communities of the Middle Ages is not likely to succeed in a more complex, interconnected, and fluid modernity.

A few things are sure:

18. See Cantor and Swetlitz, *Jewish Tradition and the Challenge*, 71–88.

19. See Slifkin, "Strange Reactions."

20. See Haidt, *Righteous Mind*, 221–23.

21. Haidt, *Righteous Mind*, 222, 364.

- Of the more than seven billion humans on Earth today, there are only about fourteen and a half million Jews. That is some two million fewer Jews than were alive in 1939, before the Shoah.[22] We Jews don't have minds to waste.
- The prophet Jeremiah was right: Those who have eyes and see not, those who have ears and hear not are foolish indeed, and without understanding (see Jer 5:21).
- Burying one's head in the sand in order to avoid seeing reality is a myth about ostrich behavior, and would be counterproductive for human beings as well. So, too, while a cultural fortress may be built on moral grounds, if the foundation is not set in hard science it will, in time, crumble.
- Claiming that biological evolution does not operate in the world and failing to encourage the serious studying of science is like placing a stumbling block before the blind. It creates a dangerous obstacle to a person's well-being. It inhibits the full development of his or her understanding of and ability to interact with the universe as it is and, further, with whatever Creative Force may exist and permeate it. And placing stumbling blocks before an individual or a community is forbidden (see Lev 19:14).

Let us hope that those feeling their way in the darkness will soon see the light.

22. See "The Continuing Decline of Europe's Jewish Population," http://www.pewresearch.org/fact-tank/2015/02/09/europes-jewish-population/.

Chapter 8

Biblical Numbers, Mathematics, and Attributed Patriarchal Ages

THE TORAH IS FILLED with numbers. There are different kinds of numbers—cardinals and ordinals, integers and fractions, even primes. And they are everywhere in the text. There are numbers for days and numbers for life spans; for populations and for the duration of events; for the measurement of quantities and for the size of objects and areas; and for a host of seemingly mundane things, such as the number of visitors and the number of palm trees.

Some biblical numbers are curiously round. For instance, Noah reportedly was five hundred years old when his son Shem was born, and six hundred when the great flood occurred. Shem was, then, one hundred years old at the time of the flood, and he died five hundred years later at the age of six hundred. Isaac married Rebekah at age forty, became a father of twins at age sixty, and was one hundred when Esau was married. He died at age 180.

Some biblical numbers are identical and strangely coincidental. The rain that created the flood for Noah lasted forty days and nights. That is the same number of days and nights that Moses reportedly spent on Mount Sinai on each of two occasions to receive the tablets engraved with the teachings and the commandments.

Other numbers appear hyperbolic and incredible. The Torah states that Jacob's descendants consisted of seventy individuals when the family entered Egypt. After centuries of involuntary servitude, the number of adult males that left Egypt with Moses is asserted to be about six hundred thousand. Including wives, concubines and children, the total number of those leaving must have been in the millions. Understood literally, the number seems absurdly large.

When we encounter numbers in the Torah, what are we to do with them? Are all or some of the numbers to be taken literally, or are one or more of them to be understood symbolically? And how can we tell which numbers fall into which category? Do we have a collection of essentially random numbers? Or, are there patterns that provide information, suggest meaning, or maybe reveal secrets?

Of all the number puzzles in the Torah, perhaps none is more intriguing than the longevity of the generations from Adam though Moses. Using a variety of approaches, scholars and others have long considered the numbers found in the Bible. They have speculated fancifully in an effort to make sense of some of them. With perhaps rare exceptions, though, these efforts have not been particularly satisfying, leaving the original problem, as the mathematician Lewis Carroll had Alice say in a different context, "curiouser and curiouser!"[1]

Let's look at the data first. According to the Torah, there were ten generations from Adam to and including Noah. There were another ten generations after Noah to and including Abraham. Coincidence?

The life spans recorded in the Torah for the first group are as follows: Adam (930), Seth (912), Enos (905), Cainan (910), Mahlaleel (895), Jared (962), Enoch (365), Methuselah (969), Lamech (777), and Noah (950). The life spans recorded for the second group, exclusive of Terah's son Abraham, are as follows: Shem (600), Arphaxad (438), Salah (433), Eber (468), Peleg (239), Reu (239), Serug (230), Nahor (148), and Terah (205).

Looking at the three primary patriarchs separately, Abraham, Isaac, and Jacob lived 175, 180, and 147 years respectively. Parenthetically, the only life span we have for a matriarch is that for Sarah, who died at age 127.

Finally, while at one point the Torah states that Jacob's descendants were in Egypt for 430 years, the book of Exodus records just four generations from Jacob to Moses. Jacob's son Levi lived 137 years, Levi's son Kohath lived 133 years, and Kohath's son Amram lived 137 years (cf. Exod 12:40–41; 6:16, 18, 20). Moses, who was Amram's son and, therefore,

1. See Carroll, *Annotated Alice*, 20.

Jacob's great-great grandson, lived 120 years (Deut 34:7).[2] What, if anything, do these twenty-eight reported ages tell us?

ATTEMPTS TO EXPLAIN BIBLICAL AGES VARY

Even a quick review suggests that the life spans of the first group of biblical humanity was reasonably consistent within a relatively narrow range of 895–969 years, with the notable exceptions of the seventh generation descendant Enoch and Noah's father, Lamech. As a general matter, the life spans of the second group, and through Moses, then continued on a downward slope. The first three individuals after Shem lived between 433 and 468 biblical years, and the next three lived in a reduced range of 230 to 239 biblical years. After Terah, no biblical personality is indicated to have lived in excess of two hundred years.

Rabbinic sages of the past accepted the reported ages as accurate. To explain the longevity of Adam, one noted that Adam was made in God's image and therefore physically perfect. Adam's immediate descendants were similarly vigorous. To explain the decline after the flood, one sage suggested that the Earth's atmosphere deteriorated. Another opined that the people faced a harsher climate after the post-Babel dispersion.[3] There is, unfortunately, no actual evidence to support such speculation, no way to test any of these propositions.

A more modern and purely mathematical analysis of the numbers by Charles A. Glatt Jr. concludes that the longevity of the individuals from Noah to Joseph (and then to Moses) declined in a manner consistent with base *e* exponential decay (where $e \approx 2.718$).[4] As we noted previously when discussing Dr. Schroeder's time dilation argument, base *e* is a rate of growth (or decay) that appears to exist commonly in nature in continuing processes. If Glatt's analysis is solid, then it would not only provide a naturalistic explanation for the reduction of the life spans studied, but might also impart some credence to the ages themselves.

Glatt uses the ages of fourteen individuals from Noah through Joseph as the raw data for his study. He claims that ten of the fourteen data points fall with one estimate of the standard deviation, and the other four come within two standard deviations. While Glatt concedes that his

2. See Plaut and Stein, *Torah*, 383n20.
3. See Scherman and Zlotowitz, *Chumash*, 25, 51n19.
4. See generally Glatt, "Patriarchal Life Span Exponential Decay."

equation does not work if extended to today, he does argue that it "does fit the data for the 1500 years from the flood to Moses." He reaches this conclusion by looking at the calculated lifespan for the time in which Moses lived (said by Glatt to be 2,580 years after Adam), which yields sixty-eight years, and he implies a consistency with the seventy-year lifespan asserted in Psalm 90, the psalm of Moses.[5]

Initially, there is a concern with Glatt and his analysis. His bona fides are not evident. He provides no background information so that his qualifications can be checked, and no contact information so that he may be called upon to respond to inquiries. His paper is not even dated. While the paper has been republished (in the *Creation Research Society Quarterly* Winter 2016 issue), it again came without any biographical information. Still, treating the analysis as it appears, is Glatt on to something, or is he, like Schroeder, merely trying to squeeze a result out of the data?

There are both general and specific questions to be asked here. As a general matter, why is decay analysis an appropriate form of investigation of population ages? Isn't this sort of exponential analysis more suited to radioactive decay, interest calculations, or the growth of crystals or populations? And if decay analysis is appropriate generally, are fourteen data points really enough information on which to base a serious study?

There are specific issues, as well. One major error appears to be in the time frame used by Glatt. He has Terah born in 1879 AC and Terah's son Abraham born in 2009 AC, when Terah would have been 130 years old.[6] According to Genesis 11:26, however, Terah was seventy years old at Abraham's birth, meaning that Abraham was born in 1948 AC. All subsequent dates used by Glatt, therefore, are erroneous as well.[7]

Two other problems relate to the predicted and the reported duration of Moses's life. First, and most simply, as noted before, according to the Torah, Moses lived to be 120 years old. Glatt makes no effort in his study to address that number or relate it to his study. Second, Glatt does not explain how he reached either the 2,580-year number for "after Adam to Moses" or the 1,500-year number "from the flood to Moses." Moreover, if Adam died in 930 AC (Gen 5:4), then according to Glatt Moses would have been alive in 3,510 AC, but if Moses were alive 1,500 years after the flood, which purportedly occurred in 1656 AC (see Gen 7:6),

5. Glatt, "Patriarchal Life Span Exponential Decay," 9.

6. Glatt, "Patriarchal Life Span Exponential Decay," 8.

7. For a traditional chronology and timeline, see Scherman and Zlotowitz, *Chumash*, 53.

then he was alive in 3,156 AC. Not only are those results inconsistent, if in most of 2019 CE the count of years after biblical creation comes to 5779 AC, then both of Glatt's timeframes place Moses much closer to the Common Era than seems justified.

Glatt's real focus, though, is from Noah to Joseph. Yet, even within that range, Glatt does not express any opinion as to why the numbers seem to fall in line with his equation. Is he simply noting or claiming implicitly that a natural phenomenon was at work. Or is he implicitly contending that supernatural activity was involved? Regardless of the answers to these questions, if one accepts the notion that base *e* decay was at work, how does Glatt explain why there are any deviations from the path predicted by the decay equation?

Finally, and despite the obvious decline in life spans, the raw numbers in the second group of biblical characters, from Shem through Abraham, continuing through Moses, still seem high. After all, the ages at the respective deaths of the fourteen identified individuals, Glatt's data points, never fall below 110. Other than an unsupported statement, that "a global longevity of around 100 years is historically recorded in Egypt around the time of Joseph," Glatt does not provide any information on life spans generally for populations three to five thousand years ago.[8]

Of course, accepting the accuracy of the given numbers, whether done so by rabbis centuries ago or mathematicians today, is problematic on two basic levels. First, it assumes the accuracy of the numbers given in the Torah. Young Earth creationists may do so, but science teaches that our species has populated the planet for much longer than six thousand years. Applying an exponential decay analysis to data derived from one limited, and presumed, time period without discussing the larger population which existed over a longer time frame is disingenuous at best. Moreover, the decay-analysis approach avoids, even precludes, the possibility that the numbers we read or hear mean something other than what they appear to mean.

In connection with the antediluvian life spans recorded in the Torah, the editors of *The Torah: A Modern Commentary*, published by the Union for Reform Judaism, refer to the Metonic Cycle, a relationship of the solar and lunar cycles bearing the name of an Athenian astronomer Meton in the fifth century BCE, but, apparently, recognized long before him. The Metonic Cycle is premised on the equivalence in time of 235

8. See Glatt, "Patriarchal Life Span Exponential Decay," 3.

lunar cycles and nineteen solar years. The editors then note that certain life spans of the ancient aged, like Methuselah's 969 years and Noah's 950, are divisible by 19. Under this approach, Methuselah lived fifty-one Metonic Cycles and Noah lived fifty.[9]

But what good is this information? If it is meant to suggest a formula by which we can devise the true age of these biblical individuals, then it does bring the numbers down to an area where we are more comfortable, but applying the formula creates other problems. Noah was five hundred biblical years old, or 26.3 Metonic Cycles, when he fathered Shem, but Methuselah was 187 biblical years old when his son Lamech was born. Was Methuselah just under ten in Metonic Cycles? What about Mahalaleel, who was sixty-five biblical years old when his son Jared was born? That comes to just over three Metonic Cycles.

Rabbi Umberto (Moshe David) Cassuto (1883–1951) was the director of the Rabbinic Seminary in Florence Italy and, starting in 1939, chair of Bible studies at Hebrew University in Jerusalem. Cassuto has argued generally that it is not possible to comprehend Torah without reference to Israel's neighbors. More specifically, he has contended that the ancient chronologies, and the use of numbers generally in the Torah, cannot be understood without reference to the sexagesimal system which prevailed in ancient Mesopotamia, often supplemented with the use of the number seven or some multiple of it.[10]

The sexagesimal system placed great emphasis on the number six and extensions of that number, including sixty, six hundred, six thousand, sixty thousand, and six hundred thousand. Fractions and multiplications of those numbers were also important. So, if sixty was a positive value, then two times sixty, or 120, was even more special.

The numbers five and seven were also important. Five years was the equivalent of sixty months. The number five factored—i.e., the product of five times four times three times two times one—turns out to be 120. The number seven was considered an ideal or perfect number, one that was a sign of totality or completeness.[11] The number seven appears by itself hundreds of times in a wide variety of contexts in the Torah, and as

9. See Plaut and Stein, *Torah*, 42.

10. See generally Cassuto, *Commentary on Genesis Part One*, 1, 7, 192–93, 258–62; *Commentary on Genesis Part Two*, 255, 257–62.

11. See Cassuto, *Commentary on Genesis Part One*, 12–17, 191–92, 243, 256, 258–62, 276; see also Gen 2:1–3.

an additional element or factor, such as seventeen or seventy-seven, in scores more.

For Cassuto, then, the years listed in Genesis for the first ten generations, with two special and explainable exceptions, can be divided into two groups, one where the ages end in five or a multiple of five and another where seven has been added to such a number. As he notes, the reported life spans of six of the first ten individuals identified from Adam through Noah end either in five or a multiple of five, which yields an end number of zero. Two more ages fit the pattern if seven is added. Methuselah's age of 969 requires the addition of two times seven to conform, perhaps further underscoring the exceptional importance of his age.

The two individuals whose ages do not fit the main pattern are, to Cassuto, the proverbial exceptions that prove the rule. Methuselah's father, Enoch, lived only 365 years. The age 365 can be viewed as special in at least three different ways. It may refer to the number of days in a solar year. It may be seen as special under the sexagesimal system, being the total of six times sixty plus five. It is also the sum of 10^2 plus 11^2 plus 12^2. We do not know what the number denotes in this particular instance, but it does not seem merely coincidental that this seventh-generation person was singled out for special treatment. Lamech's age of 777, as described in the Torah (Gen 5:31), is seven and seventy years and seven hundred years, an obvious emphasis on the ideal number seven, and again more likely symbolic than factual.

The ages of the second group of individuals, from Shem through Abraham, do not fall so neatly into Cassuto's system. At six hundred, Shem does, and so does Terah at 205, but the others do not. Perhaps, as Sherlock Holmes found with respect to the dog in the short story "The Adventure of Silver Blaze," the telling clue is in the absence, rather than the presence, of something expected. Here, perhaps, the lack of numbers ending in a five or a zero signals that the member of this group are not as special or deserving as the original group of ten.

THE ATTRIBUTED AGES OF THE PRIMARY PATRIARCHS

What happens when we reach the ages of the three primary patriarchs? As noted, the life spans recorded for Abraham, Isaac, and Jacob are 175, 180, and 147, respectively. Each number falls neatly within Cassuto's analysis. The numbers 175 and 180 end in five and zero, respectively, and 147 reflects an addition of seven to the product of two times seventy (itself the product of special numbers seven and ten, each considered a number of ordinal perfection).

Not incidentally, the approach works for the matriarch Sarah, as well. Torah tells us that Sarah died at age 127, which is two times sixty (a good sexagesimal number) plus seven. If 120 was an honorific number, then 127 was even more so.

The ages of the three primary patriarchs are, however, also part of a sequence, and a unique one at that. As illustrated by the late Nahum Sarna (1923–2005), a professor of biblical studies at Brandeis University, each number in the sequence is the product of a number squared multiplied by another number. And the numbers involved follow in two clear descending and ascending orders.[12] So:

$$175 = 7 \times 5^2$$
$$180 = 5 \times 6^2$$
$$147 = 3 \times 7^2$$

That formula can be extended in two directions, resulting in 144 for nine times four squared and sixty-four for one times eight squared. No patriarch or other biblical personage is assigned either of the extended numbers, however, underscoring how remarkable the ages attributed to the primary patriarchs are.

In his marvelous review of Genesis, then University of Chicago professor Leon Kass added that the sum of the three equations is identical. That is, five plus five plus seven is the same as six plus six plus five, which also equals seven plus seven plus three. Each total is seventeen. As to what possible significance this could be, Kass says, however: "I have no idea."[13]

And there lies the rub. We are so removed from the mathematical considerations that even when we think we can identify some of the signals, we cannot determine where they point. The number seventeen

12. See Sarna, *Understanding Genesis*, 84.

13. See Kass, *Beginning of Wisdom*, 629n. The meaning of 120, the age attributed to Moses, is also puzzling. The number 120 can be factored, as noted, but it could also be simply sixty, an important sexagesimal number, doubled. Perhaps it is the sum of the idealized age span in ancient Egypt of 110, an age attributed to Joseph and Joshua, supplemented by a perfect number (ten); in other words, more than an ideal age. We just don't know.

is, of course, a prime number—that is, a natural number greater than one that is not divisible by any number other than itself and one. Moreover, it is the seventh prime number after two, three, five, seven, eleven, and thirteen. It is also the sum of seven and ten, again, numbers of ordinal perfection.

The number seventeen occurs in interesting and varied contexts. For instance, after he was born, Joseph lived with his father Jacob for seventeen years until he was sold to the Ishmaelites or, alternatively, the Midianites. Later, Jacob lived with Joseph in Egypt for seventeen years until he died (see Gen 37:2, 25–28; 47:28). Further, the life span of Levi, another of Jacob's son, was 137, which is the sum of 120 and seventeen.

ATTRIBUTED AGES WERE NEITHER RANDOM NOR FANCIFUL

But what, if anything, is there beyond coincidences and parallelisms? We do not know for sure.[14] What we do know, though, and what is important to know, is that the attributed ages of the primary patriarchs are neither random nor mere numbers located as data points on a line. Instead, whether products or sums, they are numbers which suggest an intent by the author or editor of the text to convey something important about the three primary patriarchs, be it their distinctiveness from all others or their relationship to each other.

A few final points need be made. First, this discussion has been limited to numbers that appear in the Masoretic text. There are some different numbers that appear in the Greek Septuagint and the Samaritan Pentateuch.[15] Second, Cassuto was not a disinterested academic. In addition to desiring to show what the text of the Torah meant to those who first heard it, he was interested in demonstrating that the text of Genesis was part of a unified whole, rather than a collection of writings of different authors as postulated by the documentary hypothesis we have used previously to understand better the story of Noah and the flood.[16] Nevertheless, his insights cannot be denied.

14. Kvasnica, "Ages of the Antediluvian Patriarchs," 4, 6.

15. See Cassuto, *Commentary on Genesis Part One*, 264–65; *Commentary on Genesis Part Two*, 258–59.

16. See, e.g., Cassuto, *Commentary on Genesis Part One*, 94, 193.

Sarna concludes that "the biblical chronologies of the patriarchal age are not intended to be accurate historical records in our sense of the term." Rather, they "fall within the scope of historiosophy, or philosophy of history, rather than historiography."[17] In this view, they tell us less about the duration of the lives of real people than about what was important at the time the text was developed. In short, numerical symmetry or harmony is not a matter of coincidence among random events, but a signal of importance, and a sign of presumed divine control and direction.[18]

Lastly, the writers and redactors of the *Tanakh*, and especially the Torah, were more than drafters of national history, recorders or developers of laws and mores, and masterful storytellers. They also had a real competency with mathematics. We do not yet understand the signals, or the extent to which they might have been perceived as theologically purposeful or more mundane narrative drivers or both, but the signals are clearly there, mean something, and deserve serious and further consideration.

17. See Sarna, *Understanding Genesis*, 84.

18. See Sarna, *Understanding Genesis*, 84–85.

Chapter 9

The Myth and Function of the Passover Plagues

PASSOVER IS A WONDERFUL holiday. It is a time to gather together with family and friends. It provides an opportunity to reconnect with the millennia-old line of the Jewish people. On Passover, we reach back through the mists of time to the myths of our national origin, those ancient stories that sought to explain the nature of things. We seek to find lessons from the distant past which might guide us in our present.

The highlight of the festival is the reading of a story from the *Haggadah*, literally meaning "the story." The story tells of the enslavement of ancient Israelites in the land of Egypt and their release from bondage following a series of ten calamities, commonly understood as plagues, which devastated Egypt. Those plagues, in the order of the story in the book of Exodus, are blood, frogs, lice, insects, pestilence, boils, hail, locusts, darkness, and the death of the Egyptian firstborn (see Exod 7:14—12:30).

Today that core story, and its centuries of embellishments, is read, sung, and discussed throughout the Passover *seder* (a ritual meal, literally "order"). All along the way we are requested, challenged, even required to ask questions, to probe into the meaning of the story. The whole exercise is quite dramatic, sometimes even including costumes and choreography. No wonder Passover is an incredibly popular Jewish holiday, with more Jews participating in a *seder* than fasting on the traditional holiest of holy days, Yom Kippur.[1]

1. See Pew Research Center, "Religious Beliefs and Practices," para. 2.

The Passover story is so powerful that its magic has not been dimmed by the increasing recognition that the premise of the story lacks a solid historical foundation. The Torah states that six hundred thousand Israelite males, along with women, children, and others, left Egypt as part of a national exodus (Exod 12:37). According to the traditional timetable, this mass migration occurred around 1300 BCE.[2] That idea, however, has been largely discredited.

First, there is no evidence to date of any mass slavery of ancient Israelites during the relevant time period. Second, consider the nature of the reported biblical caravan. According to Nahum Sarna, the departure of six hundred thousand adult males "conservatively" implies that a group of "well over" two million individuals would have come out of Egypt.[3] If a population of that size marched twenty abreast, there would have been more than one hundred thousand rows of participants, exclusive of animals, carts, and other items. If those rows were separated by just ten feet, the entire entourage would have, by application of simple mathematics, extended for around 190 miles. Aside from the problems that result raises with the sea crossing tale, there is no evidence that any movement of a population of that magnitude ever occurred into the Sinai Peninsula and up to the east bank of the Jordan River. Third, there is no evidence of any new settlement patterns established west of the Jordan by a substantial influx of millions of new immigrants in the thirteenth century BCE. If the narrative were intended to be history as we moderns understand it, that is, a reasonably accurate statement and chronology of actual events, the story fails.

The view that the Torah is unhistorical is not limited to scientists and skeptics. In 2001, for instance, during Passover week, Conservative Rabbi and prolific author David Wolpe set off a firestorm when he spoke to his Los Angeles congregation about the lack of hard evidence for the exodus story. According to a writer for the Los Angeles Times, after reviewing revolutionary discoveries in then current archaeology, Rabbi Wolpe told them: "The truth is that virtually every modern archeologist who has investigated the story of the exodus, with very few exceptions, agrees that the way the Bible describes the exodus is not the way it happened, if it happened at all."[4] Rabbi Wolpe's disclosure, which met a heated visceral

2. See Isaacs, "Exodus from Egypt."

3. See Sarna, *Exploring Exodus*, 95.

4. See Watanabe, "Doubting the Exodus Story," para. 6. For a subsequent summary of Wolpe's thinking, see Wolpe, "Did the Exodus Really Happen?"

response, was preceded a few years earlier by a fuller and more detailed exploration of the unhistorical nature of the Torah by Rabbi S. David Sperling, professor of Bible and faculty chair at Hebrew Union College: Jewish Institute of Religion. Sperling concludes that many aspects of the exodus story are allegorical. For example, he sees the claim of servitude in Egypt as a portrayal of Israel's "servitude *to* Egypt" during Egypt's rule of Canaan.[5]

Now, if there were no mass enslavement of Israelites and no mass exodus, then surely there would not have been any need for liberating plagues either. Some still maintain, though, that there is significant evidence for the biblical plagues outside of the biblical text. One such advocate is Israeli Egyptologist Galit Dayan, who cites as proof of the biblical plagues an ancient Egyptian document known formally as the Admonitions of Ipuwer. The Ipuwer papyrus describes a time of considerable social and political chaos in Egypt. Dayan translates the hieroglyphs as follows: "Plague is throughout the land . . . the river is blood . . . and the hail smote every herd of the field . . . there is a thick darkness throughout the land . . . the Lord smote all of the firstborn in the land of Egypt (including) the first born of Pharaoh that sat on his throne."[6]

There are, however, a number of serious problems with the claim that the Ipuwer papyrus is evidence of the biblical plagues. One is that the Ipuwer papyrus contains a longer and more complex story than Dayan implies, and her list of events similar to certain biblical plagues amounts to a cherry-picking of like situations, while failing to explain the absence in the Ipuwer papyrus of other biblical plagues like lice, insects, and locusts.[7] Moreover, the ordeals Ipuwer describes are not seen as coming from a powerful god acting on behalf of his people, but as the result of the ineptitude of an unnamed king. The social dynamics of Ipuwer's story are also directly contrary to those in the biblical tale. Ipuwer's story concerned the immigration of foreigners into Egypt, not the emigration of slaves from it. Perhaps most importantly, while there is a debate among Egyptologists regarding the dating of the events related in the papyrus, with some setting the story in the First Intermediate Period (c. 2200–2050 BCE) and others in the late Middle Kingdom (c. 2050–1700 BCE), plus

5. See Sperling, *Original Torah*, 54 (emphasis original).
6. See Berrin, "Passover Proof," para. 4.
7. See generally "Admonitions of Ipuwer."

or minus,[8] both of those dates are centuries before the thirteenth century date traditionally assigned to the exodus.

The Ipuwer papyrus also has an extrabiblical competitor. Israeli-born producer, director, and writer Simcha Jacobovici argues that a 3,500-year-old Egyptian monument known as the Tempest or Storm Stela provides archeological evidence for the exodus. He contends that a new translation of the stela proves that a massive eruption of a volcano on the Greek island of Santorini generated a storm which flooded Egyptian temples and plunged Egypt into darkness for days. As in the biblical plague story, loud voices were heard, and the Egyptians were seized with terror (see Exod 9:29; 15:14). Jacobovici claims that the stela proves that the Pharaoh at the time, Ahmose (r. 1550–1525 BCE), the storm, and the contemporaneous expulsion of certain Asiatics known as the Hyksos are the basis of the exodus story.[9]

Jacobovici's argument displays the same defects as the claim based on the Ipuwer's papyrus. While there clearly was a massive volcanic eruption on Santorni, which scientists date to between 1645 and 1600 BCE,[10] and while that event may even have had some impact more than 450 miles away in Egypt, the explosion occurred at least half a century before Ahmose's reign. Timing aside, there is no claim, much less any proof, that the volcanic eruption generated a series of plagues in Egypt as related in the exodus story. Finally, no convincing explanation is offered to fill a long historical gap and connect the expulsion of some (but not all) Hyskos in the sixteenth century BCE with the existence of a recognizable Israelite community in the late thirteenth century BCE and a kingdom in the tenth century BCE.[11]

Not only are the attempts to establish the historicity of the Egyptian plagues wanting for lack of hard proof, there is also no basis for the initial assumption that the Passover story generally and the plagues specifically were even intended to be taken literally, to be historical statements, as we moderns understand that concept. To the contrary, both the text of the Torah we have today and other references in the Hebrew Bible strongly suggest otherwise.

8. See, e.g., Sarna, *Exploring Exodus*, 69.

9. See Jacobovici, "Proof for the Biblical Exodus!"

10. See "Santorini Volcano Eruption Date."

11. The Merneptah Stele, a stone slab of Egyptian origin which has been dated to late in the thirteenth century BCE, identifies Israel as a nation or city-state, albeit a defeated one. See "Merneptah Stele."

Exclusive of the recitation of ten plagues in the book of Exodus, plagues in Egypt are discussed two separate times in the Hebrew Bible, both in Psalms (see Pss 78:42–51; 105:28–36). In both of the recitations in Psalms, there are only seven plagues, though, and the seven are neither the same in both lists, nor is their order the same. What does this mean?

The presence in the *Tanakh* of these different accounts is actually quite instructive. First, it indicates that when elements of the text were being collected and collated, the editors were familiar with more than one tradition respecting plagues. This is no different, and therefore no more surprising, than the retention in the Torah of alternative traditions concerning such matters as creation, the flood, the Ten Commandments, and the spies, to name a few instances where different renditions of traditional stories have been maintained.

The larger story, as found in the book of Exodus, itself appears to be an edited and conflated version of several traditions. Referencing the classic biblical source criticism we have encountered before, Professor Hayes teaches that each of the primary biblical sources known as J, E, and P supplied some, but not all, of the ten plagues. Specifically, source critics claim that J provided eight plagues, while E supplied three, and P is the source of five, but there are some overlaps.[12] Significantly, Hayes does not identify D as a source for any of the plagues. In discussing the exodus in Deuteronomy, Moses merely obliquely references "trials, signs, portents and deeds-of-war," and fails to mention any specific plagues at all, save perhaps boils and locusts (or the "buzzing-cricket"; see Deut 4:34; 28:27, 38, 42).

Literary analysis of the plagues lists is also instructive. Each list in Exodus and in Psalms was written as if complete, signaled by either seven or ten components. As we have seen when considering biblical ages, in the world of biblical symbolism, those numbers indicate wholeness and perfection.[13] Further, the more extensive narrative in Exodus is structured carefully, not only as three series of three plagues each, with a stunning climax, but also including within each series repeated patterns of and phrasing for elements of the story.

In short, the theme of plagues seems to have been common during the extended time the Hebrew Bible was being formulated, but the details of the story were quite fluid. There can be little doubt, then, that

12. See Hayes, *Introduction to the Bible*, 111.

13. See also Sarna, *Exploring Exodus*, 74–75.

the story of the plagues in the Torah we have received today is a product of craftsmanship rather than reporting.

But all this begs a critical question: why include a plague story at all in the larger Exodus drama? If the authors merely wanted to convey a spectacle of the majesty and triumph of the Israelite God, they could have invoked images of God splitting the Nile, a feat more difficult than simply turning it red, as even Egyptian magicians could do (see Exod 7:22). They could have had God appear as alternate pillars of cloud and fire, as later claimed during the trek though the wilderness (see Exod 13:21). Or God could have created an oasis, a tiny Eden, or rained down quail and manna instead of hail and locusts (see Exod 16:13–15), not only to demonstrate creative and fulsome power, but to illustrate the rewards that Egypt could earn through conciliation with the Israelites. That is, the story could have offered divine carrots instead of sticks.

Clearly the purpose of this carefully designed and structured composition was not meant merely to demonstrate either awesome supernatural power generally or the control of nature specifically. It certainly was not meant to induce behavior with compassion and beneficence. Rather, the purpose of invoking plagues seems to have been an exceptionally clever use of a story that was itself dramatic and had some broad acceptability in the popular culture in order to advance a theology at least of monolatry, if not monotheism.[14]

As the Torah text explicitly states, the plagues were selected to defeat and humiliate the gods and symbols of imperial Egypt. They were aimed "*uvechol elohey Mitzrayim*," that is, at "all the gods of Egypt" (Exod 12:12). This purpose is confirmed later in the Torah. Describing the day after the first Passover, the text claims success for the onslaught: "on their gods, YHWH had rendered judgment" (Num 33:4).[15]

Who were these gods? The ancient Egyptians had many deities, and the names and roles changed over time. But it is possible to construct a list of the plausible targets of the biblical authors.

The attack begins with the lifeline of Egypt, the Nile River (Exod 7:19). Turning the river to blood would cripple all agriculture and commerce which depended on the river, which is to say most of the Egyptian economy. For the biblical authors, it also represented a multipronged

14. While monotheism is the belief that only one god exists, monolatry is the worship of one god in a recognized pantheon of gods.

15. Friedman, *Torah Commentary*, 537.

assault: the defeat of Hapi, the guardian the Nile; of Khnum, the god of the inundation of the Nile; and of Osiris, god of the underworld, for whom the Nile served as his bloodstream. The second plague was directed to a god symbolized by a frog, that is, Heqet, the goddess of fertility. The Egyptian god of the earth was Geb. Turning the dust of the earth into lice (or perhaps fleas) showed his impotence.

The war on the Egyptian pantheon continues in the second series of plagues. The definition of the fourth plague, *arov*, is uncertain. It suggests a swarm or horde of insects, often understood as flies. But the text also says that the swarm would fill not only the Egyptian houses, but the land under them (see Exod 8:17). Quite possibly the reference is to the scarab or dung beetle, as one of the most prominent Egyptian insect gods, Khepri, was depicted with the head of a scarab. Striking cattle with disease surely would have embarrassed any one of several Egyptian gods represented by animal heads, such as Apis, the bull, and Hathor, the goddess of the desert and symbolic mother of the Pharaoh. Similarly, the spread of boils illustrated the impotence of Imhotep, the god of medicine and healing.

In the third series, the rain of very heavy hail would demonstrate the weakness of Nut, the sky goddess, and mother of other prominent deities. The swarm of locusts that ate everything apparently could not be stopped by any of the agricultural gods and goddesses like Renenutet, the goddess of the harvest, or her son Neper, the grain god, or by the god of wind and chaos, Seth. The final plague in the series, that of a thick multiday darkness in all the land of Egypt, was surely an act of war (and a successful one at that) on the supreme sun god known as Ra (or Re) or Horus, and often depicted with a man's body and the head of a falcon.[16]

The import of the story so far, then, was that the gods of Egypt were incapable of protecting their respective domains, and that Pharaoh could not protect his subjects. With the final, and most devastating plague, that of the death of Egypt's firstborn males at midnight, we learn that Pharaoh could not even protect his own household or the system of primogeniture on which Egyptian law was based. Neither Renenutet, the guardian of Pharaoh, nor Selket, the guardian of protection and healing, were of any use.

As Professor Rendsburg teaches, modern readers of the Hebrew Bible, unfamiliar with the authors' society and the cultural clues contained

16. See generally Rendsburg, "YHWH's War."

in the text, will "often miss many of the nuances that make the stories so fresh and loaded with meaning."[17] That is true, of course, and important. Still, we are left with critical questions. Why was any account of plagues ultimately included in the Torah? What function did it serve? What did the final redactors want their immediate audience to learn?

Unfortunately, we cannot say with precision when particular stories were first written or when they were incorporated into the canon. Much work appears to have been done in the eighth through the sixth centuries BCE, with final revisions coming during and after the Babylonian exile. Arguably, given the inconsistencies in the biblical books of Ezra and Nehemiah about something as seemingly basic as a fall holiday, one could argue, as does University of Michigan scholar Lisbeth Fried, that the text was not even agreed and set by the end of the Persian Period, which lasted from about 639 to 332 BCE.[18] Obviously, this would stretch canonization over quite an extended time.

Moreover, this was a time of considerable turmoil, politically and theologically. Egypt's influence over the children of Israel was not as strong during this period as it once was. Rather, other powers flexed their muscles. The Assyrians crushed the Northern Kingdom of Israel around 720 BCE, and the Kingdom of Judah having barely survived a subsequent assault existed at the sufferance of the Assyrians. The Assyrians, in turn, fell to King Nebuchadnezzer and the Babylonians. Then, following a series of invasions at the beginning of the sixth century BCE, the complete destruction of the Judahite capital, Jerusalem, and the transfer of Judahite royalty and leadership to Babylon, Judah became a vassal state of Babylon. Babylon fell to the Persian King Cyrus about sixty years later. While Cyrus allowed Judahites to return home, to territory then known as Yehud, it was now a small province in the Persian Empire and ultimately subject to Persian control. Toward the end of the fourth century BCE, Persia, and with it Yehud, fell to Alexander and the Greeks.

In the midst of this lengthy geopolitical war, a multi-faceted religious battle continued as well. With the advent of the reform prophets in the eighth century BCE, polytheism came under increased attack from both those who favored either the supremacy of YHWH over lesser gods and those who recognized YHWH as the sole god, and those camps contended with each other. The monotheistic view seems to have gained

17. Rendsburg, "Reading the Plagues," 3.

18. See Fried, "Sukkot in Ezra-Nehemiah," 6.

ascendance in the seventh century with the rise of the Deuteronomistic school, but fate, in the form of the death of King Josiah of Judah and the ascendancy of King Nebuchadnezzer of Babylon, intervened.[19]

The destruction of Jerusalem could have ended this nascent monotheism. After all, if YHWH were the sole true god, he certainly did not protect his treasured people, or his promised land, or even his house, his temple, from ruin. Ironically, though, far from ending monotheism because of the impotence of the Deity, the exile from and return to Judah was understood by Judahite leadership differently. Influenced by the Deuteronomists, they argued that the people failed YHWH, not the other way around. The solution of the surviving and returning leaders, like Ezra, was a stronger commitment to what they saw as the one true God.

This was the broad context in which the contents of the Torah seem to have been finalized. And, if so, this context helps us understand how the plague story may have functioned at that time. Note that had the leaders of the community wished to restrict access to the text, they could have done so. In that case, a member of the educated class, with access and attuned to cultural cues, still might well have recognized the Egyptian motifs referenced in the story of the plagues. If he did, then he would also know that the story was not an eyewitness account of events, but a symbolic war between the then dominant Israelite god, YHWH, and the gods of Egypt, headed by the sun god Ra. But the substance of the plague story was designed, like that of many of the stories and regulations that surround it, to be told widely, to become part of accepted lore. And so the process for its dissemination was open and multifaceted. The *Tanakh* contains several models for the reading of and discussion about this story, and they all involve the public, broadly defined, regardless of gender or age.[20]

So, rather than merely directed to a sophisticated reader in Judah, or even in the established community of Judahite emigres in Egypt (who would understand the references to the Egyptian pantheon), the story of the plagues may well have been intended to underscore for postexilic Judahites who had returned, or were thinking of returning, that the worship of false gods of any kind, whether of Canaanite, Babylonian, Persian, or other origin, was improper and destructive. That is, beyond the explicit

19. See Hayes, *Introduction to the Bible*, 230–35.

20. See generally Demsky, "Historical *Hakhel* Ceremonies."

message, lay an implicit lesson: just as the gods of Egypt were no match for the Israelite God, neither are any of the current local gods.

In a time that required nation-building, the story served, then, to provide a unifying theological feature within the larger text which functioned as a unifying statement of a people's creation and history and a unifying anthology of its traditions. And the public readings of this larger text, were, in turn, reinforced by the insistence on diligently teaching the children about these "instructions," talking about them at home and while walking on one's way, upon lying down and rising up, and even posting the teachings on one's doorposts and gates (see Deut 6:7–9).

Myth-based Judaism has generated compelling stories and, at its best, a compassionate culture. The story of the Israelites' escape from bondage serves as a beacon to all who are oppressed and as a prompt to those of us fortunate enough to be free to imagine what slavery might have been like, how it must have felt to have been a stranger in a strange land. During the *seder*, as the plagues are mentioned, we remove some wine from our cups to diminish the sweetness of the Israelite's escape, to recognize the suffering of others, and to temper our joy. These are worthy lessons, and there are others, too.

But myth-based Judaism has its limits, and pretending that myths are reality is not only intellectually indefensible, it can be counter-productive, even self-destructive, as well. If we take Bible stories as statements of historical truth when they are not, and if we purposefully avoid trying to understand the authors' agendas and what the authors intended their audience to learn, we act as nothing other than illiterate literalists.

When Judaism meets science on Passover, reality-based Judaism acknowledges that neither the exodus nor the plagues occurred as depicted, that the plagues are a myth within a larger myth, set in a time when humankind often identified each aspect of nature, of life itself, with a separate god. But reality-based Judaism also rejects the nihilism of some superficial contemporary readers who may erroneously conclude that such an acknowledgement means that the Jewish freedom narrative lacks not only foundation, but merit. Instead, reality-based Judaism accepts the challenge of Passover to dig deeply into the tale, to ask a question, and then another and yet another. It seeks to come to grips with both the original intent and redeeming transcendent value of the story, to wrestle with the text, and extract both truth and wisdom from this powerful story. When we do so, when we engage the broader myth, and struggle with the more troubling one contained in it, we may well recognize that the authors had matured enough to grasp how flimsy is a foundation premised on false gods. That alone is a lesson well worth learning.

Chapter 10

The Camel's Nose in the Torah's Tent

The Torah is the foundational text of the Jewish people. Initially, it asserts a prehistory of and a purpose for the ancient Judahite kingdom to which contemporary Jews trace their emotional and often actual genetic origin, incorporating the kingdom's legends and lore, its poetry and prose, its customs and commitments.

For those who hold that the Torah is also the word of God, without flaw and inerrant, the last few hundred years have been very frustrating. The development of the documentary hypothesis, the idea that the Torah was a compilation of works from several discrete sources, was and—despite scholarly challenge—remains a formidable obstacle to the claim of unitary and divine authorship. But the documentary hypothesis is, for all its power and value, just that: a hypothesis. Similarly, an alternative idea that much of the Torah text is pretext—i.e., a series of allegories designed to enhance the image of one or more Kings of Judah—is another provocative and persuasive concept, but again, just a concept.

What if we were to dismiss such broad theories as too sweeping and not definitive? We would still be confronted with textual problems, which, though small, are both numerous and stubborn. Nor can they be easily refuted or disregarded. One sign that the Torah is not the work of a single writer, much less a divine one, is the presence of anachronisms in the text.

An anachronism is a word or reference that is temporally out of place. It may be a person who is named, but was not yet born at the time in which his identification was set. Or, it may be a location or thing or event which is mentioned, but which did not exist or had not occurred when the story was placed. In such instances, the presence of the word both counters the claim of inerrancy and, conversely, helps to show when and where the passage in question may really have been drafted. For instance, if the Torah had said that Moses turned on electric lights the night before the exodus from Egypt so that he could review a map of his escape route, we would know that the text was flawed because electric lights were not invented until about thirty-one centuries after Moses supposedly lived. Moreover, the reference would help place the writing of the passage to a time in or after the nineteenth century of the Common Era when electric lighting became a reality.

Consequently, in order to determine whether a text actually includes an anachronism, you need to know at least two things. The first is the time in which the story is set. The other is the time when the person, place, or thing mentioned first existed or occurred.

Sometimes, an anachronism is obvious from the text itself. For instance, in Genesis 34:7, we read that Shechem committed a "disgrace in Israel" by lying forcibly with Jacob's daughter, Dinah. The narrative, however, has not yet identified any people known as Israel. There was no nation, nor any group, by that name around at the point in the Genesis saga to be disgraced (cf. Deut 22:21). Similarly, in Exodus 19:22 and 24, we read that the priests must stay pure. But the priesthood had not yet been established, and would not be until after the revelation of Sinai and the subsequent consecration of Aaron and his sons described in chapters 28 and 29 of the book of Exodus. In each of the foregoing instances, the author or editor seems to have made reference to a circumstance that his audience would have understood—i.e., Israel, priesthood. But each reference was also internally inconsistent with the chronology of the story.

Sometimes, discovering an anachronism requires knowledge outside of the text at issue. The book of Genesis, for instance, is set in parts of what today we call the Middle East, a wide swath of territory which does not have precise borders, but conventionally extends from present day Egypt and Israel east to Iran, from modern Turkey south to Yemen and Oman. After the initial primeval stories of creation, wickedness, flood, hubris, and Babel, the main story begins to unfold, focusing on the journey of Abraham, his family, and his descendants. Tracing the dates for Abraham and his family in a traditional manner solely from the biblical text itself, we see that Abraham lived between 1948 and 2123 AC (After Creation), his son Isaac lived between 2048 and 2228 AC, his grandson

Jacob lived between 2108 and 2255 AC, and his grandson Joseph lived between 2199 to 2309 AC.[1] Assuming solely for the purpose of this exercise that 5779, the Jewish year in which this book was published, is the actual number of years since creation (and we know that it is not), then Abraham lived between 1812 and 1637 BCE, Isaac lived between 1712 and 1532 BCE, Jacob lived between 1652 and 1505 BCE, and Joseph lived between 1561 and 1451 BCE. From an archeological viewpoint, the patriarchs lived in the Middle Bronze Age, and Joseph in the Late Bronze Age, all well before the start of the Iron Age, which is conventionally dated to 1200 BCE.

At Genesis 47:1–3, we read about Joseph introducing his father and brothers to an unnamed pharaoh. The brothers request permission to stay in the Nile Delta area known as Goshen. Pharaoh grants their wish, and allows the family to settle in "the goodliest part of the land," in the "region of Goshen" (Gen 47:6). The story concludes with a note that Joseph settled his father and brothers "in the goodliest part of the land of Egypt, in the region of Ra'meses" (Gen 47:11). The problems here are twofold. First, according to Nahum Sarna, the reign of Rameses the Great did not begin until about 1290 BCE. It lasted until about 1224 BCE.[2] Consequently, the area at issue was not named for Rameses until the thirteenth century BCE or subsequently, but at least two hundred years after the initial settlement of Jacob's family according to Genesis. Moreover, the name Goshen may be related to an Arabic tribe whose domination of the area did not occur prior to the sixth or fifth centuries BCE, or almost a millennium later than Joseph.[3]

At Genesis 26:1, we read that at a time when famine forced him to move, Isaac traveled to the king of the Philistines. The story seems perfectly reasonable, until one realizes that the Philistines, as part of the Sea Peoples migration, did not arrive in Canaan until after 1200 BCE, centuries after Isaac died.[4]

There's more.

At Genesis 11:28, we are told that Haran, brother of Abram (as he was then named), died in his native land, called Ur of the Chaldeans. Ur, located in what is now Iraq, was an ancient city, once the capital of

1. See Scherman and Zlotowitz, *Chumash*, 53.
2. See Sarna, *Exploring Exodus*, 10.
3. See Finkelstein and Silberman, *Bible Unearthed*, 67.
4. See Sarna, *Exploring Exodus*, 106.

Sumer. But the Chaldean Empire existed only relatively briefly. It began with the ascent of King Nobanidus and ended with the Chaldeans being conquered by Cyrus, essentially from about 626 to 539 BCE.[5] That is, there were no Chaldeans in control of the territory around Ur until the late seventh century BCE, and they were gone by the late sixth century BCE, perhaps a thousand years or more after the reported death of Haran. The author of the text could not have inserted the reference prior to the rise of Nobanidus.

Similarly, after the death of the matriarch Sarah, the story relates that Abraham sought an appropriate burial site for his wife (see Gen 23:7–18). Instead of simply saying that Abraham purchased a cave near Hebron, the text describes in considerable detail the negotiations, and even the choreography of the negotiations, for the transaction. The entire drama, including the honorific language exchanged between principals, the opening bid, and appropriate bowing, reminds some of a type of negotiation evidenced by a much later neo-Babylonian "Dialogue Document." While the conclusion of the transaction, Abraham's successful acquisition of real property in exchange for valuable consideration (four hundred shekels of silver) was important to the Israelite claim to and relationship with the land, and indeed remains important to this day, what is important for present purposes is that the first evidence for the dialogue format does not appear prior to the eighth century BCE.[6]

One more example: in chapter 28 of Exodus, the Torah includes a detailed discussion of the vestments that are to be made for and worn by Aaron and his sons in their capacity as priests. After the robe, tunic, breastplate, sash, and other items are described, verse 42 states, "You are to make them breeches of linen to cover the flesh of nakedness; from the hips to the thighs shall they extend." These trousers or undergarments were to be worn as the priests entered the Tent of Meeting or approached the altar. Bible professor Rabbi S. David Sperling has demonstrated, however, that trousers were invented by the Persians around the sixth century BCE. The sartorial direction at Exod 28:42 could not, therefore, have been written prior to then, certainly not during any fourteenth- or thirteenth-century BCE Exodus.[7]

5. See "Chaldean Empire"; see also Finkelstein and Silberman, *Bible Unearthed*, 311.

6. See Plaut and Stein, *Torah*, 155n15.

7. See Sperling, *Original Torah*, 116.

In short, there are a variety of anachronisms in the text of the Torah which indicate, first, that the author of those passages lived long after the time in which his story was set and, second, that he retrojected commonly understood circumstances back into an era that had no connection to them. Why he did that is another topic, but the fact that he did cannot be disputed credibly.

Moreover, at least some passages of the Torah can be no older than the sixth century BCE. That is, not only were they not written at Mount Sinai just after the exodus, they were not written prior to the alleged entry from the wilderness into Canaan. Indeed, they were not written before the time of Joshua, Judges, or Kings David and Solomon.

Of all the possible anachronisms in the Torah, perhaps none has caused as much controversy as the references in it to camels. The Torah contains just over two dozen such references, and the entire Hebrew Bible contains no fewer than fifty references to camels, extending from mentions in the stories of the patriarchs to the travels of Ezra and Nehemiah to Jerusalem from Babylonia at the very start of the Persian Period, around 538 BCE.

The first reference is at Genesis 12:16, where Abram and Sarai (as she was then known) were well-received in Egypt, especially Sarai, and Abram is reported to have acquired sheep, oxen, asses, slaves, and camels. Camels are also mentioned with respect to Isaac, Jacob, and Joseph (see, e.g., Gen 24:61–64; 31:17, 34; 37:25).

These references, and others, all seem to make perfect sense within the storyline—except for the camels. The history of the camel, it turns out, is rather unusual, complex, and not well-detailed or understood. The ancestors of modern day camels, by which we really mean the dromedary or one-humped camel, originated in North America, and then about two million years ago, at the end of the Pliocene Epoch, traveled north and west to the Asian land mass, ultimately reaching Mesopotamia, and even what is now the Saharan desert. While there is sporadic evidence of the presence of camels in Syria and the Dead Sea area well over hundreds of thousands of years ago, former Missouri Southwest University professor Juris Zarins reports that wild camels "seem to have disappeared or to have been driven out of their natural habitat into the more inhospitable reaches of the Arabian peninsula" by about 3000 BCE, near the beginning of the Bronze Age.[8] Based on the existence of jars and figurines that

8. See Zarins, "Camel," 1.824.

are said to be camels, various individuals have proposed a wide range of dates for the domestication of the camel, including prior to 2000 BCE. Ancient records of the Egyptian Nile Valley, however, while depicting a broad menagerie, including all of the larger mammals, do not have a word for the camel. Moreover, there is a thousand-year gap between about 2180 and 1170 BCE in representations of camels in pottery.[9]

Columbia University professor Richard Bulliet states that "[h]istorically, the earliest explicit indications of camel use in northeastern Africa date back to the sixth and seventh centuries B.C. and are related to Assyrian and Persian invasions of Egypt across the Sinai peninsula."[10] Archaeologists Israel Finkelstein and Neil Asher Silberman effectively concur, noting both that Assyrian texts from the seventh century BCE are the first to refer to camel trade caravans in Canaan and that archeological excavations have revealed a noticeable increase in camel bones discovered from that period.[11]

So, while camels may have been domesticated (meaning that they may have been used as a source of milk and meat in the second millennia BCE in other locations such as Persia, or present-day Iran), there does not appear to be any serious evidence discovered to date that camels were domesticated in Egypt prior to 800 BCE. Thus, the timing of the stories of Rebekah riding a camel (Gen 24:61–64), of camel caravans to Egypt (Gen 37:25), and of camels as part of Pharaoh's livestock herds (Exod 9:3) appear to be, as Hamlet had it, "out of joint."[12]

All this talk about camels and the Middle East naturally reminds us of the ancient teaching that you should not allow a camel to put its nose in your tent, lest you will soon have the entire camel in there with you. The lesson is a metaphorical warning that permitting a small act can lead to greater and quite undesirable consequences. Thus, the anachronistic camel could serve for some as a reason, or yet another reason, to dispute the divine origin and, therefore, the importance of the Torah.

It is true that the mention of the camel, and the presence of all of these anachronisms, those mentioned here and others, challenges the idea that the Torah was written by a Divine Author. It undermines both the character and the credibility of the presumed supernatural Author.

9. See generally Saber, "Camel in Ancient Egypt."

10. See Bulliet, *Camel and the Wheel*, 116; Saber, *Camel in Ancient Egypt*, 209.

11. See Finkelstein and Silberman, *Bible Unearthed*, 37.

12. Shakespeare, *Hamlet*, 1.5.190.

Any objective analysis of the resulting text demonstrates a lack of the primary "omnis" traditionally attributed to God, omniscience and omnipotence, as an all-knowing and all-powerful Deity, or even just a good editor, would not have written a text so chock-full of factual errors and internal inconsistencies.

One response to all of this criticism is, of course, to claim that it is incredibly pretentious: humans cannot know the mind of God, and if the Author of All provides us with a text with puzzles, then the Author must have had a reason to do so. The argument, in short, is that God gave the Torah to Moses at Sinai and, as the Psalmist said long ago, "The teaching of the LORD is perfect, . . . The decrees of the LORD are enduring" (Ps 19:8).[13] But for that argument to be credible, its primary assumption must be tested. That is what we shall do in the next chapter.

13. The JPS *Tanakh* uses "the LORD" instead of the personal name transliterated in the tetragrammaton, YHWH.

Chapter 11

Is This Really the Torah that God Gave Moses at Sinai?

THE TORAH IS MORE than the purported history contained in it. When its contents were reduced to writing, text trumped tradition as the source of both political and religious authority in the Judahite world.[1] That result initiated nothing less than a textual revolution. In the words of Israeli writer Amoz Oz and his daughter historian Fania Oz-Salzberger, a "lineage of literacy" followed.[2] Transmitted over millennia and eliciting commentary which itself then begot more commentary, the written Torah has bound and continues to bind the Jewish people together over space and across time as they read it, study it, participate in its interpretation and organic growth, and act out its lessons.

Jewish tradition ascribes the highest honor to the Torah. Honoring one's father and mother, performing deeds of kindness, and making peace between one man and another are all deserving of the greatest reward, but, according to the ancient sages, the study of Torah is equal to all of them.[3] Why? In part because the tradition sees Torah as the word of God, and that teaching has been transmitted to the present day.

The traditional view has been that the Israelites left Egypt in the spring of the year 2448 AC, which corresponds to 1312 BCE.[4] The *Tal-*

1. See generally Schniedewind, *How the Bible Became a Book*, 91–117.
2. Oz and Oz-Salzberger, *Jews and Words*, 15.
3. m. Pe'ah 1:1; b. Šabb. 127a, https://www.sefaria.org/Shabbat.127a?lang=bi.
4. See "Exodus."

mud relates that Moses received the Torah at Sinai, and then gave it over to Joshua. Joshua gave it to the elders, the elders to the prophets, and the prophets to the men of the great assembly.[5] The latter were understood as a group of 120 prophets and sages who, beginning in the fourth century BCE during the Second Temple period, were the final religious authority for the reconstituted Jewish community in Judea.

The notion that contemporary Jews are the inheritors of this transmitted Torah is still expressed widely today. Take a look at a *siddur* (prayerbook) in any Orthodox, Conservative, or Reform (though not Reconstructionist) congregation, and turn to the conclusion of the service for the reading of the Torah. That is the moment when the Torah is lifted and the community joins in reciting the words "*V'zot haTorah asher sam Moshe lifnei b'nei Yisrael al pi YHWH b'yad Moshe*"—that is, "This is the Torah that Moses set before the children of Israel, from God's mouth through Moses's hand."

Based on passages in the Torah and commentary in the Talmud, Orthodox Rabbi Gil Student argues not only that God dictated the Torah to Moses, but also that Moses wrote thirteen copies of the text.[6] There was one copy for each of the twelve tribes of Israel, plus one which was maintained and safeguarded by the priests and ultimately deposited in the First Temple. Student concedes that what happened to the priests' copy is not clear, but contends that after the exiled Judahites in Persia were allowed by Cyrus, around 538 BCE, to return to Yihud, they found three Torah scrolls. Ultimately led by the priest and scribe Ezra in the middle of the fifth century BCE, they reconciled any differences in the texts by majority vote of the scrolls. The resulting text was then protected by the priests of the Second Temple.[7]

Of course, the premise of Student's argument—that God dictated the entire Torah to Moses—is a claim that the Torah itself does not make explicitly, as Rabbi Daniel Gordis has recognized in his discussion of revelation in the Conservative movement's edition of the Torah text.[8] Moreover, even if the Torah did make a claim of divine authorship, the problem inherent in a text (self-)serving as its own proof text is obvious.

5. Pirqe 'Abot, 1:1.
6. See Student, "On the Text of the Torah," 3.
7. See Student, "On the Text of the Torah," 5.
8. See Gordis, "Revelation," 1394.

Nor are references to statements of Talmudic sages necessarily persuasive to any except those predisposed to accept their authority. The statement at the beginning of Pirqe 'Abot upon which Rabbi Student builds his case was written not by eyewitnesses, but well more than a thousand years after the original purported transmission and, further, by men who, however well-meaning, had an interest in presenting themselves as the primary interpreters and adjudicators of Jewish law and practice.

A Talmudic debate not referenced by Rabbi Student, however, is instructive. Considering what God gave Moses and when, one rabbi stated that the entire Torah as we know it was given scroll by scroll. Another claimed that it was disclosed in its entirety. Yet another thought that certain passages were revealed before others, as needed.[9] The rabbis, of course, were trying to fill the gaps in the sparse history of the development of the Torah text. Ironically, and no doubt unintentionally, this debate underscores the lack of certainty about what God gave Moses and when.

As we have discussed previously, recent archeological investigations and new understandings of literature from the Levant strongly suggest (1) that a key story in the text, the exodus from Egypt, did not occur as represented, and (2) that some other, smaller details are anachronisms which demonstrate that the Torah could not have been written at the turn into the thirteenth century BCE. Documentary analysis also argues persuasively that different sources or versions of the same stories were interwoven into the final text. Thus, the traditional view that a single author, whether directed or inspired, wrote the entire text of the Torah at or shortly after some revelatory event at Mount Sinai about thirty-three centuries ago is undermined by different but consistent streams of evidence which agree that the Torah we have today was written and edited much later and over a number of centuries by several individuals or schools.

Nevertheless, the idea that the Torah we possess today, what Rabbi Student calls the *textus receptus*, is identical to that given by God to Moses is a powerful one, and persists to this day, as evidenced by its promotion by Rabbi Student and its acceptance in contemporary prayer texts and elsewhere. So let's revisit the claim, this time taking the contents and physical nature of the Torah we have today as we find them and ask how, if at all, the received text compares to the presumed original Torah text

9. See b. Giṭ. 60a–b; "Talmud—Mas. Gittin 2a," 169–73.

that a scribe such as Moses might have written over three millennia ago. As we proceed, we are faced, analytically, with at least five issues which must be addressed: contents, language, script, security, and transmission.

THE CONTENTS OF MOSES'S SEFER HATORAH

The only mention of sacred writings in our Torah, other than the two sets of tablets on which God and Moses respectively wrote, occurs at the end of Deuteronomy when the story relates that Moses, then one hundred and twenty years old and impaired, wrote down "*haTorah*" (see Deut 31:2, 9, 24). The term "Torah" could mean "law," but is better understood here as "teaching" or "instruction."[10] Moses then gave the "*sefer haTorah*" (the document of instruction) to the Levitical priests, who were in charge of the ark of the covenant (see Deut 31:26). The contents of the *sefer haTorah* are not expressly delineated.

Other passages surrounding Moses's inscription and conveyance may provide a clue, though. In anticipation of crossing over the Jordan into Canaan, Moses previously had directed the people to erect large stones after their passage and to write on them all the words of "*haTorah*" (see Deut 27:3–8). Subsequently, Joshua reportedly wrote on stones a copy of the "*Torat* [*sic*]" (the "Teaching") that Moses had written (see Josh 8:32). Joshua then read all the words of "*haTorah*," the "blessing and the curse," as written in the "*sefer haTorah*" (see Josh 8:34).

What could Joshua have written? Consider the logistics. If two tablets could hold only ten commandments (see Exod 20:1–14; 31:18), how many would be necessary to contain all 613 in the Torah we have today, plus the many stories and genealogies? In addition, today a trained scribe requires about a year to write a Torah on parchment.[11] An effort to chisel an entire modern Torah on stone would require even more time. But time was not available, as neighboring kings were preparing to attack the invading Israelites (see Josh 9:1–2). All this suggests that the "*sefer haTorah*" Joshua chiseled was limited in scope, perhaps to the terms of another covenant that we find in Deut 27:11—28:68. And that, in turn, suggests the same for what Moses may have written in the *sefer haTorah* he handed to the priests. But we simply cannot know for sure.

10. See Berlin and Brettler, *Jewish Study Bible*, 2.

11. See Zaklikowski, "Torah Scroll Facts," para. 6.

THE LANGUAGE AND THE SCRIPT OF MOSES'S TORAH

If we are unsure about the words Moses wrote, what do we know about the language and script he used? The language of the Torah we have is Hebrew, but it is not contemporary Hebrew. Rather, the language of our Torah is what scholars call Biblical Hebrew, or Classical Biblical Hebrew. Whether Moses and the biblical Israelites actually spoke Hebrew of any sort is, however, doubtful.

According to the biblical story, Moses's writing came at the end of a forty-year journey which followed over two hundred years of life in Egypt for the descendants of the patriarch Jacob, much of which was spent in slavery.[12] Taking the story as true for present purposes, what was the language Moses likely used while writing his *sefer haTorah*? Again, the text we have today is conspicuously silent on the subject, but one fact and two assumptions relate to this issue.

The fact is that whatever language Jacob and his immediate family may have spoken, there is no evidence that during the time tradition ascribes to Jacob's life (c. 1653–1506 BCE) that language was Hebrew. According to Bible scholar Jan Joosten, the Canaanite language used in the territory in which Abraham, Isaac, and Jacob are said to have traveled was "a local Northwest Semitic dialect spoken in the land long before there was any mention of a 'people of Israel.'"[13] Given the American experience with both slaves and immigrants, it does not seem unreasonable to assume that when Jacob's family arrived in Egypt under the protection of Jacob's son, Joseph, who was acting as governor of Egypt, the family began to learn the language used by their Egyptian hosts. Further, when Jacob's descendants became enslaved, it is similarly reasonable to assume that they used the language of their overseers. Of course, it is possible that the descendants of Jacob were isolated sufficiently from the Egyptian population so that they could have maintained their original language, whatever it was, but that is just more speculation. The following question remains: if Moses intended that the recently freed Israelites understand the words he wrote, did he write it in the language of Egypt, or in some other tongue?

Intimately related to the question of the language Moses used is the issue of the script he wrote. As might be expected, the rabbis in the

12. See Scherman and Zlotowitz, *Chumash*, 359n40.

13. Joosten, "How Hebrew Became a Holy Language," 46.

Talmudic period speculated about the script used in the original Torah. And, as might be expected, they disagreed with each other.

For these rabbis, the possibilities involved two types of lettering, and the dispute was about which was used and when. The two candidates, both West Semitic scripts, were *Ivri* and *Ashuri*. One of the rabbis, Rav Elazar HaModal, contended that the script was always *Ashuri*. Another, either Mar Zutra or Mar Ukva, opined that the original lettering was *Ivri*, but that in the time of Ezra it changed to *Ashuri*. A third argued that it was originally in *Ashuri*, later changed to *Ivri* and then reverted to *Ashuri* once again.[14]

The factual bases, if any, for the rabbis' opinions are unknown. What is known is that they did not have the benefit of recently developed information. What modern archaeology and studies of Near Eastern literature, linguistics, and lettering teach us is that the *Ivri*, or Paleo-Hebrew, script was an offshoot of Phoenician. This Paleo-Hebrew script was used in the First Temple period. It can be seen, among other places, in the inscription carved in the Siloam Tunnel at the end of the eighth century and in a wide variety of seals used in the sixth century.[15] According to present evidence, prior to the destruction of the First Temple, "the paleo-Hebrew script was the only alphabet used by the Israelites."[16]

A distinctive square script, known as *Ashuri*, originated in Aramean kingdoms, evolved, and was promoted by the Assyrians as they expanded their empire. It was maintained by the Persians after they conquered the Assyrians. When the exiled Judahites returned from Babylon, during the early years of the Persian Period, they brought back with them the distinctive square script.[17]

Paleo-Hebrew continued to be used along with Ashuri into the second century of the Common Era, ending with the Second Jewish Revolt (132–35 CE). After that time, the *Ashuri* script was the only script used by Jews, ultimately becoming the square Hebrew lettering in use today.[18] Consequently, when Moses reportedly wrote the *sefer haTorah*, there is

14. See b. Sanh. 21b, 22a, http://halakhah.com/sanhedrin/sanhedrin_21.html.

15. See Siegel, "Evolution of the Hebrew Scripts," 28; see also Jewish Virtual Library, "Hebrew: History of the Aleph-Bet," 4–5.

16. See Siegel, "Evolution of the Hebrew Scripts," 28.

17. Siegel, "Evolution of the Hebrew Scripts," 28, 30; see also Schniedewind, *How the Bible Became a Book*, 176

18. See Siegel, "Evolution of the Hebrew Scripts," 30, 33.

no evidence that the script used for today's scrolls even existed, and no such evidence would arise for over half a millennium.

So what script might he have used? To try to answer that question we have to go back in time, initially to the mid-ninth century BCE, and start with the Mesha Stele, also known as the Moabite Stone. Discovered in the last half of the nineteenth century, in what is present-day Jordan, the Mesha Stele is a basalt slab about a meter tall which contains one of the earliest and clearest examples of writing with some Hebrew characteristics. The stele describes the exploits of King Mesha of Moab, who is also discussed in the Hebrew Bible (see 2 Kgs 3:4–27). The script here displays no evidence of square letters, however, and would be unrecognizable to any Hebrew school, or perhaps even any rabbinic student. It more resembles Phoenician writing of the same period. Some would characterize the letters as Paleo-Hebrew.

An older inscription, being a record of agricultural events, was found in 1908 in the old biblical city of Gezer, located between Jerusalem and what is now Tel Aviv. Known as the Gezer Calendar, this eleven-by-seven-centimeter limestone tablet dates to about 925 BCE, the time of the biblical King Solomon. While scholarly views on the script and language used have shifted, the consensus seems to be that there is no feature of the Gezer Calendar that is particularly Hebraic.[19] The structure of the letters is certainly nothing like that of contemporary Hebrew.

Apparently older still is a fragment of a storage jar, discovered in Jerusalem in 2012 at the southern wall of the Temple Mount. The wall in which the fragment was found has been tentatively dated to the tenth century BCE. The fragment itself contains writing in a Proto-Canaanite script which has been dated to the eleventh or tenth century BCE.[20]

From these three bits of evidence, we can draw a tentative conclusion. Even two to three hundred years after an often-proposed Exodus date, the script used in Canaan was barely, if at all, recognizable as Hebrew. Presumably, therefore, whatever script Moses may have used must have been even further removed from yet-unborn Hebrew and closer to the then current script used in Egypt or in international commerce at the beginning of the thirteenth century BCE.

19. See Biblical Archaeology Society Staff, "Oldest Hebrew Script and Language," https://www.biblicalarchaeology.org/daily/biblical-artifacts/inscriptions/the-oldest-hebrew-script-and-language/.

20. See Ngo, "Precursor to Paleo-Hebrew Script."

A prime example of the kind of lettering which might have been used at that time can be seen in the El-Amarna Tablets, a collection of hundreds of clay tablets first discovered in 1887 in Amarna, located between Cairo and Luxor in Egypt.[21] The messages are mostly from kings in communities throughout the Middle East and are dated to 1350–1300 BCE, around the traditional time calculated for the exodus.

Despite the location of their discovery, however, the tablets were not written in Egyptian, but in Akkadian infused with various Canaanite dialects. The script was not Egyptian hieroglyphics, but cuneiform wedges. The quantity and origins of these tablets suggest that there was, at least for commercial purposes, a language and a script that was commonly used at the time not only in Egypt and Canaan, but also northeast to Assyria and east to Babylon. If not in the hieroglyphics of their despised Egyptian taskmasters, and not in the as yet un-invented square lettered Hebrew, might Moses have written the *sefer haTorah* in something like the cuneiform found in the El-Amarna Tablets? If he did, that raises a question regarding the sufficiency of the translation from the original text into what we now call Biblical Hebrew. If he did not, what language and script did he use that would have been comprehensible by the Israelites?

SECURING THE TRANSMISSION OF THE ORIGINALLY INSCRIBED TEXT

There are other problems. Let's start with the medium of Moses's inscription of the *sefer haTorah* that our Torah says Moses wrote just before he died (see Deut 31:9, 24–26) and the security afforded the resulting work. Our Torah does not say precisely whether Moses chiseled the words into stone, wrote them with a stylus in wet clay, or used a quill on parchment or papyrus. If the entire Torah as we know it was inscribed on stone or clay tablets, there must have been many of them to include almost eighty thousand words containing over three hundred thousand letters.[22] If one or more scrolls were used, the material involved must have been sizable as well. In any event, it is certainly hard to imagine the 120-year-old Moses chiseling, pressing, or writing that much text as he was about to die.

The word *sefer* is used in the Hebrew Bible over 150 times. The meaning of the word changes depending on the context of the usage.

21. See Knott, "Amarna Letters."

22. See "Sefer Torah."

At one time, it may indicate a letter, while at another time a legal document. While the material on which Moses wrote was not bound on a side edge like the books of our day, as the binding of paper into books did not begin until soon before the beginning of the Common Era, the word *sefer* seems more suggestive of writing on parchment, or similar material, than it is of chiseling in stone or impressing in clay. In other words, *sefer haTorah* here appears to indicate a scroll.

That conclusion is buttressed by a consideration of how the *sefer haTorah* was to be handled for the journey into Canaan. Our Torah states that Moses directed the Levites to take the *sefer haTorah* and place it by the ark of the covenant, there to remain as a witness (see Deut 31:26). Later commentators speculated about this placement. Each assumed that the item consisted of a single scroll. Under Rabbi Judah's theory, that scroll was placed near the outside wall of the ark on a shelf or ledge projecting from and attached to the ark, while Rabbi Meir thought that it was placed inside the ark between the tablets inscribed with the Ten Commandments and the inner wall of the ark.[23] Both explanations are problematical. For present purposes, though, the main point is the general consensus among early rabbis that the *sefer haTorah* was a scroll. Professor Richard Elliott Friedman agrees, and translates *sefer haTorah* as a "scroll of instruction" in his commentary on the Torah.[24]

Whether the *sefer haTorah* was, in fact, a scroll or not, its subsequent history is quite mysterious. As the Hebrew Bible relates, after Moses's death, the ark, with the scroll presumably either inside or on a protruding ledge, was taken on a long and difficult journey. Taking the story as true for present purposes, over the course of more than two hundred years, the ark was, among other things, carried across the Jordan River, paraded around Jericho, set in a tabernacle in Shiloh, brought into the field during battles with the Philistines, captured by the Philistines and taken to various cities, returned to the Israelites, moved by King David to a private house and later put in a tabernacle, and then placed by King Solomon in the temple in Jerusalem. There is no further mention of the ark in the historical narrative of the Hebrew Bible, except for a report in Chronicles that several hundred years after Solomon, King Josiah ordered the Levites to place the ark back in the temple (see 2 Chr 35:1–3). This report is not corroborated by any similar statement in the story of Josiah as told

23. See b. B. Bat. 14b, http://halakhah.com/bababathra/bababathra_14.html.

24. See Friedman, *Commentary on the Torah*, 664.

in 2 Kings, but if given credence certainly raises the question of when the ark had been removed from the temple and where it was in the interim.

In any event, biblical history as related in Kings (and Chronicles) is rather clear in its description of what the Babylonian King Nebuchadnezzar took as his army destroyed the temple in Jerusalem around 586 BCE. The *Tanakh* relates that the Babylonians carried off valuables that they did not otherwise destroy or burn (see 2 Kgs 24:8–13; 2 Chr 36:18–19). There is, however, no specific mention of the ark, or any similar container, being confiscated. Nor is there any claim or even hint expressed that the royal family or any priest or anyone else either took with them into exile or hid or otherwise protected or preserved the *safer haTorah* that was, at Moses's direction, to be located by the side of the ark. This does not mean that those transported to Babylon did not take any documents or records, but merely that a specific ancient treasure was not specified in the national record, as one might have expected it to be had the object actually been (1) extant and (2) moved.

Is it reasonable to believe that, throughout almost seven hundred years, a scroll exposed to the enemy in battle and to the elements in peace could have survived intact? Given its importance as a sacred writing by Moses himself, the fact that the historical sections of the Hebrew Bible after the book of Joshua, with one exception, do not mention the *sefer haTorah* is telling. And that exception, the story of King Josiah's surprise at a newly discovered scroll by the priest Hilkiah, seems to confirm at the very least that control over and protection of sacred scrolls was not well managed (see 2 Kgs 22:3–20).

Similarly, as an object, the existence of the scroll on which Moses wrote God's words seems to have been of no concern to the authors of the Prophets (the *Nevi'im*) and the Writings (the *K'tuvim*), not to mention the editors of the *Tanakh*. Again, the silence suggests strongly that if a *sefer haTorah* ever existed, it was lost long ago.

VERSIONS AND CHANGES FROM TRANSCRIPTION

Given the many problems inherent in the claim of an unsullied transmission of the *sefer haTorah* authored by Moses, we should not be surprised to learn from Duke Bible professor Marc Zvi Brettler that a number of versions of the Torah existed in the Second Temple period (c. 538 BCE—70

CE).[25] The evidence comes, in part, in the form of pre-Common Era texts like the Samaritan Pentateuch, the Greek Septuagint, and the Dead Sea Scrolls, among others, which contain different wording than we find in the Torah we have today.

Rabbi Student argues that these other documents are not indisputable proof, or even good witnesses, against the assumption that the "text that is agreed upon by the Jewish community—the *textus receptus* that is claimed to be the Masoretic text—is correct."[26] Student is certainly persuasive when discussing a translation like the Septuagint.[27] The philosophy, principles, and methodology at play in any translation can be quite complex. For instance, did the first translation attempt to be true to the literal nature of the subject text word by word? Did it seek to convey the meaning of that text thought by thought? Did it seek to balance those two approaches? Did it seek to mimic the sound and cadence of the original language, or to appeal to certain linguistic or philosophical sensibilities of the readers by using, among other things, colloquial or gender neutral expressions? And was the first translation consistent? The same questions, and more, would apply to any effort to translate the first translation back independently in order to determine an original text. In short, reverse engineering a translated text compounds the inherent complexity of the translation exercise.

Rabbi Student is on considerably less solid ground with other variants, like the Samaritan Pentateuch.[28] By one count, the Samaritan Pentateuch contains not only over three thousand differences in spelling when compared to a contemporary Torah, but it also contains over three thousand words and phrases which clarify or change the meaning of the story found in current text.[29]

Sometimes the changes are small, but even those small changes can be important, as when dialogue is added in the Samaritan version of the familiar story of Cain and Abel. The story in the book of Genesis reports that Cain said something to Abel before they went into a field where Cain killed Abel, but it does not disclose what Cain said (see Gen 4:8). The Samaritan version supplies dialogue, which indicates that Cain enticed

25. See Brettler, *How to Read the Jewish Bible*, 22.

26. See Student, "On the Text of the Torah," 1.

27. See Student, "On the Text of the Torah," 6–12.

28. See Student, "On the Text of the Torah," 1–2.

29. See generally Lieber, "Other Torah."

his brother to accompany him and, therefore, supports a determination that Cain's murder of his brother was premeditated.

Then, too, sometimes the changes are substantial. For example, the Samaritan text, in addition to the well-known Ten Commandments, includes an additional commandment to establish an altar on Mount Gerizim.

Student acknowledges the obvious, which is that there are "many differences" between the Samaritan Torah and the *textus receptus* we have, but concludes that these differences "can be due to the free hand the Samaritan scribes exhibited in developing their Torah." Consequently, for him, "the Samaritan Torah fails our test of being a reliable witness."[30]

The differences, however, individually and collectively, are more probative than Rabbi Student allows. In his argument, Student concedes that he is departing from the approach of modern scholars and admits that he is assuming that today's standard text, his *textus receptus*, is "correct," unless "categorically disproven" otherwise.[31] But giving something a Latin label does not make it sacrosanct, and placing a thumb on the scale of evidence will always skew the analysis and never lead to an accurate reading of that evidence, even without the imposition of a stringent burden of proof. Why, for instance, assert that Samaritan scribes "developed" their holy text with a "free hand" and not allow for the possibility that Judahite scribes did so similarly? If we want to evaluate the totality of circumstances objectively, as opposed to proving a point, we cannot proceed under any, much less unwarranted, assumptions. More specifically, given what we know today, "we cannot," according to Professor Brettler, "assume that the text . . . as we now have it is the same as the text . . . when it was originally written."[32]

In addition, if there were, once, one text written by Moses, or even one redacted by Ezra, surely that text was modified over the centuries. As another Hebrew Bible scholar, University of Pennsylvania emeritus professor Jeffrey H. Tigay, has discussed—albeit in the context of his analysis of Bible codes from a textual perspective—the ideal of an unchanged Torah text "was not achieved in practice as far back as manuscripts and other evidence enable us to see."[33] According to Prof. Tigay, the "manual

30. See Student, "On the Text of the Torah," 2.

31. Student, "On the Text of the Torah," 1.

32. Brettler, *How to Read the Jewish Bible*, 22.

33. See Tigay, "Bible 'Codes,'" 5.

copying of texts naturally created variants."[34] In addition, other changes to ancient texts involved the spelling system for those texts, including, specifically, the use or non-use of vowels. Tigay has looked at the Gezer Calendar, discussed earlier, and noted that the letters which were inscribed on the tablet only represented consonants. At some time in or after the tenth century BCE, Hebrew began to use a limited number of consonants as vowels. The current system of marks placed above and below letters to indicate vowels was not adopted until sometime between the sixth and eighth centuries of the Common Era.[35]

Scholars today can trace the changes in biblical manuscripts through the ages. They can identify the sources available and used by different scribes, as well as the editorial choices they made. They can see how a particular manuscript once considered to be definitive was corrected later and supplanted. This was true prior to the advent of machine printing in the fifteenth century of the Common Era and also true after.[36]

The end result of this textual history is inconsistency in the manuscripts we have and use today. Contemporary Hebrew Bibles are based primarily on thousand-year-old manuscripts known as the Aleppo Codex (c. 930 CE) and the Leningrad Codex (c. 1010 CE), the latter being the oldest complete *Tanakh* existing today. Both are part of the Masoretic Text tradition, but agreement among modern works is still lacking. For instance, the well-regarded Koren Jerusalem Bible, first published in 1962, is based on the Leningrad Codex, but according to Professor Tigay, the text of the Torah contained in it has forty-five more letters than does the Michigan-Claremont Westminster computerized text of the Leningrad codex which is used by most scholars.[37] Similarly, a text provided to Israeli soldiers is also based on the Leningrad Codex, but reportedly it, too, contains spellings different than those in the Koren.[38]

The illusive nature of the original Torah has not deterred researchers from seeking to find it. Currently, there are at least two academic efforts aimed at producing a scholarly edition of the *Tanakh*, one known as the Hebrew University Bible Project and the other as the Hebrew Bible Critical Edition Project, formerly known as the Oxford Bible Project. Yet,

34. Tigay, "Bible 'Codes,'" 6.
35. Tigay, "Bible 'Codes,'" 5–6, 19–20.
36. See generally Penkower, "Development of the Masoretic Text," 2077–84.
37. See Tigay, "Bible 'Codes,'" 5, 19.
38. See Zeligman, "List of Some Problematic Issues," 2–3.

even as one may anxiously wait to read of new developments, perhaps Professor Brettler is correct in concluding that "[i]t is naïve to believe that we may recover the Bible's original text (what scholars call the 'Urtext'), namely the text as penned by its original authors."[39]

THE WORTH OF TORAH

The traditional claim that the Torah we have today is identical to a text authored in the thirteenth century BCE strains credulity. Regardless of whether the author was divine or human, for the traditional view to be valid, not one but a series of improbable events would have had to occur in the creation and multiple transmissions of that text over more than one hundred generations. The evidence, internal in the *Tanakh* and external on hard stone and clay and soft manuscripts, says those events did not occur. Analyzing the content, language, script, security, and transcription of a proposed original Torah demonstrates why such a document, if it ever existed, must have been different, perhaps considerably so, from the Torah we have today.

There is no reason to stress or strain over either the evidence or the result of our inquiry. The Torah we have might not have come from Moses, and certainly it has not arrived unimpaired from some original manuscript, but it remains a very special document, one that holds appeal for a wide variety of individuals.

Implicitly conceding the lack of historicity for an original thirteenth century BCE Torah from Moses, at least for the purpose of making a greater point, University of Illinois at Chicago philosophy professor Samuel Fleischacker, himself Orthodox, has argued that what is important is the authority rather than the authorship of the Torah, whether the Torah "represents a supremely good ("divine') way for us to live."[40] It is an interesting and significant argument, one that (perhaps ironically) parallels positions taken by less ritually observant individuals.

Putting aside the problematic issue of "authority," we can agree that the Torah plays and deserves a unique role in the Jewish civilization. Those who appreciate miracles (whether supernatural or not) can marvel at the Torah's continued existence and power. We still have it, and we continue to read and study and even wrestle with it. And atheists like

39. Brettler, *How to Read the Jewish Bible*, 22.

40. Fleischacker, "Making Sense of the Revelation," 6–7.

Amoz Oz and his daughter Fania Oz-Salzberger also can recognize that honor is due the Torah as literature that "transcends scientific dissection and devotional reading." In their words, "no other work of literature so effectively carved a legal codex, so convincingly laid out a social ethic."[41]

With its depth, its breadth, and its reach, the Torah may well have been intended initially as a vehicle to hold and share the sometimes inconsistent stories, codes, and customs of the residents of ancient Judah, perhaps before but certainly after their exile from and return to their homeland. In this light, its role was to bind the people together with a common created history and collective purpose. The Torah that we have today serves a similar function, as transnational, transgenerational glue for the Jewish people. And, against all historical odds, it remains, in words from Proverbs, an *etz hayim*, a tree of life to those who hold fast to it (Prov 3:18).

Consequently, if the Torah is a text written by mere mortals, a work of human minds struggling to understand their place in a presumed grand heavenly scheme, a work written by people dared by their geography and history to survive on their earthly abode, trying to figure out how to live day by day, week to week, season to season, and year to year, we still have a work worthy of continuous study. For here are stories of a people seeking to distinguish themselves from their neighbors, and, on their better days, choosing, rather than being chosen, to live an ethical life, to love each other, to treat the stranger with compassion, and to become a holy nation. Here are stories, sometimes written in frank and salty language, and sometimes with puns, sarcasm, and humor, that are both rooted in reality and aspirational, and because of that duality so challenging and inspirational for us.

A Torah less of divine decree and more of human hand does not make it less worthy of reverence. It is, rather, a deep well of endless worthwhile lessons about life, a fitting foundational text for Western Civilization, and a work to treasure.

41. Oz and Oz-Salzberger, *Jews and Words*, 5–6.

Chapter 12

The Battle for Jerusalem and the Origin of Fake News—2700 Years Ago

How can you tell what is true and what is not in the Hebrew Bible? How can you separate fact from fiction and fable? In some instances, science can help. As we have seen, both geological and archaeological records confirm that the whole planet was not submerged in flood waters during the last six thousand years, and evolutionary biology demonstrates that all land animals and birds do not owe their existence to creatures that were on a vessel floating on those mythical waters. Similarly, we know that the Sun did not stop in the sky for twenty-four hours during a battle at Gibeon, for that would have meant that the Earth ceased to rotate during that period of time, which, in turn, would have caused cataclysmic consequences neither reported in the story nor elsewhere (see Gen 7:6–8; Josh 10:12–14).

From a modern perspective, then, it is reasonably easy to identify some biblical stories that are not factually accurate. They may well contain worthy moral or other lessons, but as factual recitations of actual occurrences, they fail.

At the same time, there are other stories in the Hebrew Bible that seem quite plausible, even contemporary in their nature. How can we tell if they are historically true, historical fiction, or simply imagined? One such story concerns the siege of Jerusalem by the Assyrian king Sennacherib (pronounced Seh-*nack*-er-ib) during the reign of the Judahite king Hezekiah about 2,700 years ago.

THE SETTING AT THE CROSSROAD OF CONTINENTS

Seemingly from its origins, millennia ago, Jerusalem (like much of the rest of the ancient land of Canaan) was blessed and cursed by being located near the southwestern portion of a strip of land known as the Fertile Crescent. Beginning in present-day Iraq, the Crescent extended north and west along the great rivers, the Euphrates and the Tigris, into what is now Syria and Turkey, and then descended south and west, encompassing land on both sides of the Jordan River. Further to the south and west lie Egypt and the Nile Delta.

The location of the Fertile Crescent was critical because, among other things, it supported agriculture, trade, and the development of great civilizations. It was also the passage way between the continents of Asia and Africa. Not surprisingly, it was the scene of constant conflict, as empires rose and fell, elevating and devastating smaller kingdoms, city states, and tribes along the way.

Assyria asserts itself

One of the cycles in that waxing and waning of empires began in the thirteenth century BCE with the demise of Pharaoh Ramses II. The Egyptian Empire, which previously reached up into and beyond Canaan, began to decline. Over a period of time, the withdrawal of Egyptian presence and influence allowed the people native to the land of Canaan and those who migrated there to begin to establish their own cultural, tribal, and national identities. Then, and there, a confederation of tribes emerged under the banner of Israel, ultimately claiming a common history and destiny, a common religious tradition, and a common polity. Egypt's withdrawal, and the demise by the twelfth century of a power in the North known as the Hittites, also allowed an opportunity for another great northern power, the Assyrians (who had their own periods of strength and weakness) to expand to the South.

Assyria's recovery began in the latter third of the tenth century BCE. By the middle of the ninth century, Assyria under Assur-Nasipal II and Shalmaneser III became a fierce and dominant power. Assyrian writings contained in artifacts like the Monolith of Shalmaneser (also known as the Kurkh Stela) and the Black Obelisk of Shalmaneser, both now at the British Museum in London, seem to some to indicate the engagement of the Assyrians with the kingdom of Israel. The Monolith, a six-foot limestone writing, arguably identifies King Ahab of Israel in northern Canaan as a major participant in a coalition that fought with the Assyrians in the Syrian city of Qarqar. Similarly, the Black Obelisk, a black limestone sculpture also about six feet high, has been read as depicting the Israelite King Jehu paying tribute to Shalmaneser III. The biblical book of Kings ("BOK") also notes the invasion of Israel by the Assyrians and the payment of tribute during this time period (see, e.g., 2 Kgs 15:17–20). But then, toward the end of that century, Assyria, too, began to decline again.

With the rise of Tiglath-Pileser III in the eighth century BCE, Assyria became strong again, so much so that it was able to initiate a fatal fight with Israel. BOK mentions the situation briefly, noting the capture by Assyria of substantial territory and the deportation of its inhabitants to Assyria, such a dispersal being consistent with Assyrian policy toward defeated peoples (see 2 Kgs 15:29).

To the south of Israel lie the Kingdom of Judah, and Tilgath-Pileser exerted influence on Judah as well. King Ahaz assumed the throne of Judah between 743 and 727 BCE at the age of twenty (see 2 Kgs 16:2). Early in Ahaz's reign, King Pekah of Israel allied with the Aramaen King Rezin to march on Judah's capital, Jerusalem. The prophet Isaiah counseled Ahaz to be "firm and calm" and rely on YHWH, for then the enemy would not succeed (see Isa 7:4–9). Instead, according to the book of Kings, Ahaz, claiming to be his "servant" and "son," sent gold and silver to Tilgath-Pileser and pleaded for his assistance. The Assyrian king did, in fact, come to the aid of Ahaz. He not only took Damascus, the capital of Aram, from Rezin, but he killed him as well (see 2 Kgs 16:1–9). Jerusalem was saved for the moment.

Subsequently, Tilgath-Pileser's successor, King Shalmaneser V, "marched against" a new king of Israel, Hoshea, and forced him to pay tribute to and become a vassal of Assyria. Hoshea then apparently tried to appeal to Egypt for help and stopped paying tribute to Assyria. From the Assyrian perspective, this was an act of treachery, and Sargon II completed the work of his two predecessors. Samaria, the capital of the Kingdom of Israel, fell to Assyria c. 721–720 BCE. Again, the book of Kings records that Israelites were deported to various areas controlled by Assyria, even beyond the Assyrian capital in Nineveh, near present-day Mosul, Iraq, to Media, now in northwestern Iran (see 2 Kgs 17:1–6).

Hezekiah assumes the throne of Judah

Not long after Israel fell, Ahaz died and his son, Hezekiah, was anointed king of Judah. The dates of his ascension (whether 727 BCE or 715 BCE) are in some dispute, and we will not attempt to resolve that issue now. For our purposes, what is crucial is that Hezekiah differed significantly from his father in his approach. For example, where Ahaz accepted his position as a vassal of Assyria and paid tribute to it, Hezekiah sought national independence.

Hezekiah recognized that the destruction of the Northern Kingdom created several challenges and opportunities for Judah. Each was sharpened by the inflow of tens of thousands of refugees who moved from Israel into Judah and, especially, into Jerusalem. Archaeologist Israel Finkelstein and author Neil Asher Silberman have written that the population of Judah, which had consisted of a "few tens of thousands" dispersed across only some "villages and modest towns," blossomed to 120,000 individuals in "about three hundred settlements of all sizes."[1] The increased population affected Jerusalem in particular, which grew from "about one thousand to fifteen thousand" in a few decades and expanded from a small town of "about ten or twelve acres to a huge urban area of 150 acres, of closely packed houses, workshops, and public buildings."[2]

Ironically, an expanded Jerusalem was especially vulnerable to any attack by Assyria. If it was to be able to defend itself, the capital needed both to strengthen its walls and secure its water supply. According to Near Eastern archaeologist and anthropologist William Dever, in the mid-to-late eighth century Judah undertook both tasks (see also 2 Chr 32:1–5). He confirms that an "enlargement of the city wall," called the Broad Wall, has been found and dated to that time period.[3] And broad it was, reaching twenty-three feet wide in places.

Even more impressively, Hezekiah is credited with directing the construction of a tunnel project designed to bring water, which was then sourced at the Spring of Gihon, located outside of the existing walls of the city, to a reservoir within the city walls. Starting from both ends of the tunnel, workers dug through over 1,700 feet of bedrock in the most primitive of conditions. Yet they met as planned in the middle! And water could flow from the Gihon Spring to the Siloam Pool. It was, as Dever

1. Finkelstein and Silberman, *Bible Unearthed*, 245.

2. Finkelstein and Silberman, *Bible Unearthed*, 243.

3. Dever, *Lives of Ordinary People*, 351.

has described it, a project "engineered with amazing expertise."[4] Some doubts have been raised as to whether the tunnel was in fact begun by one of Hezekiah's predecessors, but the book of Kings does not address such matters.

The book of Isaiah suggests that the prophet both opposed rebellion against Syria (see Isa 20:1–4) and the efforts to strengthen the city. Presumably speaking of Hezekiah, Isaiah noted that he identified deficiencies in the city's fortifications and acted to improve them, but also that the king did so by pulling down existing houses, no doubt disrupting the lives of numerous residents. Moreover, while the king "constructed a basin" for the waters newly directed into the city, he "gave no thought to Him who planned it, . . . took no note of Him who designed it long before" (see Isa 22:9–11). In other words, Isaiah berated Hezekiah as insufficiently confident in God's plan and protection.

The influx of people with similar (but not the same) traditions as the Judahites created another predicament, one which necessitated, spurred, or at least provided an excuse for the initiation of a wide range of reforms which could unify the now enlarged and more religiously diverse population. Where Ahaz had made offerings "under every leafy tree," and even erected a copy of an altar he saw in Damascus, Hezekiah abolished rural shrines and posts and centralized worship in the temple in Jerusalem (see 2 Kgs 16:4, 10–13; 18:4). Hezekiah was, essentially, a political and religious nationalist.

THE REBELLION AGAINST ASSYRIA

In or around 705 BCE, Sargon II was killed in battle away from Nineveh. It was the first time that an Assyrian king had met such a fate.[5] The demise of Sargon yielded two related consequences: a power struggle in Nineveh and restlessness and rebellion in Assyria's vassal states. From Babylon in the southeast to numerous kingdoms to the west and southwest, kings were ceasing to pay tribute to Assyria and talking about forging alliances against the Assyrians. Hezekiah, despite the debt Judah owed Assyria for saving it from the joint attack of Israel and Aram, was one of those kings.

Hezekiah knew that he would need allies, or at least would have to neutralize his neighbors. BOK reports that Hezekiah "overran Philistia

4. Dever, *Lives of Ordinary People*, 349.

5. See Stern, "Assyrian March against Judah," 1.

as far as Gaza and its border areas, from watchtower to fortified town" (2 Kgs 18:8). The conquered communities included Ekron, one of the five capitals of the greater Gazan pentapolis (see Josh 3:3). Apparently, Padi, the King of Ekron, remained loyal to Assyria, so Hezekiah removed him and brought him in chains to be imprisoned in Jerusalem.

Sennacherib and the Assyrian Empire fight back

Within a year or so, Sennacherib emerged as the successor to Sargon II and launched a series of military campaigns designed to reassert Assyrian hegemony. Eight military campaigns of Sennacherib are reported on each of three virtually identical six-sided clay artifacts, called prisms. Each prism was about thirty-eight centimeters high, and each of the six panels on each prism was between seven and eight centimeters in width. Each prism contains about five hundred lines written between 691 and 689 BCE in the Akkadian language with cuneiform script. The first of the prisms to be discovered was found in Nineveh in 1830. Known as the Taylor Prism, it is currently in the British Museum. A second version, called the Sennacherib Prism, is housed in the Oriental Institute of the University of Chicago. A third is in the Israel Museum in Jerusalem.

In 1924, Daniel David Luckenbill, a professor of Semitic languages and literatures at the University of Chicago, translated the text of the Sennacherib Prism. That translation, along with Luckenbill's historical analysis and related sources and records, was published under the title "The Annals of Sennacherib."[6]

The first of Sennacherib's campaigns was directed against Babylon, Assyria's most formidable foe and the one of any consequence closest to Nineveh. Over the course of several years, Sennacherib defeated the Babylonian king and his allies in Chaldea (near the Euphrates delta and along the Persian Gulf, in today's Iraq) and Elam (now part of southwestern Iran).[7] Around 701 BCE, in his third campaign, Sennacherib turned west and attacked kingdoms in "Hittite-land" and along the Mediterranean coast. In each instance, these were areas that were in rebellion against Assyria or Judah's allies or both. Sennacherib started in Phoenicia by removing the king of Sidon (in present-day Lebanon) and replacing him

6. As of this writing, the entire text of the Annals is available online: see Luckenbill, "Annals of Sennacherib."

7. See Luckenbill, "Annals of Sennacherib," 38–39.

with a friendlier ruler.[8] Then, and although the order of the campaign is recited summarily, he seems to have headed both south along the Mediterranean coast through land formerly held by Israel and into Philistia, as well as east beyond the Jordan River and the Dead Sea to pacify Ammon, Moab, and Edom.[9]

Hezekiah's influence over Philistia proved short-lived as Sennacherib secured promises of fealty and tribute from numerous local kings, including rulers of Ashdod and Joppa. When the king of Ashkelon would not kiss his feet, literally or figuratively, Sennacherib simply took him and his family to Assyria, and, as was the usual custom, replaced him with a more pliant sovereign.[10] When Ekron not only resisted, but called on reinforcements from Egypt and Ethiopia, Sennacherib defeated them all and hung the bodies of rebellious Ekronite governors and nobles on stakes spread around the city. He then then sent a messenger to Jerusalem demanding the release of former King Padi. Hezekiah, now effectively isolated, complied.[11] And nothing stood in the way of an Assyrian attack on Judah and its capital, Jerusalem.

Sennacherib's assault on Judah

According to the Sennacherib Prism, the Assyrian king entered Judah, laying siege to and taking "46 of (Hezekiah's) strong, walled cities, as well as the small cities in their neighborhood, which were without number, by leveling with battering-rams, and by bringing up siege-engines, by attacking and storming on foot, by mines, tunnels and breaches." Sennacherib claims to have driven out "200,150 people, great and small, male and female, horses, mules, asses, camels, cattle and sheep, without number, (which he) brought away from them and counted as spoil."[12]

The archaeologist Dever doubts that there were forty-six walled cities in Judah to be destroyed, but even if Sennacherib's exploits were exaggerated, the dramatic destruction wrought by his campaign is amply evidenced by the archeological excavations that Dever has identified.[13]

8. See Luckenbill, "Annals of Sennacherib," 43.
9. See Luckenbill, "Annals of Sennacherib," 44.
10. See Luckenbill, "Annals of Sennacherib," 44.
11. See Luckenbill, "Annals of Sennacherib," 45–46.
12. Luckenbill, "Annals of Sennacherib," 46–48.
13. See Dever, *Lives of Ordinary People*, 360.

Finkelstein and Silberman concur, observing that "[t]he devastation of the Judahite cities can be seen in almost every mound excavated in the Judaean hinterland."[14] Even the biblical account acknowledges, tersely but without dispute, that "King Sennacherib of Assyria marched against all the fortified towns of Judah and seized them" (2 Kgs 18:13).

Before confronting the capital, however, Sennacherib decided to attack Judah's second strongest fortification, Lachish, about thirty-five miles to the southwest of Jerusalem. The *Tanakh* relates nothing of the battle at Lachish, but large gypsum wall reliefs from Sennacherib's palace, which were discovered and excavated around 1845–47 by British archeologist Sir Austen Henry Layard, tell a detailed story. Currently housed in the British Museum, the wall reliefs depict a clash between the well-armed Assyrians, supported by battering rams, and their outmanned opponents, who were mainly archers, torch hurlers, and stone slingers. The battle ended with the comprehensive rout and ruin of Lachish by the Assyrians. The enormity of the Assyrian victory can be seen on wall reliefs that depict Judahites alternatively impaled on spears or spikes, bowing in submission to their captors, or being carted off to Assyria.[15] Lest anyone think that the tale told in the Sennacherib Prism and vividly illustrated in the wall reliefs is fanciful exaggeration, archaeologists have found corroborating evidence, not just in the ashes of a city set on fire, but in caves near Lachish which contain mass graves for "about fifteen hundred people—men, women, and children."[16]

With the surrounding territory now firmly in his control, Sennacherib was now able to initiate Assyria's siege of Jerusalem. Of Hezekiah, Sennacherib claimed, "like a caged bird, I shut [him] up in Jerusalem, his royal city. Earthworks I threw against him,—the one coming out of the city gate, I turned back to his misery." And, having "cut off [Hezekiah] from his land," Sennacherib then returned territory Hezekiah had conquered to neighboring kings.[17]

14. Finkelstein and Silberman, *Bible Unearthed*, 260.

15. For more details, see generally Amin, "Siege of Lachlish."

16. Finkelstein and Silberman, *Bible Unearthed*, 262–63.

17. See Luckenbill, "Annals of Sennacherib," 47.

The terms of settlement

Having confined Hezekiah, Sennacherib also claimed that the "terrifying splendor of [his] majesty overcame" the Judahite king and that mercenaries which Hezekiah had procured to defend Jerusalem "deserted him." Consequently, Hezekiah "dispatched his messengers" in order to "accept servitude" and "pay tribute." The tribute, by Sennacherib's reckoning, included "30 talents of gold and 800 talents of silver, gems, antimony, jewels, large sandu-stones, house chairs of ivory, elephant hide, elephant teeth [tusks], ebony, boxwood, all kinds of valuable treasures, as well as his daughters, his harem, [and] his male and female musicians."[18]

The BOK similarly states that Hezekiah sent a messenger to Sennacherib who was reportedly camped in Lachish. The message contained a concession that Hezekiah had "done wrong" and an acceptance in advance of "whatever" Sennacherib would impose on him. According to Kings, the penalty assessed was "a payment of three hundred talents of silver and thirty talents of gold." It states that Hezekiah delivered "all the silver that was on hand in the House of [YHWH] and in the treasuries of the palace." Moreover, Hezekiah "cut down the doors and doorposts of the Temple of [YHWH], which [he] had overlaid [with gold] and gave them to the king of Assyria" (see 2 Kgs 18:14–16).

While different than the tally recited in the Sennacherib Prism, notably in the omission of the transfer of daughters and other personnel, this is still corroboration of a substantial payment by Hezekiah. Determining the true value of weights and measures in ancient Judah is problematic, of course, but respectable scholars believe that a talent weighed just over 75 pounds.[19] If so, this would mean that Hezekiah sent over a ton of gold and ten tons of silver to Sennacherib. Exchange rates vary over time, but clearly this was an enormous transfer of wealth from Judah to Assyria.

A dramatic change

At this point in its telling, though, BOK interjects a new and fascinating story. It suggests that Sennacherib was not satisfied with his booty and sent two high-ranking officials, supported by a "large force," to meet with Hezekiah. The king, in turn, sent the palace executive officer, a scribe,

18. Luckenbill, "Annals of Sennacherib," 47–48.

19. See Berlin and Brettler, *Jewish Study Bible*, 2105.

and a recorder to meet Sennacherib's emissaries. The Assyrians, speaking not in Aramaic—which was then the international language of commerce and diplomacy—but in Yehudit, the vernacular of the Judahite common folk who were listening to this conversation, then threatened war. They taunted the Judahites about their lack of lack of military strength, asserted that any reliance on Egyptian protection was misplaced, and boldly claimed that the Judahite God, YHWH Himself, told the Assyrians to destroy Judah. Reaching directly to the public, the Assyrians further argued that the Judahites should not listen to Hezekiah, lest the Assyrians destroy the city and take the people away (see 2 Kgs 18:17, 19–35).

The story in BOK takes pains to relate how seriously the establishment took these threats. Hezekiah's messengers reportedly tore their cloths before they returned to the king. So did Hezekiah, who also "rent his clothes" and even "covered himself with sackcloth" (2 Kgs 19:1).

Hezekiah also sent his team to Isaiah for counsel. Once again, the prophet urged his king not to rely on foreign forces, this time the Egyptians, and, instead, to have faith in YHWH, who would shield and protect Jerusalem (see Isa 30:1–7; 31:1–3). Then, claiming to speak the word of God to the messengers, Isaiah said "I will delude him; he will hear a rumor and return to his land, and I will make him fall by the sword in his land" (2 Kgs 19:7).

According to BOK, further negotiations ensued. Sennacherib's officials learned that their king had left Lachish and attacked Libnah, just north of Lachish. At the same time, reportedly, the Nubians had also decided to engage Sennacherib (see 2 Kgs 19:8–9). Perhaps they believed that Assyria was overextended. Perhaps Hezekiah had rejected Isaiah's advice and made a side deal with an outside force.

As told in BOK, Sennacherib directed that one more message, another threat, be extended to Hezekiah (see 2 Kgs 19:10–13). Hezekiah took the letter from the messengers and, seeking divine guidance, "went up to the House of the LORD and spread it out before the LORD" (2 Kgs 19:14). Subsequently, he received a lengthy message from Isaiah. Isaiah assured Hezekiah, in the name of God, that Sennacherib would neither enter Jerusalem nor "shoot an arrow at it, Or advance upon it with a shield, Or pile up a siege mound against it." Instead, he would return by "the way he came; He shall not enter this city" (see 2 Kgs 19:32–33).

Sennacherib does not enter Jerusalem

The story in BOK concludes in remarkable fashion. It relates that the night following the transmission of Isaiah's message to Hezekiah, 185,000 Assyrians camped outside of Jerusalem died, "struck down" by "an angel of the LORD" (see 2 Kgs 19:35). This allegedly causes Sennacherib to retreat and return to Nineveh, where he stayed, and was then killed by two of his sons and succeeded by a third son, Esarhaddon (see 2 Kgs 19:36–37). If true, this story would seem to confirm the viability of YHWH's promise to maintain forever a Davidic dynasty in Jerusalem (see, e.g., 2 Sam 7:12–16; 1 Kgs 11:32, 36; 2 Kgs 8:19).

For its part, Assyrian records contain, not surprisingly, no mention of any mass slaughter of troops during the campaign against Judah. And, in stark contrast to the situation at Lachish, there has been no discovery to date of any mass grave or other indication of the demise of such an army near Jerusalem. Indeed, there has been no report of the recovery of a single sword, shield, or bow or arrow, much less the tens of thousands of them, which might have been left by the dead Assyrian troops and which, if recovered, could have served to enhance a weak Judahite military.

Moreover, while the *Tanakh* would have you believe that the total destruction of the Assyrian army led to the assassination of Sennacherib, there is no record of any of the subjugated peoples under the Assyrian thumb rising at the time to take advantage of a depleted and defeated Assyrian military. To the contrary, the Sennacherib Prism and other Assyrian records indicate that, while Sennacherib did not return to Judah again, he did engage in multiple military campaigns elsewhere after leaving Judah.[20] And, while it is true that he was killed by two of his sons, and succeeded by Esarhaddon, the assassination did not occur until about 681 BCE, some twenty years after the siege of Jerusalem.

THE MINGLING OF FACT, FICTION, AND FAITH

What are we to make of these two versions of a significant historical event? Clearly, the main features of the invasion and siege by Assyria are corroborated by the reports in the *Tanakh*. A larger, stronger power threatens a smaller, weaker polity, secures substantial treasure from a cowed king, but does not enter the gates of the targeted capital and ultimately departs.

20. See, e.g., Luckenbill, "Annals of Sennacherib," 48–61.

The reports differ on some details, such as with respect to the nature of the tribute paid by Hezekiah to Sennacherib, but the primary differences relate to the reasons for the Assyrian failure to attack Jerusalem and the fate of the invading army and the king of Assyria himself.

Today, we are all too familiar with the phenomenon sometimes called fake news. While it comes in various flavors, at its core fake news is the dissemination of information which is either false or substantially misleading, but still designed to appear as accurate and trustworthy, all for the purpose of influencing the person who receives the information.

We also tend to think that fake news is a recent invention. But, in fact, fake news with respect to political (and other) events has a long history. The differing versions of the siege of Jerusalem around 701 BCE may be two of the, if not the, earliest examples. Reviewing the reports in this light may help us understand why they differ as they do, because the Sennacherib Prism and the stories in the *Tanakh* were written for quite different purposes and handled in dramatically different fashion.

The Sennacherib Prism and its siblings were monuments to the vanity of the king of Assyria, written presumably at his behest shortly after the events described in them. In contrast to the wall reliefs depicting the victory at Lachish, the prisms were not intended for public consumption, either by distribution or display. Rather, as was the custom, they were placed in areas thought to be protected and secure, like a vault or foundation block, within a government building. While they were artifacts extolling the king, and surely were written in a manner to please his ego, they were to be preserved for posterity and available for reading only by subsequent kings or the Assyrian gods.[21]

The timing of the creation and authorship of the story in the book of Kings regarding the siege of Jerusalem is less clear. Some contend that the biblical story is a mix of two (or more) invasions by Sennacherib. Perhaps. Clearly, certain features of the story could have been written at or near the time of the main events, but the conflation of the story of the siege and Sennacherib's assassination could not have been written, or added, until he was murdered decades after the siege. And at least two hands are visible in the writing. As Professor Brettler points out, just after the segment about the booty paid by Judah to Assyria (2 Kgs 18:13–16),

21. See "Hezekiah's Defeat: The Annals of Sennacherib on the Taylor, Jerusalem, and Oriental Institute Prisms, 700 BCE," para. 2, http://cojs.org/hezekiah-s_defeat-_the_annals_of_sennacherib_on_the_taylor-_jerusalem-_and_oriental_institute_prisms-_700_bce/.

not only does the "tone and content" of the material change for the duration of the story, but so does the Hebrew spelling of the name Hezekiah, indicating that a new author was involved in drafting the text.[22]

Moreover, the siege story was just part of a larger collection of the history of the kings of Israel and Judah. As such, it was undoubtedly at least edited over a long period of time up to and even after the fall of Jerusalem to the Babylonians in around 586 BCE. The current scholarly consensus places the BOK together with the books of Deuteronomy, Joshua, Judges, and Samuel in a collection called the Deuteronomistic History, works bound together by similar language and reflecting a singular point of view. Rather than attempting to provide a factual recitation of actual events, the authors were looking back, probably from the time that the kingdom of Judah was near collapse or had collapsed, and trying to respond to the catastrophe.

Professor Hayes teaches that the Deuteronomistic History is an attempt to "respond to the first major historical challenge to confront the Israelite people and the Hebrew religion."[23] The destruction of Jerusalem, and the temple within it, as well as the forcible exile of the royal family to Babylonia, raised fundamental theological as well as political problems. Divine promises respecting an eternal Davidic monarchy, the inviolability of Jerusalem, and control of the greater homeland itself had seemingly all been broken. Was it because YHWH, the Creator God and the God of History, was now impotent or faithless?

The genius of the Deuteronomistic Historian (a term meant to include all who contributed to the final text) was an interpretation of events that preserved the integrity of YHWH by characterizing the demise of Israel and Judah as punishment for the sins their disobedient, primarily idolatrous, kings. It was, in short, not just historical fiction, but purposeful historiosophy. The political aim, according to historian Yoram Hazony, was to "embrace and heal a broken people after the loss of its land and freedom," first by "staunch(ing) the hemorrhaging of the community of exiles as they sought solace in foreign ways," and then preparing them "for the long struggle to return to their land and restore their city and their kingdom."[24]

22. See Brettler, *How to Read the Jewish Bible*, 125–26.

23. Hayes, *Introduction to the Bible*, 231.

24. Hazony, *Philosophy of Hebrew Scripture*, 41, 58, 63.

Consequently, clearly, and crucially, the Deuteronomistic History was, as Ziony Zevit of the American Jewish University writes, "relevant to the post-destruction communities in Judah and Babylonia," not only explaining their situation, but demonstrating "the power of God, the validity of His covenant with Israel, and His meticulousness in maintaining it."[25] This reconciliation raised its own problems, to say the least, but as Hayes points out, it allowed the exiles in Babylonia to avoid both "despair and apostasy," and even to view their captors as God's agents, so that they could look forward to a day of return and redemption.[26]

In sum, just as the Torah is not a science book, neither is it or the larger *Tanakh* a history book. Rather, facts were mixed with fiction and faith in order to serve the authors' agenda. Hezekiah—though he failed utterly to protect any part of Judah outside of the walls of Jerusalem, saw his lands reallocated to kingdoms he had conquered previously, and ransomed away what gold and silver was in the capital, even stripping the temple—was characterized as doing "what was pleasing to the LORD." "He trusted only in the LORD the God of Israel; there was none like him among all the kings of Judah after him, nor among those before him" (see 2 Kgs 18:3, 5; but also see 23:25).

In the years to come, others would try to explain how Jerusalem managed to avoid destruction. Luckenbill, for instance, notes that several hundred years after the event, the Greek historian Herodotus wrote that field mice devoured the quivers, bowstrings, and shield thongs of the Assyrians, and much later the British poet Lord Byron attributed the matter to some unspecified Assyrian discomfiture.[27] We may never know for sure.

Perhaps some skirmish distracted Sennacherib. Or, given the enormity of the booty gained, and the logistical challenges in transporting it over seven hundred miles back to Nineveh, he simply took yes for an answer and calculated that there was nothing more to be gained that would be worth the loss of an Assyrian soldier. If Sennacherib thought so, his prudence benefitted Assyria, at least for a while. After Hezekiah's death, Manasseh, Hezekiah's son, reverted to his grandfather's approach, both in terms of domestic religious practice and foreign policy. Judah once again became a vassal of Assyria. BOK provides a litany of Manasseh's

25. Zevit, "First Kings: Introduction," 670.

26. See Hayes, *Introduction to the Bible*, 233.

27. See Luckenbill, "Annals of Sennacherib," 26.

outrageous apostasies, but makes no mention of either war or economic duress, suggesting that his fifty-five-year reign was one of relative peace and prosperity. Archaeologist Finkelstein goes further. He argues that, under Manasseh, Judah became a main supplier of olives for the Assyrian oil industry, increased its trade economy, grew its population, became more literate, and "reached unprecedented prosperity."[28] Of course, BOK gives Manasseh no credit for such results, only finding his conduct "displeasing to the LORD" because he followed "the abhorrent practices of the nations that the LORD had dispossessed before the Israelites" (2 Kgs 21:2).

TWO LESSONS FROM A PRISM AND A SCROLL

The story of Hezekiah and Sennacherib may be considered more historical fiction than fake news, but regardless of label, it was an early example of intentional textual manipulation of recorded reality and creative copy, of inclusion and omission. And it serves to teach us at least two important lessons.

First, as the impact of *Uncle Tom's Cabin* dramatically demonstrated over 165 years ago, historical fiction can be very powerful and persuasive. Surely it was in the case of the Deuteronomistic History. The theocratic system advocated by the Deuteronomistic Historian may lack present appeal, as very few Jews today would like to see a third temple built and sacrifices reinstituted, but it received a positive reception among at least a critical mass of fifth century BCE Judahites.

Moreover, if it is questionable whether Judaism as we know it would have developed had not Cyrus allowed the exiled Judahites to return to their ancestral home, and it is,[29] surely proto-Judaism would not have had any chance to survive and grow had not the Deuteronomistic Historian provided the intellectual and emotional literature of the Deuteronomistic History. His interpretation both rationalized the national catastrophe which occurred and the exile that followed, and, ultimately, laid the groundwork for a program for the future. And that program, to paraphrase words attributed to the poet Issac Leib Peretz (1852–1915), was to be built for one people on one land under one God with one law.[30] With-

28. Finkelstein, "Digging for the Truth," 16.

29. See generally Price, "What If."

30. Baron, *Treasury of Jewish Quotations*, 246.

out the rationale the Historian offered, who is to say that the defeated and dispersed Judahites would have been able to maintain their cohesiveness and their faith such that they would be motivated to return when the opportunity to do so arose?

Further proof of the power of the story is that we are still thinking about it today. We do so not because the dream of the Historian was fully realized, because it was not. For over four hundred years after the return of Judahites and the building of a Second Temple, with the exception of a period of troubled years of rule by the Hasmonean family, Judah (subsequently called Yihud and, later, Judea) remained controlled by foreign powers: Persians, Greeks, Seleucids, and Romans. Yet the underlying theme of the Deuteronomistic History resonated. Indeed, the Deuteronomistic History became part of the foundational history of the Jewish people, canonized in the *Tanakh*. That history, in turn, sustained the people for well over eighteen centuries in their various settlements and migrations following the destruction of the Second Temple in 70 CE. It supported a Jewish desire to reclaim and rebuild an ancient homeland. One result of that dream is that today Hezekiah's Jerusalem is again a vibrant community and the capital of the state of Israel, while Sennacherib's Nineveh lies in decaying ruins near troubled Mosul, and his Annals are hardly read at all. Sometimes, it seems, history is written (and read) not just by the victors, but by the survivors.

A second key lesson of the story, and one applicable to both fake news and historical fiction, is that the essence of a story is not necessarily to be found in the names and activities and dates discussed in the text, but in discovering the purpose with which the author wrote as he did. So, to understand and evaluate what is being said, we do need to know (or try to know) who said it and why. We need to consider the source and the author's agenda. This approach was at the heart of the prime, and capitalized, directive at the old City News Bureau of Chicago: IF YOUR MOTHER SAYS SHE LOVES YOU, CHECK IT OUT.[31] This rule is true for tales in the *Tanakh* and, as it is sometimes written, it is true "until this [very] day" (see, e.g., Deut 2:22).

Consequently, when Jewish texts meet science, we should accept the challenge to seek the truth behind the tale. Whether that challenge comes from the classical sciences or the social sciences, and whether by use of instruments or analysis, we should be prepared to meet the texts

31. Minow, "If Your Mother Says ."

we encounter with critical thinking. This means testing premises, noting inconsistencies and anachronisms, and seeking confirmation, corroboration, or contradiction. The credibility of the text, and our own integrity, demand and deserve no less.

Part Three

RELEVANCE FOR THE PRESENT

Chapter 13

Jews, Genes, and Genetics

IN TRADITIONAL THOUGHT, JEWS are part of two covenantal communities. One has its origins in the covenant the Torah asserts was established at Mount Sinai. The Sinaitic or Mosaic covenant, formulated like Ancient Near Eastern treaties, established mutual terms and obligations by and between the biblical Israelites and their God, YHWH. Should the Israelites obey the terms of the covenant, YHWH agreed to bless them and treat them as a "special treasure," "a kingdom of priests, a holy nation" (Exod 19:5–6). This was so attractive, we are told, that even before hearing the terms of the arrangement, the Israelites in unison recognized YHWH as their deity, pledged their loyalty and love, and agreed to follow commandments, laws, and ordinances set forth in YHWH's name (see Exod 19:8). Later, they reaffirmed their commitment, twice (see Exod 24:3, 7).[1]

By the time the Israelites are ready to enter the Promised Land, though, all of the adults present at Sinai (save for Joshua and Caleb) have died. This presents a problem about the continuing viability of the Mosaic covenant. The author of Deuteronomy resolves this dilemma, though, with an exceedingly clever recapitulation of the story of the revelation at Sinai. Addressing this new group of Israelites, the Deuteronomist has Moses telling them that the covenant at Sinai was entered into by those present at Sinai and those not present (see Deut 29:13–14). To be sure

1. Should the Israelites not obey the rules and regulations, YHWH reserved the right to draw upon a lengthy list of penalties expressed as curses (see Lev 26:14–38; Deut 28:15–68).

that there can be no doubt that the agreement is in full force and effect, though, Joshua reenacts the covenant ceremony after the Israelites cross into the Promised Land (see Josh 24:1–28).

Prior to the covenant at Sinai, though, there was another covenant. This one was neither mutual nor conditional. To the contrary, rather than resembling an ancient suzerainty treaty, it was more of a royal grant. In this scenario, YHWH commands Abram to leave his home for a "land that I will let you see" (Gen 12:1). If he obeys, YHWH will provide Abram not only with land, but will bless him and make of him a "great nation" (see Gen 12:1–3). This arrangement is then modified a few verses later (with the deity known as El-Shaddai), and the ritual of circumcision is introduced a sign of an enduring "covenant for the ages" with Abram's progeny (see Gen 17:12–13).[2]

Both of these communal covenants resonate with Jews today, but in different ways. Membership in the Mosaic covenantal community requires affirmative conduct, a willingness to engage as if one were at Sinai, a voluntary commitment to observe codes, rites, and rituals. Membership in the Abrahamic covenantal community does not. That is, in the United States today, an open and fluid society characterized by opportunity and mobility, most adult Jews are not forced to act Jewishly, i.e., to engage in conduct commonly understood to be specifically Jewish, such as attending *shul*, keeping kosher, or studying sacred texts. Nor are they forced to believe in a particular collection of ideas or ideals, including whether God exists; or, if they think that God does, what attributes or aspects God may or may not have. The reality is that most Jewish Americans are Jews, if at all, by choice.

GENES AND GENETICS

There is, of course, one matter that is not open to choice, much less dispute or revision, and that is one's genetic structure. And here, the Abrahamic community meets science, for here, as elsewhere, advances

2. The *Tanakh* discloses three other covenants. The Noahide covenant is God's unconditional pledge to all humanity to never again destroy the world by a flood (see Gen 9:8–11). The Davidic covenant, articulated by the prophet Nathan, asserts an unconditional and eternal commitment by YHWH to maintain the house of King David as Israel's royal ruler (see 2 Sam 7:8–17). A final, promised "new" covenant advocated by Jeremiah was aimed at the entire community, but was imprecise and has never gained much traction (see Jer 31:31–34).

in science in the relatively recent past have allowed us to investigate, to probe, to attempt to provide science-based perspectives, if not answers, to the most basic questions: "Who am I?" and "Where did I come from?".

Much as astrophysicists and cosmologists can look at electromagnetic radiation in various forms such as invisible cosmic microwaves or visible light that reaches Earth each day and use that information to look back in spacetime to the origins of the known universe, geneticists study the mechanisms of inheritance. They can look deep into the cellular structure of organisms and derive information which allows them to understand the historical ancestors of the organisms they are studying, both individually and in relation to other organisms.

The astrophysicist and cosmologist can assert that we are made of stardust, and that is true. But it is also true that that stardust has been fashioned by biochemical mechanisms that we do not fully understand. And the geneticist knows that, with respect to the human species, our evolution has been both effected and affected by a process of genetic shuffling.

The mechanisms which are the focus of genetic study are found in the cellular structure of organisms. The body of every living organism on Earth is made up of one or more cells. The number of cells an average-sized human male may have is not known, but seemingly reasonable estimates range from about thirty to forty trillion cells, some for brain and blood, others for skin and bone, and still others for the myriad functions that cells perform.[3] Inside each of these cells is an enclosed portion called the nucleus.

The nucleus of a cell houses genes which consist of molecules of deoxyribonucleic acid, or DNA, a molecule characterized by two long, intertwined, and interconnected helical shaped strands of sugar and phosphate units.[4] DNA is associated with the transmission of genetic information from parent to child. This information is contained in a sort of code which can be understood as consisting of four nucleotide bases, generally known by the letters A, T, C, and G. In the DNA, the letters are arranged in various combinations of three letters to form the words of the genetic code. Each of the sixty-four possible triplets relates to a specific amino acid, the building block of a protein. Each protein, in turn, plays a specific role in the cell of the host organism. The double-stranded

3. See Zimmer, "How Many Cells"; see also Hewings-Martin, "How many cells are in the human body?"

4. See Whitfield, *From So Simple A Beginning*, 61.

DNA molecule is, then, a chain of information (call it a roadmap) which governs the development of the organism.[5] The totality of the genetic information of an organism is known as its genome. The genome for the human species is about three billion base pairs of DNA.[6]

In humans, as in other organisms, the DNA is divided into packages called chromosomes. There are forty-six such segments, combined into twenty-three pairs of chromosomes in a human's body.[7] In the process of human procreation, each mate contributes half of its chromosomes, which combine to form the genetic structure of a new individual. With two exceptions, the bits and pieces of each parent are shuffled, i.e., essentially sliced and diced and recombined.

One of the exceptions to the process of recombination relates to the male sex chromosome, known as the Y chromosome. With respect to sex chromosomes, while women have two X chromosomes, men have one Y chromosome in addition to one X chromosome. Therefore, each sperm will contain either an X or a Y chromosome. Because human males have them, and females do not, if the successful sperm carries a Y chromosome, the resulting embryo will be male. Consequently, the Y chromosome is transmitted from father to son, but never from father to daughter. Moreover, the Y chromosome "carries few genes [and] possesses little variation."[8] While "small changes in chromosomal DNA do occur," a son receives an "essentially unchanged" Y chromosome from his father.[9]

In theory, one could study genetic information in the Y chromosome of one human male and see whether there were any similarities to another human male such that the two males might be related. In theory, with sufficient data, one could see whether male members of a particular social, ethnic, national, or other group were related. By virtue of their history, both mythic and real, Jews should be ideally suited to such a study. Until the 1980s, however, the scientific tools to undertake such a study had not yet been developed.[10]

5. Whitfield, *From So Simple A Beginning*, 61–62.

6. See Goldstein, *Jacob's Legacy*, 128.

7. Goldstein, *Jacob's Legacy*, 127.

8. Goldstein, *Jacob's Legacy*, 25.

9. Goldstein, *Jacob's Legacy*, 25; see also Entine, *Abraham's Children*, 55.

10. See Entine, *Abraham's Children*, 54–55.

THE QUEST TO FIND A JEWISH CHROMOSOME

The story of the first Jewish Y chromosome studies is fascinating on several levels, and provides a wonderful example of how scientific discoveries are often the result of circumstance and serendipity. In 1994–95, within a very brief period of time, not one but two men independently came up with the same idea of testing the Y chromosomes of a limited subset of Jews. They arrived at the same place from different paths and with different perspectives, but they joined together and embarked on a remarkable journey.

In late 1994, Neil Bradman, a successful middle-aged businessman, was in the midst of leaving his publishing and real-estate ventures in England and returning to his youthful love: science. His son Robert was then pursuing a graduate degree in genetics, and looking for a research project to be undertaken in Israel where his girlfriend lived. Bradman is a Levite by tradition, i.e., a theoretical descendant of the biblical Levi, third son of the patriarch Jacob. He recalls that he and his son quickly arrived at the concept of "'testing the story of the Jewish priests.'"[11]

In January, 1995, clinical nephrologist and university professor Karl Skorecki was attending a Shabbat morning service in Toronto, Canada. Like Bradman, Skorecki was in his mid-fifties and also planning a major change in his life. In Skorecki's case, the plan was to move to Israel.[12] At this particular service, the rabbi was looking for a *Kohen* to read from the Torah. A *Kohen*, according to Jewish tradition, is a descendant of the priests who served in the temples in Jerusalem. The biblical narrative sets the inception of the Jewish priesthood to a date several months after the exodus from Egypt, or, by some accounts, just over 3,300 years ago, when Moses presided over the investiture of Aaron as the first *Kohen Gadol*, or high priest, and his sons as *Kohanim* (plural for priests; see Exod 29:9). Once established, the priests were assigned specific unique duties and granted special privileges (see, e.g., Lev 6:16, 29; 22:10–16; Num 6:22–26; 27:21; Deut 21:5; 31:9–13; 33:10). The appointment was to be everlasting, extending throughout all the generations (see Exod 40:15; Num 18:19; 25:13).

The rabbi's search was answered by a Moroccan Sephardic Jew. Skorecki, of Polish descent, and the sole child of Holocaust survivors, considered himself to be a *Kohen* as well. As he looked at the Jew of North

11. See Entine, *Abraham's Children*, 76–77.

12. See Entine, *Abraham's Children*, 65–69.

African ancestry, whose physical traits were so different than his, he wondered whether they were really descended, as the Torah and subsequent tradition would have it, from the same person, including possibly the first high priest, Aaron. If they were, and more broadly, if all *Kohanim* were as well, Skorecki hypothesized, they should share common genetic markers, or changes or mutations in a gene. Testing this idea would be "a great summer project for a medical student," he thought.[13] Bradman and Skorecki independently reached out to Batsheva Bonne-Tamir, a professor of genetics at Tel Aviv University, to inquire about the possibility of locating distinguishing markers in the genes of Jewish men. Bonne-Tamir encouraged both men to contact Professor Michael Hammer, a genetic anthropologist and an expert on the Y chromosome, at the University of Arizona in Tucson. Ultimately, the Bradman-Skorecki-Hammer team was formed and a study plan adopted. The purpose of the exercise was not to prove the historical truth of the biblical account of the priesthood, but rather to see whether there was any genetic basis for the *Kohen* narrative of common ancestry extending back to Aaron.[14]

Later in 1995, between Rosh Hashanah and Yom Kippur, Bradman's son Robert collected almost two hundred vials containing the saliva of Jewish males, split evenly between those of Ashkenazi (Central and Eastern European) descent and Sephardi (Iberian) descent. About one-third of these men also claimed to be *Kohanim*.[15] When the data were analyzed, the researchers discovered an amazing fact: regardless of the national origin of each of the participants, regardless of their designation as Ashkenazi or Sephardi, they almost all (98.5 percent) had a relatively unique mutation, called a haplotype, one shared with only 3–5 percent of the world Jewish population.[16] This result suggested at least two important things: First, there was a high correlation between the cultural and genetic records and, second, the original possessor of the haplotype predated the millennia-old separation of Jews into Ashkenazim and Sephardim.[17]

The researchers were reasonably, but not always, cautious when disclosing their results. Their circumspection apparently did not matter. Some in the press exuberantly declared that science had traced the Jewish

13. See Entine, *Abraham's Children*, 69–70.
14. See Entine, *Abraham's Children*, 85.
15. See Entine, *Abraham's Children*, 78.
16. See Entine, *Abraham's Children*, 79; see also Goldstein, *Jacob's Legacy*, 27.
17. See Entine, *Abraham's Children*, 79.

priestly line back three thousand years to Aaron, the first high priest.[18] The research team, however, apparently understood better than others the limits of their initial study, and the need for more sampling and further analysis. Enter David Goldstein, an American transplant to England. Goldstein was a geneticist who had done research on microsatellite markers, variations that exist in more than one form, and used a process of linkage analysis to identify genes susceptible to disease.[19] To further their research, the expanded team collected samples of 306 Jewish men who self-identified as *Kohanim*, Levites, or Israelites.[20] What they found was 109 different types of Y chromosomes, which is a substantial variety of such chromosomes. But there was a significant difference between the kind of Y chromosomes found in the Israelite and non-Israelite populations. Only 12 percent of the Israelite group shared a common type of Y chromosome, but over half of the *Kohanim* had the same type of Y chromosome.[21] Moreover, the substantial prevalence of that particular Y chromosome was found in *Kohanim* of both Ashkenazi (45 percent) and Sephardi (56 percent) origin.[22] The scientists named that frequently appearing chromosomal type the Cohen Modal Haplotype ("CMH").

Accounting for chromosome types with small mutations in very short base pairs, they then found other types "clearly related" to the CMH. Together, these accounted for almost two-thirds of the chromosomes studied in the *Kohanim*. Again, the prevalence of a shared haplotype within the modal cluster was similar in the Ashkenazim (69 percent) and the Sephardim (61 percent). By clear contrast, only 14 percent of Jews identified as Israelites have chromosomes within the modal cluster.[23] These findings were undoubtedly not the result of random behavior.[24] The second research team therefore not only confirmed, but extended, the findings of the first team. It identified the CMH, and refuted the notion that there was much outside interference with the line of male ancestry. What the results did not prove, though, was how old that line

18. See Entine, *Abraham's Children*, 81.
19. See Entine, *Abraham's Children*, 86; see also Goldstein, *Jacob's Legacy*, 128.
20. See Goldstein, *Jacob's Legacy*, 30.
21. See Goldstein, *Jacob's Legacy*, 31.
22. Goldstein, *Jacob's Legacy*, 31.
23. Goldstein, *Jacob's Legacy*, 31–32.
24. See Entine, *Abraham's Children*, 87.

is, and specifically whether it traced back continuously to the time of the biblical priesthood.

How far back could this line be traced? Hammer had thought the first study showed lineage extending fifty generations in the past, itself a stunning result.[25] Goldstein's novel approach of dating by mutations held the promise of more precision. He developed a computer model that considered the time necessary to develop the differences in the observed chromosomes from the Cohen Modal Haplotype.

As Neil Bradman in Israel provided Goldstein with raw data telephonically, Goldstein could see where the data was leading. The results of the second study were eclipsing those of the first. According to Goldstein, he then inputted the numbers into his computer for refinement. After entering the information for the microsatellites, the computer model indicated that the founder of the *Kohanim* line lived about three thousand years ago, give or take a few hundred years.[26] At thirty years per generation, that would cover one hundred generations. And, even at the lower end of the range, the start date for the Kohanim line would have preceded the date of the destruction of the First Temple in 586 BCE by at least a century.[27] Goldstein concedes that the "range of possible dates was and is very broad" and that "the origin of the line cannot be assigned with precision to the time of Aaron, Solomon, or any other named individual."[28] Still, Goldstein cannot resist the suggestion that *if* the biblical Exodus occurred at all, and *if* it can be dated to the late thirteenth century BCE, then the start date for the CMH haplotype would be "well within the interval we predict for the origin of the [K]ohanim."[29] His more modest and "best guess" is that the priestly line was founded prior to "the time of the Romans and perhaps before the Babylonian conquest in the sixth century B.C.E."[30]

What does all this mean? Well, paradoxically, quite a bit and not all that much, both with respect to the *Kohanim* and Jews generally. The science is more than intriguing. For the first time, we have data which seem to be consistent with extensive oral and written traditions and

25. See Entine, *Abraham's Children*, 79.

26. See Goldstein, *Jacob's Legacy*, 37–38.

27. See also Entine, *Abraham's Children*, 89.

28. See Goldstein, *Jacob's Legacy*, 38.

29. Goldstein, *Jacob's Legacy*, 39.

30. Goldstein, *Jacob's Legacy*, 39; see also Entine, *Abraham's Children*, 90.

appear to show extraordinary, perhaps unique, continuity of lineage over two to three millennia in time and across national boundaries, despite countless wars, persecutions, famines, diseases, and other scourges and challenges, not the least of which is random happenstance. Even if one is not religious, the breadth and depth of this lineage and its implied fidelity to family and tradition is astonishing, and gives new resonance to the ancient reference in the Jewish prayer books *l'dor vador* (to all generations).

At the same time, CMH is not limited to Jews. It has been found, albeit in lesser concentrations, in numerous non-Jewish populations from Iraq to Italy, including Palestinian Arabs.[31] What are the practical consequences of what we have learned? The Reform, Reconstructionist, and Humanist approaches to Judaism are already disinclined to recognize distinctions between classes of Jews. They are unlikely to reinstate favored status for *Kohanim* based on this genetic information. Over the last seventy years, the Conservative movement has modified its views on several previously understood priestly privileges.[32] Though persuasive, the lack of certainty attendant to genetic testing today does not seem sufficient to cause a reversion to prior practice. Because the Orthodox community's commitment to traditional rituals was not dependent on genetics in the first place, the *Kohanim* studies should not have much affect at all, unless there were a move to formalize recognition under Jewish law, which seems doubtful. Similarly, Israel's Law of Return for Jews is unlikely to be amended to provide for genetic "proofs."

The Center for Kohanim in Israel promotes study of the biblical role and history of *Kohanim* and *L'viyyim* (the Levites). While it was formed over a decade before the studies on the *Kohanim* yielded the results discussed above, the Center features these results, seeing the findings as supportive of the statements in the Torah to the effect that not only will the Aaronide priesthood be everlasting, but the Torah is "truth." It encourages all *Kohanim* and Levites to register with it, and looks forward to a day when the Israelite temple is rebuilt and "Kohanim [are] at their service, Levites on the Temple platform Israelites at their places."[33] Aside from the massive political problems associated with the rebuilding of the temple, the narrative of migration and conquest is highly questionable.

31. See Entine, *Abraham's Children*, 91.

32. See Frydman-Kohl, "Priests and Levites," 1444–46.

33. See "Kohanim Forever."

Similarly, the historicity of Aaron and his immediate descendants has not been proven by either of the *Kohanim* studies mentioned here.

Further, the story of an unbroken line of priests is internally contradicted by the Torah text itself, as well as other biblical and historic writings. For example, Deuteronomy refers to *L'viyyim* generally as well as descendants of Aaron being eligible for the priesthood (see Deut 17:18; 10:8–9; 33:8–10). And the Maccabean revolt in the second century BCE, celebrated in the Hanukkah holiday, was not simply against the Greek king Antiochus IV, but also against the Jewish priests who, whether or not in the Aaronide line, supported the king's efforts to Hellenize Judaism. Matityahu's son Shimon (Simon) may have assumed Aaron's role as *Kohen Gadol*, the high priest, but his doing so was a stark departure from the theme of direct lineage, whether justified or not. Finally, it is not at all clear that Aaronide lineage was restored during the Roman occupation of ancient Palestine.

In short, as resonant as are the results of the *Kohanim* studies, they do not appear likely to make much of a difference, except perhaps to the psyche of some self-identified *Kohanim*.

What of the 95–97 percent of Jews who are not, or do not believe themselves to be, *Kohanim*? There are a number of broader lessons here. First, science is not static. The *Kohanim* studies were ground-breaking, but also limited in scope due to subject matter, sampling size, and the sophistication, or lack of it, of the researchers' tool kit. Moreover, while patrilineality was key to Jewish identification in biblical times, it no longer is today, as a general matter, and has not been for fifteen to twenty centuries. Tracing Jewishness along the maternal path obviously requires investigating something other than the Y chromosome studied for the *Kohanim*. It means focusing on mitochondrial DNA, or mtDNA, which, while present in both males and females, is transmitted between generations only from mother to daughter.[34]

The *Kohanim* studies, and more advanced technology, did spur a spate of investigations into Jewish population genetics. Collectively, these studies did several things that the *Kohanim* studies did not do. They looked at a wider variety of Jewish populations, presumably non-Jewish populations for comparisons and control, greater numbers of participants, and, by no means least, one more very important gender.[35] The

34. See Goldstein, *Jacob's Legacy*, 80.

35. Information concerning some of these more recent projects may be found at "Jewish Genetics: Abstracts and Summaries."

results were not entirely consistent, however. For instance, studies led by Naama Kopelman and Harry Ostrer separately do confirm that Jews in different diaspora groups share a common Middle Eastern deep ancestry, and "show a high level of genetic similarity to each other."[36] While each of the Jewish groups studied demonstrated its own distinctive genetic traits, there remained a "genetic coherence," a "high level of genetic relatedness" among the group.[37] Steven Bray's study, though, while agreeing that Ashkenazi Jews share a common Middle Eastern origin with other Jewish populations, also suggests that there is more genetic diversity in the Ashkenazi population than previously thought. His results show a pattern "*subtly more similar to Europeans than Middle Easterners*" and "*aligning closest to Southern European populations*."[38] Nevertheless, the ability of geneticists to specify when and how populations intermixed remains limited.

Geneticists have identified eighteen principal Y-chromosome haplogroups. A haplogroup is a population which shares the same collection of markers. Each of these major lineages bears an alphabetical designation from A through R. Ashkenazi Jews generally, but not always, fall into either the E or J haplogroup.[39] For instance, the CMH haplotype is a subset within the J haplogroup. Parenthetically, the twin ironies here cannot go unnoticed. Eighteen is the numerical equivalent of *chai*, the Hebrew word for life. And, as we have seen, higher biblical criticism posits that the Torah is an amalgam of four principle collections of writings, designated E, J, D, and P, with E referring to an author who preferred to refer to God as *Elohim* and J referring to an author who favored YHWH.

The J1 (M267) haplogroup emerged some one hundred thousand years ago in the southern Fertile Crescent, possibly where Iraq is located today. Over time, and through major immigrations, members of the group moved to Europe, Ethiopia, and later to North Africa and the Iberian Peninsula. All of this is also consistent with the biblical story that places the patriarch Abraham initially in present-day Iraq and then traveling to Canaan and Egypt.[40] It in no way proves the existence of Abraham or that he engaged in the conduct described in the Torah, but it is in harmony

36. "Jewish Genetics: Abstracts and Summaries," para. 13.

37. "Jewish Genetics: Abstracts and Summaries," para. 12.

38. "Jewish Genetics: Abstracts and Summaries," paras. 2–3 (emphasis original).

39. See Entine, *Abraham's Children*, 88, 359–62.

40. Whether northern or southern is in dispute. See Lieber, *Etz Hayim*, 62n28.

with the notion of a Middle Eastern origin for the Jewish people. More than half of J1 (M267) in one database are Ashkenazi Jews, genetically connected to historic Jewish origins in the Middle East.

JEWISH PEOPLE AND JEWISH PEOPLEHOOD

Long before genetic anthropology was a field of study, Mordecai Kaplan tried to resolve the question of the nature of Judaism by positing that Judaism was the evolving, religious civilization of the Jewish people. But that definition, while comprehensive, is still somewhat circular. Who or what are the Jewish people? Clearly, Jews are more than merely coreligionists, because there are sharp disagreements among religious Jews about the most basic of religious principles, and there are many nonreligious Jews. Equally clearly, all Jews are not members of the same race, as that term is conventionally understood. Rather, Jews exhibit a wide variety of physical traits and come in many colors. You can be a Jew in any hue.

Similarly, based on the information collected to date, no one can argue reasonably that all Jews belong to a distinct genetic group. Genetics is not a conclusive basis for confirming or denying anyone's Jewishness. No haplotype is either necessary or sufficient for Jewish identity, whether viewed internally or externally. Consequently, those who look to genetics, either to promote Jews or a Jewish cause or to reject Jews or a Jewish cause, are doing so without a firm scientific basis.[41]

In the summer of 2018, the unique character of Jewishness was considered in a federal lawsuit that alleged discrimination by Louisiana College against an applicant for a position as an assistant football coach who claimed that he was deprived of his desired job because he had "Jewish blood." The plaintiff's mother was Jewish, and he grew up attending a synagogue. Later, though, he converted to Christianity and, while a student-athlete at Louisiana College, led the football team in prayer services. Because he did not claim to adhere to the Jewish religion, the issue of religious discrimination was not before the court. Rather the question was whether Judaism could be considered a race for purposes of Title VII of the 1964 Civil Rights Act, which forbids employers from discriminating based on race, among other categories. US Magistrate Judge Mark

41. On this, Jews across the religious spectrum agree. "Judaism is not rooted in race or clan or in a genetic matter but [in] a religious tradition of choice" (Schulweis, "Keruv, Conversion and the Unchurched," para. 31). See also "Racial Theory," http://international.aish.com/seminars/whythejews/yjd05600.htm.

Hornsby held that it could. In response, Rabbi David Wolpe, of Sinai Temple in Los Angeles, said most rabbis would argue that neither race nor religion is accurate as an appropriate box into which to fit Judaism. "We're not a race because you can't convert to a race. You can't decide to be black tomorrow," he said. "On the other hand, it's not a religion because you're not born into a religion. It's a category that does not fit neatly into 21st century American ways of thinking."[42]

So, despite this judicial opinion and the recent studies, and the sensationalized headlines that accompanied them, Jews remain something of an enigma, a people bound by a variety of common threads to each other, related by one or more of faith and culture, language and literature, desire and circumstance, and custom and experience. Today's Jews are heirs not just to tales, texts, and traditions that began thousands of years ago, but are in some cases literal descendants of individuals who helped shape the collective history. Some consider themselves chosen, and others have made a choice to be part of the Jewish people. Either way, both now share not only a common history, but a common destiny. The result is a multiracial, transnational, and sometimes fractious extended family, and though an exceedingly small segment of the greater human family, an exceptionally interesting one at that.

42. See Natanson, "Louisiana Judge Says." The magistrate's opinion is subject to further review in due course. Left open by Rabbi Wolpe's observations is the troubling question of what legal protection against employment discrimination might be available for Jews if the traditional categories of race, religion, and national origin are not appropriate.

Chapter 14

Ginger Jews

What a Single Gene Tells Us about Jewishness

In the Summer of 2014, about two hundred red-haired Israeli Jews gathered for a conference at Kibbutz Gezer in Israel.[1] While that is a nice size for a group, there were, apparently, many hundreds who were interested in attending but unable to do so. Those who attended the conference shared stories, sang a popular children's song called "I am a Redhead," and reportedly had a good time. *Gezer*, by the way, is Hebrew for carrot.

And then there is Stav Shaffir, the thirty-something member of the Knesset whose hair is vibrant red. *Stav*, by the way, is Hebrew for autumn.

There is even Hebrew slang for redheads: *gingi* (Jeenji) for a male and *gingit* (Jeenjit) for a female, both Hebraicized corruptions of the English ginger.

What's with Jews and red hair?

The Jewish connection to red hair turns out to be quite complex. The first possible references to redheaded Jews appear, not surprisingly, in the *Tanakh*, the Hebrew Bible. Two well-known personalities, Esau and David, are described as *admoni*, meaning red or ruddy (see Gen 25:25; 1 Sam 16:12; 17:42). Some commentators leap quickly from *admoni* to red hair, but there is both more and less here than meets the eye. With respect to Esau, the text suggests that he was red and hairy all over, but *admoni* also serves as a pun for Edom, whose residents were said to be descendants of Esau (see Gen 36:9). When used regarding David, the

1. See Arad, "Finally, a Red Alert."

reference is even more obscure, and does not clearly involve hair. Rather, it is more suggestive of a ruddy complexion.

Even if the two references were to red hair, they provide no consistent signal or message or even information, because the two men are viewed quite differently. David, the poet-warrior and king of united Israel, has been idealized as the founder of a royal dynasty that God promised would last forever (see 2 Sam 7:12–16; 1 Kgs 9:4–7; 11:36; 15:4; 2 Kgs 8:19). By contrast, Esau has been marginalized in the Torah. Tricked by his younger twin brother Jacob, the patriarch-to-be, Esau traded his birthright for a bowl of red lentil stew, later married two Canaanite women and a daughter of Ishmael, and is viewed as the progenitor of the Edomites (see Gen 25:30–34; 26:34; 28:9; 36:1–3, 9).

In any event, these two instances of red hair, if that is what they are, seem isolated instances in the *Tanakh*, and not particularly indicative of anything worthy or special. To the contrary, black hair was viewed as normal, even idealized (see Eccl 11:10; Song of Songs 5:11).

Of course, whether Esau and even David actually existed is open to question, and the description of hair as red may have been more a literary device than actual reporting. Or not. In 2000, Dr. James Tabor, a professor of Religious Studies at the University of North Carolina, entered a recently broken entrance to a first century trilevel tomb south of Jerusalem. Inside, he found not only skeletal remains of a Jewish male, but a preserved sample of his hair. And the color of the hair was "reddish."[2] So the notion of red haired Jews during biblical times may not have been entirely fanciful.

Many centuries later, in English drama and literature, two Jews were portrayed with red hair, and quite unfavorably so. In productions of *The Merchant of Venice*, William Shakespeare's Shylock was frequently costumed with red hair, really a fright wig.[3] Similarly, Charles Dickens's Fagin, the manipulative criminal in the novel *Oliver Twist*, was adorned with red hair, albeit natural.

While the anti-Semitism prevalent in England during these times (at least) cannot be denied, one should be cautious about drawing a connection between the red hair and antipathy toward Jews. Shakespeare himself did not portray Shylock in unmitigated bad light. To the contrary, the bond story in which Shylock is prominent is of a piece with the

2. For more, see Tabor, "Only Ancient Jewish Male Hair."

3. See Jonah, "Shake-Up of Shakespeare's Shylock," 7.

two other principal themes in the play, the casket story and the ring story. In each and all, the playwright literally asks about form and substance through his characters, as well as uniqueness and commonality. So Shylock famously inquires, "If you prick us, do we not bleed?" And later, Portia, guised as a young lawyer, wonders, "Which is the merchant here and which the Jew?" As the marvelous early-twentieth-century Shakespeare scholar Harrold C. Goddard observed, this ironic play is about "what is within and what is without," and we are often more like the other than we might wish to recognize.[4]

Nor is it clear or even likely that Shakespeare would have used red hair as a sign of malevolence. After all, when *The Merchant of Venice* was first mounted in 1596, the reigning monarch was Queen Elizabeth I. As the queen had red hair,[5] there was nothing to be gained by antagonizing her.

Dickens's treatment of Fagin was quite different. The original version of his novel, published in 1837, contained over 250 references to Fagin as "the Jew,"[6] and Dickens did not mean it in a nice way. Nor did he temper his portrayal with humanistic utterances. Still, his selection of red hair for Fagin was not necessarily part of the seemingly anti-Semitic package. It surely could have been the literary equivalent of the fright wig associated with Shylock, but it may also or additionally have been meant to be one in a series of characteristics like old age, ugliness, criminality, and suggested child predation that marked Fagin as an archetypal villain.[7] That is, the hair choice could simply have been a conventional, accessible, and easily understood "marker of low moral character, of fiery hot tempers, of violence, of suspiciousness."[8]

Jews have had their own postbiblical fictional redheads, too. In Yiddish folklore, *di royte yidn* were redheaded Jewish fighters who were strong, brave, independent warriors and could rescue their fellow Jews from whatever was the persecution of the day.[9]

Esau and David, Shylock and Fagin, and *di royte yidn* notwithstanding, red hair was and is not a predominant trait of Jews. Hair color, like

4. See Goddard, *Meaning of Shakespeare*, 82.

5. See "Queen Elizabeth I: Biography, Facts, Portraits & Information."

6. See Walsh, "Dickens' Greatest Villain," para. 4.

7. See Mergler, "Dickens' Great Dark Villain."

8. Mergler, "Dickens' Great Dark Villain"; see also Johnson, "Dickens, Fagin and Mr. Riah."

9. See Philologos, "Redheaded Warrior Jews."

other traits, is determined genetically, of course, and, in this instance, by the production and regulation of two pigments by the melanocortin 1 receptor ("MC1R") gene found in chromosome 16. But MC1R is not like the Cohen Modal Haplotype that seems to allow certain male Jews to trace their lineage back 2,500 years to the Second Temple period. Nor is it found disproportionately among certain Jews as are certain BRCA1 and BRCA2 mutations which are markers, or genetic signals, for early onset breast cancer.[10]

The MC1R gene appears to be recessive. Typically, for an individual to be born with red hair, both parents must be carriers of an MC1R gene and the MC1R gene from both must combine in the fertilized egg. MC1R operates on two pigments, eumelanin and pheomelanin, and, in general, the more of the latter, the redder the hair. But MC1R is also quite variable, and may be subject to being influenced by modifiers. In fact, according to University of Delaware professor John H. McDonald, the genetics of hair color is "complicated."[11]

While human hair color varies enormously, from the lightest blonde to the darkness black, red hair manifests itself only in about 1 percent of humans worldwide. Individuals with red hair can be found around the globe, but the greatest concentrations are in Northern European populations, and in particular, in Scotland and Ireland where, respectively, 13 percent and 10 percent of the population are redheads. One theory is that genetic material for red hair was favored in such areas because it would allow for the production of Vitamin D in circumstances of low sunlight and ultraviolet radiation. The prevalence in the United States is 2 percent.[12]

Data on the percentage of Jews with red hair is uneven and questionable. Hair color does seem to have been a topic of considerable interest at the end of the nineteenth century and early in the twentieth. An article in the Jewish Encyclopedia published in 1906 contains several tables which collect various observations of hair color in Jews around the world.[13] One table concerns Jewish school children in Central Europe (Austria, Bavaria, Germany, and Hungary). While most of the children are indicated to have brown or black hair, approximately one quarter to

10. See Rubenstein, "Hereditary Breast Cancer in Jews."

11. See McDonald, "Red Hair Color."

12. See "Red Hair Gene."

13. See Hirsch et al., "Hair."

one-third of these children are said to have blond hair. The incidence of red hair is less than 1 percent. A second table concerns hair color among Jews in selected countries throughout Europe. While dark hair again predominates, the frequency of red hair often appears to be 2 percent or higher, reaching more than 4 percent in Poland, Galicia, and Russia. The information is of doubtful value, however. Among other problems, the size of the sample populations differs greatly from country to country and the method for selecting the individuals is unknown.

A somewhat similar review occurred in New York City, the results of which were published in 1903 by Maurice Fishberg, a physician and anthropologist, in the *American Anthropologist*.[14] With a sample size of almost 2,300 Jews twenty years old and older, and reasonably split between males and females, Fishberg found that about 82 percent of Jews studied had dark hair—meaning black, brown, or dark chestnut—while about 15 percent had fair hair—that is, light chestnut or blond—and about 3 percent had red hair.[15] The precise percentage of male redheads was 2.53 percent, and the percentage for females was 3.69 percent. Fishberg characterized the percentage of red-haired Jews to be "high."[16] And he stated, without reference to any authority, that erythrism (i.e., a prevalence of red pigmentation) "has been regarded as characteristic of the European Jews."[17] Similarly, he contended that that the condition "appears not to be of recent origin," referring to the biblical descriptions of Esau and David.[18]

Fishberg also observed the color of beards on 587 Jews and found that 10.9 percent of them were red. From this, he concluded that "red hair is nearly three times as common in the beard as in the hair of the head." The basis for his calculations is not clear. If the percentage of male redheads was 2.53 percent, then a 10.9 percent red-beard observation would indicate that red beards are more than four times as common as red head hair. In any event, Fishberg characterized the frequency of red beards as "not at all surprising" because "anyone who has observed Jews closely" would know that "the beard is quite frequently red."[19]

14. See Fishberg, "Physical Anthropology of the Jews."
15. Fishberg, "Physical Anthropology of the Jews," 92.
16. Fishberg, "Physical Anthropology of the Jews," 97.
17. Fishberg, "Physical Anthropology of the Jews," 98.
18. Fishberg, "Physical Anthropology of the Jews," 98.
19. Fishberg, "Physical Anthropology of the Jews," 99.

The percentage of red-headed Jews today is not at all certain. What is more apparent and important is that Jews come in all shapes and all sizes and all shades, with different aptitudes, attitudes, and orientations. Whether Jews started as one wandering family that settled in Egypt, grew over time, and were forged into a nation in the wilderness, or emerged from Canaanite tribes, the Jewish people today is truly a mixed multitude. As we have seen, Jews are a multinational, multiracial people whose members are bound together in different ways and to different extents by an uneven mix of religion and culture, language and literature, history and choice, and, yes, genetic material too. For some, that genetic material includes the MC1R gene that may make for redheads.

Red hair is not, however, a marker of Jewishness. Red hair is neither restricted to Jews, nor is it predominant among them. Natural hair grows on Jews in many colors, maybe not as many as the colors on Joseph's coat (see Gen 37:3), but more than enough to dispel unwarranted stereotypes. Literary and artistic conventions aside, the incidence of red hair among Jews evidences that Jews are just like everyone else. As Shylock might say today, "Swab our cheeks. Do we not share the same chromosomes?" While that may not be as tightly nor as sharply put as what Shylock said in his soliloquy, the point is the same.

Ginger Jews remind us of how varied Jews are. Ironically, the most important thing about redheadedness in Jews may well be that it is really not that important at all.

Chapter 15

Judaism and Nuts

Ethics and Allergies

It is one of the most dramatic moments in the entire Torah. There is no lightening or thunder, no plagues or parting of the sea, just an elderly statesman appearing before his people one more time, to teach one more lesson before they part from each other, the people to cross the river and the old man to enter eternity. Having led for so many years with the assistance of signs and wonders, now he simply speaks words, hoping to refresh their recollection and inspire them. He reminds them of their history in order to set the stage for their future. He tells them again what they should and should not do, emphasizing that they will have to make choices that will lead to prosperity or adversity, choices that will enhance life or bring death. This leader, this teacher, this Moses urges them to "choose life, that you may stay alive, you and your seed" (see Deut 30:19; see also Lev 18:5). Not for nothing is the Torah known as an "*Etz Chaim*," a tree of life (see Prov 3:18; Ezek 20:11).

This reverence for life is more than some gauzy good feeling. Judaism at its best is grounded in experience, rooted in reality. Centuries after the biblical authors first put quill to scroll, the rabbis in the Talmudic period considered situations where observance of biblical ordinances on the sanctity of the Sabbath might adversely, perhaps fatally, affect real people—a wall that had collapsed on a child but could be removed, a fire that could be extinguished.[1] Referring to an obscure statement in the

1. See b. Yoma 84b, https://www.sefaria.org/Yoma.84b?lang=bi; see also, b. Yoma

Holiness Code which seems to prohibit standing by or upon the blood of your neighbor (Lev 19:16), the rabbis formulated the doctrine of *pikuach nefesh* (the preservation of human life), the principle that all of the laws, all of the rules, and all of the regulations which are in Torah can be abrogated to save a life. There are three major exceptions, essentially related to the cardinal sins of idolatry, murder, and improper sexual behavior such as adultery and incest, but the bias is otherwise comprehensive in favor of saving the life of another: "Whoever saves a life is considered to have saved the entire world."[2]

And if someone should fall ill, Jewish tradition seeks healing. In the Torah, God was conceived as *Rofeh Cholim*, the Healer of the sick (see, e.g., Gen 20:17; Num 12:13; see also 2 Kgs 20:1–5; Jer 17:14). Not surprisingly, the traditional prayer service contains a prayer for the sick, the *Mi-Shebeirach*. Even for those who cannot accept the notion of an intervening Divine Doctor, the expression of concern, of desire, of hope for a *r'fua shleima*, literally a complete cure, resonates with great power.

The biblical authors also knew that it was not enough just to revere life or treat illness. Prevention of harm was seen as crucial. So the Torah warns us not to place a stumbling block in the path of the blind (Lev 19:14), and we understand that we are responsible for the welfare of others—especially those whose circumstances or condition place their health or safety at risk.

And what does all of this have to do with nuts, the delicious treat that can lower bad cholesterol and provide heart healthy nutrients? The short, if not simple, answer is that for many people, and an increasing number of them, nuts can be deadly. For instance, as *Tablet* columnist Marjorie Ingall has written, they can turn *charoset* into the Mortar of Doom.[3]

Researchers estimate that fifteen million Americans have food allergies.[4] These allergies affect one in thirteen children under the age of 18 (about two children in the average classroom). And the problem is getting worse. As the National Center for Health Statistics has reported, food allergies in general are increasing rapidly among children.[5] More-

83a, https://www.sefaria.org/Yoma.83a?lang=bi.

2. See b. Sanhedrin 37a, https://www.sefaria.org/Sanhedrin.37a?lang=bi.
3. See Ingall, "Going Nuts," para. 3.
4. See Food Allergy Research and Education, "Facts and Statistics."
5. See Blue Cross Blue Shield, "Childhood Allergies in America."

over, children in large urban centers have significantly greater incidences of food allergies than do children in rural communities. Summarizing the statistics, *ScienceDaily* reports that almost 10 percent of urban children have food allergies.[6]

Ninety percent of allergic reactions to food are caused by eight particular allergens. Two of the primary triggers are peanuts and tree nuts.[7] Rates of peanut allergies, specifically, tripled between 1997 and 2008.[8] The results of a multicenter study published in 2014 show that today about 6 percent of children in urban areas are allergic to peanuts.[9]

Distinguishable from food intolerance, a food allergy involves the immune system. When an allergic person eats an offending food, that person's immune system perceives the food as a foreign invader and attacks it, releasing a number of chemicals in the process. Symptoms may be relatively moderate, such as a tingling sensation in the mouth, hives, or cramps, but also may include swollen lips, difficulty in breathing or swallowing, reduced blood pressure, and loss of consciousness.[10] When a person's blood pressure drops to a dangerously low level, anaphylactic shock occurs. Even a very small amount of allergen can result in life-threatening anaphylaxis.

Injection of epinephrine (adrenaline) is both the first line of defense and the only available treatment for anaphylaxis, but its use is not really a cure. Rather, if administered quickly enough, it merely buys a few minutes of time to get to a hospital emergency room for further treatment. To really avoid allergic reactions, a person with a food allergy must avoid the allergy-causing food.

So, as important as it is to be able to recognize and treat an allergic reaction, prevention in the form of avoidance is truly the best medicine. The good news is that congregations, temples, synagogues, and *shuls*, as well as Jewish federations, centers, and other organizations, can respond to the growth of food allergies by adhering to the principle of *pikuach nefesh*.

One such policy is to become a nut-free facility—that is, to prohibit any food containing peanuts or other nuts from being being cooked or

6. See Gupta et al., "Geographic Variability."

7. See American Academy of Allergy, Asthma, and Immunology, "Food Allergy Overview."

8. See McCoy, "Peanut Allergies on the Rise."

9. See Wood, "Food Allergies More Widespread."

10. See Mayo Clinic, "Food Allergy."

served in the facility, or even brought inside. This approach is especially helpful in protecting young children who cannot read food labels or otherwise guard against their own allergic mishap. Allergies aside, some organizations already ban any food that is not prepared in their own kitchens or under certain religious supervision. Groups that do allow food to be brought in, regarding such events as potluck meals, need to recognize that for some people with food allergies, luck is not a good policy.

A number of congregations across the country already strive to be nut-aware and nut-free—for example, Woodlands Community Temple in White Plains, New York (Reform) and Congregation Beth Shalom in Seattle, Washington (Conservative). For other organizations, here is a protocol based on what some groups are doing now to implement the principle of *pikuach nefesh*.

1. *Pikuach nefesh*, the saving of life, is a core Jewish value. We as a community take this value seriously and are committed to making our home a safe environment for our members and guests.
2. We recognize the danger that certain foods may pose to those who would pray, study, or just gather together with us. So we seek to remove the stumbling blocks.
3. Peanuts and tree nuts and foods containing or derived from nuts (including peanut paste, peanut butter, Nutella, Bamba, cooking oil containing peanut oil, as well as nut granola bars, Reese's candies, or food processed on equipment that process nuts) are not allowed anywhere in our facility.
4. This policy applies to all meals, *Ongei Shabbat*, snacks, in-school events or parties in our facility, and on field trips and congregational outings.
5. Congregants are and will be advised to read labels on all foods in order to identify peanut and tree-nut ingredients "hidden" in foods and to avoid bringing or sending prohibited foods to our building.
6. Notices of potluck and other similar events will include a reminder that no peanuts or tree nuts or any food containing peanuts or tree nuts are allowed to be brought to the building.
7. Caterers may not serve peanuts or tree nuts or foods containing or derived from them in the building.

8. The school noard and principal will develop procedures for implementing this policy in the religious school. As an additional precaution, teachers will be trained in preventing and recognizing allergic reactions and in the use of injectable epinephrine.
9. This peanut and tree nut policy will be communicated to the congregation each year, and the policy will be posted on the community website.

Yes, yes, some will feel that such a policy inconveniences them. But *L'Chaim* is not just a toast with a click of the glass. It is a core value and a promise. And so Jewish tradition has not recognized inconvenience as an exception to the principle of *pikuach nefesh* at any time or in any place. Rather, across time and space, Jews have consistently opted to protect those in need and preserve life. Here, clearly, when Judaism meets science, it should know how to respond: *L'Chaim! To life!*

Chapter 16

A Nice Jewish Shot

Why Vaccinations are Kosher and Required

Let's face it. Sometimes you can deny certain reasonably established scientific truths and it does not make much difference. You can, for instance, believe that the Earth was created about six thousand years ago, and life as we know it will still go on. Okay, maybe college professors and television comics will make fun of you, but sure as you are reading this page, the Sun will appear to rise in this east tomorrow, and later it will set over the western horizon. Yes, another day. Life will go on.

If, however, you deny the safety and efficacy of approved medical vaccinations designed to prevent harmful, debilitating, even deadly diseases, such as polio, measles, hepatitis, and tetanus, your belief may well make a great deal of difference to you, your family, your community, and, indeed, all of humanity.

And yet, there are those who for a variety of reasons refuse to inoculate themselves or their children, or both, even when established governmental authorities require such action. While it is tempting to stereotype all such persons as undereducated or acting out of ignorance, some are not. Aside from the rare situations based on the medical condition of a child, some people object to a particular vaccine or procedure. Others have broader religious, philosophical, and personal beliefs that militate against inoculations. Some even may be part of an otherwise-socially conscious community.

"Jewcology" is a website that aims to be a resource for the "entire Jewish-environmental community."[1] One of the bloggers that Jewcology hosts is Raphael Bratman, and one of Bratman's posts is entitled "Water Fluoridation and Vaccinations are Contrary to Torah Principles."[2] Bratman's article is filled with so many erroneous statements that one is tempted to let this dog lie and hope that it gets lost in the tangle that is the World Wide Web. The problem, however, is that people visit the Jewcology site, and several hundred of them have already indicated that they "like" the anti-vaccination post. While undoubtedly some people get their kicks from clicks, and will "like" anything, one cannot discount the possibility that some readers actually believe the nonsense contained in the piece.

Moreover, as we know, Jewish tradition teaches that we may not exploit our neighbors by placing a stumbling block in their path, and neither may we stand idly by the blood of our neighbor (see Lev 19:14, 16). To the contrary, we are obligated to remove the impediment, to protect the neighbor. So let's remove this particular stumbling block, piece by piece. Let's set the record straight.

Bratman's post asserts three main arguments in opposition to vaccinations. The first is that they don't work, that the benefits are "unclear and unproven." The second and third relate to Jewish law. Bratman argues that vaccines are *treif*, that is, they contain ingredients forbidden to observant Jews, and also that their use violates a primary Jewish principle we have discussed before, that of *pikuach nefesh*, the preservation of life. None of the arguments is supported by any credible authority. The first is contrary to well-established and well-documented medical science. The second and third thoroughly misstate Jewish tradition, as understood by a broad spectrum of scholars.

VACCINATION PROGRAMS ARE INDISPUTABLY SAFE AND EFFECTIVE FOR SUITABLE PATIENTS

As explained by the US Department of Health and Human Services ("HHS"), a vaccine is a chemical compound that contains "very small amounts of a weak or dead germs" and "prepares your body to fight [a]

1. See generally http://jewcology.org/about/.
2. See Bratman, "Water Fluoridation and Vaccinations."

disease faster and more effectively so you won't get sick."[3] The diseases for which vaccines are administered include measles, mumps, pertussis (whooping cough), hepatitis, and smallpox, among more than a dozen. If you are not familiar with one or more of these diseases, thank a vaccination program.

According to the National Institute of Allergy and Infectious Diseases ("NIAID"), a part of the US National Institute of Health, vaccines work because they stimulate the human immune system to attack the invading microbes present in the vaccine.[4] That is, they simulate a disease and trick the body into fighting it resulting in immunity to the real disease.

When introduced into a community, vaccination programs have been extraordinarily successful in preventing cases of disease and related consequences. The long view is instructive. In 1900, 53 percent of the causes of death in the United States were infectious diseases. By 2010, as a result of various vaccination programs, infectious diseases were responsible for only 3 percent of deaths in the United States.[5]

The data are even more compelling when one looks at the results of introducing a vaccination program for a particular disease. For instance, prior to the widespread use of measles vaccines in the United States in 1963, the annual number of reported measles cases exceeded five hundred thousand. The drop-off after the introduction of the program was swift and dramatic. In 2009, there were only seventy-one reported cases, for a decline from base of 99.9 percent. Diphtheria, with a former baseline of 175,000 annual cases, and smallpox with a base line of forty-eight thousand cases, were nonexistent in 2009, thanks to vaccination programs. Even where diseases have not been eliminated, e.g., mumps and pertussis, the reduction of cases has been dramatic.[6]

Over and over, facts demonstrate that vaccines work. And facts matter today in this situation, and in others, for the same reason that a young John Adams taught us long ago when defending eight British soldiers accused of murder during a riot in Boston. Facts, he argued, are "stubborn things." They protect us from the unfiltered application of "our wishes,

3. See https://www.vaccines.gov/basics/index.html.

4. See National Institute of Allergy and Infectious Diseases, "How Do Vaccines Work?"

5. See Jones et al., "Burden of Disease."

6. See National Institute of Allergy and Infectious Diseases, "Vaccine Benefits," https://www.niaid.nih.gov/research/vaccine-benefits.

our inclinations, or the dictums of our passions."[7] Amen. Moreover, if enough people in a particular community are vaccinated, a process known as community or herd immunity develops, offering protection to the few who, due to special circumstance, cannot be vaccinated.

While Bratman acknowledges that the incidence of polio has declined over the years, he attributes that result to "improvements in sanitation and hygene [*sic*]." Of course, Bratman fails to cite a single such improvement, much less one that occurred throughout the United States at exactly the time in 1955 when the newly approved Salk vaccine was first being administered and which would account for a rapid 93-percent decline in cases of paralytic polio. What hard data shows is that incidents of infectious diseases, including polio, did not decline slowly and steadily over time due to continuous environmental improvements, but dramatically and directly as a consequence of the administration of a vaccine program. Not surprisingly, and despite Bratman's suggestion to the contrary, the savings in direct medical costs and in indirect social costs resulting from fewer cases of infectious disease is substantial.[8]

Conversely, where participation in a vaccination programs has declined or been refused, we have seen a resurgence of diseases like measles, mumps, and whooping cough, which some thought were contained, if not eradicated.[9] In early 2018, for instance, over sixty individuals reportedly contracted measles. The majority of these individuals were not vaccinated against the disease.[10] Outbreaks typically occur in close communities whose members fail to act vigilantly to secure comprehensive compliance with vaccination programs. In 2017, for instance, a measles outbreak arose in an Orthodox Jewish community in Los Angeles.[11] A few years earlier, there was a mumps outbreak in New York and New Jersey. An investigation found that 97 percent of the case patients were Orthodox Jews.[12] Even more recently, measles has spread to upper Manhattan and the Bronx and has been reported in other areas around Boston, San Francisco, and Los Angeles.

7. See McCullough, *John Adams*, 68.

8. See Remy et al., "Vaccination: The Cornerstone"; see also Centers for Disease Control, "Achievements in Public Health, 1900–1999 Impact of Vaccines Universally Recommended for Children—United States, 1990–1998."

9. See Shim, "3 Diseases."

10. Centers for Disease Control, "Measle Cases and Outbreaks."

11. Karlamangla, "Measles Outbreak Grows."

12. See Barskey et al., "Mumps Outbreak."

The Centers for Disease Control and Prevention ("CDC") is part of the US Department of Health and Human Services. It has found is that more than 80 percent of the reported cases occurred in persons who were not vaccinated and about four of every five such individuals *purposefully chose* not to be vaccinated![13] So, what could happen if people stopped getting vaccinated is not a matter for conjecture. Rather, the cause and effect relationship between infectious diseases and inoculations against them is clear. When vaccination programs are introduced, the incidence of disease falls sharply. When participation in such programs falters, purposefully or by negligence, disease returns and spreads, and not just individuals but whole communities suffer. Quite simply, viruses spread, and people get sick and even die.[14]

ACCEPTING A NON-ORAL ADMINISTRATION OF A VACCINE WITH "TREIF INGREDIENTS" IS NOT PROHIBITED

Referring to certain ingredients found in vaccines, Bratman asserts that for anyone who takes *kashrut* seriously, essentially the rules and regulations regarding the fitness of food for consumption, "it would seem a no brainer to avoid vaccinations." On analysis, however, this argument fares no better than his first claim.

Vaccines, like other medical preparations, contain a variety of ingredients for a variety of purposes. Some of these ingredients are the active components in the vaccine, necessary for it to perform its task. Others are used as stabilizers for live vaccines or as preservatives. For vaccines available in the United States, CDC provides a comprehensive list of ingredients contained in those vaccines, as well as ingredients used in the manufacturing process but removed or remaining only in trace amounts.[15] There is no doubt that *some* formulations of *certain* vaccines contain products that, if consumed, would possibly or certainly be considered *treif* by most authorities. These include not only chick embryo cell cultures and bovine muscle tissue, but such items as Vero (monkey kidney) cells, Madin Darby Canine Kidney cell protein, hydrolyzed por-

13. See Hiltzik, "Measles Is Spreading."

14. See Centers for Disease Control, "What Would Happen If We Stopped Vaccinations?"

15. See "Vaccine Excipient & Media Summary."

cine gelatin, embryonic guinea pig cell cultures, and human diploid cells such as lung fibroblasts.

For many, whether those items are more or less appetizing than other ingredients like formaldehyde, thimerosal, hexadecyltrimethylammonium bromide, and sodium taurodeoxycholate is a matter of, well, taste. But the question here is whether they are *treif* when received by way of inoculation or spray or some other non-oral application.

Israeli Orthodox Torah scholar Rabbi David Samson summarizes the legal situation as follows: "There is no prohibition in using medicines which contain forbidden ingredients if they are administered by injection, suppository, enema, medicated bandage, and the like, since they are not eaten."[16] Bratman seems to have missed that distinction.

The general principle announced by Rabbi Samson received specific application in the United Kingdom when Fluenz, a new flu vaccine containing a hydrolyzed pork gelatin, was introduced. Fluenz is administered as a nasal spray. When asked whether usage was permissible, British *kashrut* authority Rabbi Abraham Adler from the Kashrus and Medicines Information Service confirmed that there was "no problem" with pork or other animal-derived ingredients "in non-oral products."[17] Concurring, Rabbi Yehuda Brodie, registrar of the Manchester Beth Din, noted that the vaccines were not taken orally and were taken for medicinal and not nutritional purposes. Moreover, the gelatin was a "denatured product" which was "very far from the original piece of pork it came from," and, further, a "small part of a complex formula," all of which made it "absolutely acceptable." Said Rabbi Brodie, anyone saying otherwise is "acting in the extreme."[18]

Rather than simply conclude that the issue is a "no-brainer" after all, let's just say that some folks with very fine Jewish brains, who actually took the time to study the tradition and apply it to scientifically established facts, uniformly have agreed that the original source or nature of certain ingredients does not render non-oral applications of vaccines as *treif*.

16. Samson, "Are Vaccines Kosher," para. 4.

17. See Sheinman, "Fluenz Flu Vaccine," para. 6.

18. See Brodie, "Why the Fluenz Vaccine Is Not Treif," paras. 4, 5, 6.

JEWISH TRADITION FIRMLY SUPPORTS COMPULSORY VACCINATION PROGRAMS

Bratman's final argument is that vaccinations violate the Torah principle of *pikuach nefesh*. His contention could not be more wrong.

Obviously, there was nothing that we would recognize as resembling medical science in the days of the authors of the Torah or those of the sages whose interpretations of Torah were collected in the Mishnah and the Talmud. The rudimentary understandings of the great twelfth-century physician scholar Maimonides did not encompass vaccinations, either. Nevertheless, far from condemning vaccinations, the Jewish tradition contains principles that are both well-established and almost universally understood to be applicable today, and they support vaccinations against infectious diseases.

The analysis begins, as it usually does, with the Torah. In addition to the two statements in Leviticus mentioned before, Deuteronomy contains a number of lessons that relate to a person's obligation, his duty, to care for himself and his neighbor. In that last book of Torah, Moses is portrayed as summarizing the history of the Israelite experience from Mount Sinai to the Jordan River and giving three farewell orations to the assembled people. In the first of these speeches, Moses urges the people to watch themselves scrupulously, to "take exceeding care" (see Deut 4:9, 15). Later, in his second discourse, Moses reminds the people that indifference is prohibited, in favor of, for example, an obligation to make a parapet for a roof, so that the owner would not "put blood-guilt" upon his house should anyone fall (see Deut 22:8).

Subsequently, rabbis interpreted Moses's statements as God's commandments to protect one's health, to guard against disease. The Talmudic concern with the well-being of a child was broad and deep. For instance, a parent was obligated to teach his child how to swim, lest the child be in danger if he was traveling by boat and the boat began to sink.[19] The overarching concern was the child's needs.

The rabbinic discussion also considered whether the requirement to protect health was a negative or a positive mandate. The former would only require staying away from danger, but the latter would require affirmative conduct to protect one's health. In Rambam's summary of Jewish

19. See, e.g., b. Qidd. 29a, https://www.sefaria.org/Kiddushin.29a?lang=bi. As set forth in Qidd. 30b, the recorded reason was to provide the child with a potentially life saving skill. See https://www.sefaria.org/Kiddushin.30b?lang=bi.

law, the *Mishneh Torah*, he concluded that the duty was a positive one and required proactive measures to mitigate a foreseeable harm.[20] Similarly, the *Shulchan Aruch*, Joseph Caro's mid-sixteenth century restatement of Jewish law, reaffirms that there is a religious obligation to take affirmative steps to prevent an anticipated danger to oneself or to others.[21]

The rabbis' teachings were not merely academic, as became clear not long after Edward Jenner introduced the first effective vaccine against smallpox in England in 1796. In Eastern Europe, the Chassidic master Rabbi Nachman of Bratslov wrote in his *Kuntres Hanhagot Yesharot* that one "must be very careful about the health of children . . . [and] inoculate every baby against smallpox before the age of three months, for if he does not do so, he is like one who sheds blood."[22] Sadly, in 1810, Rabbi Nachman died at age thirty-eight of tuberculosis, a century and a half before the development of a preventative vaccine.

More recently, the Reform and Conservative movements in the United States have issued formal commentaries on the issue of vaccination. The Reform analysis was published in 1999 by the Responsa Committee of the Central Conference of American Rabbis ("CCAR"), Rabbi Mark Washofsky, Committee Chair.[23] The Reform *responsum* on vaccinations was written with respect to a challenge raised by some parents to a congregation's policy that required immunizations as a precondition to a student's attendance in the congregational school. Apparently, the parents thought the required vaccinations were "excessively risky" and elected not to immunize their children.[24]

The CCAR began its analysis by noting the "fundamental fact" that Jewish tradition considers the practice of medicine to be a *mitzvah*, that is, a religious obligation. Moreover, it is an aspect of the dominant principle of *pikuach nefesh* (the preservation of human life), the precise principle Bratman thinks prohibits vaccinations. The CCAR then reviewed the history of infectious diseases in the United States, including the reduction and even elimination of certain diseases due largely to vaccination programs. The result, it concluded, was "surely one of the great success

20. See m. Torah, *Hilchot De'ot* 4:1; *Rotzei'ach* 1:6, 14.

21. See *Shulchan Aruch, Chosen Mishpat* 427:8.

22. See Prouser, "Compulsory Immunization in Jewish Day Schools," 14.

23. See generally Washofsky, *Reform Responsa*, 107–20.

24. See Washofsky, *Reform Responsa*, 107.

stories of the twentieth century."[25] Without discounting the reality that vaccines are not "completely safe," based on credible scientific evidence, the CCAR agreed that any risks posed were "very much outweighed by the public health benefit of" vaccination programs.[26] Similarly, such programs generated enormous cost savings "in terms of direct medical costs alone."[27] Finally, the CCAR recognized the benefit of community or herd immunity created when a sufficient percentage of a particular community are immunized. This immunity provided additional protection especially to those who were, for unique reasons, unable to be vaccinated. Conversely, it determined that refusal to vaccinate a child not only created a dangerous situation for the unimmunized child, but also for the community at large.[28] Concluding that "there are no valid Jewish religious grounds to support refusal to immunize as a general principle," the CCAR found the congregation entitled to require immunization of students in their congregational school as a prerequisite to attendance.[29]

The Conservative Movement's view was expressed six years later by its Rabbinical Assembly's Committee on Jewish Law and Standards, also with respect to school immunizations. The Committee's statement "Compulsory Immunization in Jewish Day Schools" was authored by Rabbi Joseph Prouser. It begins with a review of state immunization mandates and various objections to vaccinations. In its review of Jewish law, not surprisingly, it places a greater emphasis on a larger number and wider variety of traditional sources than did the CCAR. The results were, however, the same.

Prouser's review of halachic literature begins with the telling observation that "[e]nthusiastic halachic support for immunization protocols emerged even before" Jenner's development of a smallpox vaccine, at a time when the best practice, called variolation, consisted of deliberately infecting patients with smallpox or cowpox. While the risk of contracting the disease from that technique was about one in a thousand, at that time such a risk was considered negligible, especially given the potential benefit.[30]

25. Washofsky, *Reform Responsa*, 109.
26. Washofsky, *Reform Responsa*, 112.
27. Washofsky, *Reform Responsa*, 110.
28. See Washofsky, *Reform Responsa*, 112–13.
29. See Washofsky, *Reform Responsa*, 113–15.
30. See Prouser, "Compulsory Immunization in Jewish Day Schools," 12.

The analysis also references three contemporary Orthodox authorities, each with a different (though consistent) approach. One argues that society has a right to compel "life-sustaining treatment" even when a parent is opposed to it, and even when that opposition is religiously motivated. Another supports vaccinations mandated by the state under the principle of *Dina d'Malchuta Dina*—that is, the law of the land is the law. A third finds that the laws of Shabbat may be set aside in order to avoid a "life-threatening situation."[31] The conclusion, based on hundreds of years of experience with vaccinations, is clear to Prouser. There is a "well established preference for preventive medicine as a religious mandate."[32] Consequently, "[u]nless medically contraindicated for specific children, in extraordinary and compelling cases, parents have an unambiguous religious obligation to have their children immunized against infectious disease." By doing so, parents fulfill a "religious obligation to remove hazardous conditions which imperil the public's health and safety." Conversely, failure to do so is "a serious, compound violation of Jewish Law."[33]

And let's not forget the Reconstructionists. Concerning a theoretical decision by parents to avoid vaccinating their children, Rabbi David Teutsch writes succinctly: "Since the duty to heal is communal, this choice is ethical only when parents believe that no one should be vaccinated or when the vaccine poses a grave health risk to a particular child."[34]

Needless to say, Bratman does not discuss numerous pertinent passages in the Torah and *Talmud*, the various codes of Jewish Law, or contemporary commentary from Orthodox, Conservative, Reform, and Reconstructionist perspectives, each of which leads to or reaches the same conclusions. Vaccinations are not contrary to Torah principles. Indeed, vaccinations are mandated, because Jewish tradition places the highest value on preventing foreseeable damage to individuals and the community.

31. See Prouser, "Compulsory Immunization in Jewish Day Schools," 15.
32. Prouser, "Compulsory Immunization in Jewish Day Schools," 15–16.
33. Prouser, "Compulsory Immunization in Jewish Day Schools," 29.
34. See Teutsch, *Guide to Jewish Practice*, 505.

WE MUST PROTECT OUR COMMUNITY

A recent and decent estimate suggests that the Jewish population on this planet does not exceed fifteen million.[35] And we cannot afford to lose a single one to a preventable disease. As Rabbi Prouser reminds us, in prior times, in places where Jewish law governed, those who endangered the health and wellbeing of the community could be lashed or excommunicated.[36] In America today, lashing and excommunication are not likely, or even desirable, remedies to achieve compliance with Jewish norms. But we can still speak, and still insist: For the sake of the children and in the interest of the Jewish people (indeed all humanity), we must do all we can to educate against debilitating ignorance and inoculate against deadly disease. On this platform, Judaism and science can meet and stand as partners.

35. See Maltz, "World Jewish Population." para. 1.

36. See Prouser, "Compulsory Immunization in Jewish Day Schools," 22–25.

Chapter 17

The Curious Consensus of Jews on Abortion

THAT DIFFERENT JEWS HAVE disparate views is not news. What is news is when most Jews agree on a particular idea or approach. And so it is with the curious consensus of Jews on abortion.

In mid-2012, the Public Religion Research Institute ("PRRI") published its findings from a 2012 survey of Jewish values.[1] The survey sought to measure the opinions of Jewish Americans on a wide variety of economic and political issues, both domestic and foreign, as well as with respect to certain religious beliefs and practices. Some of those opinions were analyzed internally by Jewish denomination and externally by comparison to those of other faith or ethnic groups.

While Jews varied considerably in their views of a wide range of topics, on one—abortion—they were not only reasonably cohesive in their attitude, but strikingly different from other groups surveyed. Given the emphasis in the Jewish tradition on valuing life, on equating the preservation of one life with the preservation of a world and, conversely, the destruction of one life as the destruction of the world,[2] this result, on its face, seems as anomalous as it is clear.

First, let's look at the PRRI data. Essentially regardless of denominational affiliation or demographics, American Jews think abortion should be legal in all (49 percent) or almost all (44 percent) cases. That is, fully 93

1. Jones and Cox, "Chosen for What?"

2. m. Sanh. 4:5.

percent of all American Jews support legalized abortion in some fashion. Even political leanings, while influential, are not determinative. Among Jewish Democrats, support is 95 percent, but 77 percent of Jewish Republicans also favor legalized abortion in all or most cases, far exceeding the rate of other groups studied.

The comparable numbers for other faith groups are quite different, not only in their overall support or opposition to legalized abortion, but in the internal differences within each group. Jews are the only group surveyed in which a plurality support abortion in *all* cases. While about half of all Jews support abortion in all cases, in no other faith group surveyed does such support exceed 25 percent of the population. Moreover, in comparison to the 93 percent total of Jews who support legalized abortion in all or most cases, the only other group surveyed that showed clear majority support for legalized abortion was white mainline Protestants (59 percent). The comparable numbers for black Protestants and Catholics are 50 percent and 48 percent. Just one-third of white evangelicals support abortion in all or most situations.

Further, while the survey found that just 6 percent of Jews oppose legalized abortion in most cases and 1 percent did in all cases, the other groups surveyed were much more diverse in their views. For instance, while 19 percent of Catholics thought abortion should be illegal in all cases, 31 percent said only in most cases. Similarly, 21 percent of white evangelicals opposed legal abortion in all cases, but 44 percent only opposed it in most cases.

One problem with the Jewish Values Survey, and indeed with the public debate about abortion generally, is the lack of precision with definitions. When asking about "most cases," the 2012 Jewish Values Survey did not specify what "most cases" meant. This may not matter to the half of all Jews who support legalized abortion in all cases or to the roughly one of five Catholics, white evangelicals, and black Protestants who oppose legalized abortion in all cases, but for the rest of America it appears to make a difference.

Pew Research Center conducted a similar survey in 2017, asking whether abortion should be legal in all or most cases.[3] The results of the 2017 Pew Survey, reported in January 2018, were reasonably consistent with PRRI's Jewish Values. Pew found that 83 percent of Jewish adults said that abortion should be legal in all or most cases. To be clear, there

3. See Masci, "American Religious Groups."

are Jews and Jewish organizations (e.g., the Jewish Pro-Life Foundation) that oppose legalized abortion, but their numbers are small and their influence on the general Jewish population seems limited.

In the 2017 Pew Survey, the percentage of Jews favoring legality in all or most cases was exceeded only by Unitarians, atheists, and agnostics, and was slightly more than Buddhists and Episcopalians. While 55 percent of Muslims and 48 percent of Catholics held the same view, only 27 percent of Mormons and 18 percent of Jehovah's Witnesses did. Taking all US adults together, 55 percent shared the overwhelming consensus of Jewish Americans.

While the Pew results are less than those found in the PRRI survey, they seem to confirm that the overall position of Jewish Americans is more uniform than that of almost all other religious groups. So, why are Jews so much different from others on this issue? Is there something in the Jewish tradition which leads inexorably to the overwhelming consensus most Jews have reached?

ANCIENT CONCEPTIONS

The Torah itself, indeed the entire *Tanakh*, is silent on the topic of abortion. A passage in the Torah, however, does reflect a biblical view of a fetus. The passage concerns an injury to a pregnant woman which causes a miscarriage of her fetus. The Torah states that such conduct warrants financial compensation but nothing greater, specifically not the same penalty that would be imposed for murder (see Exod 21:22–23).[4] In other words, this passage considers the fetus as not fully a *nefesh*, a person, but more akin to personal property. It treats the incident, therefore, not as a criminal matter (for involuntary manslaughter), but more like a tort. This view was similar to other laws in the Ancient Near East, such as the eighteenth-century-BCE Babylonian Code of Hammurabi (nos. 209–10).[5]

When the ancient sages talked about abortion, they did so in the context of the knowledge of their day and with at least one eye on the Bible. Consequently, as a matter of principle, abortion was generally

4. A mistranslation in the Septuagint of this passage arguably has contributed to an understanding in the Catholic Church different than the traditional Jewish view. See Bronner, "Is Abortion Murder?"

5. See King, "Hammurabi's Code of Laws."

prohibited because, for example, it destroyed something created in God's image (see Gen 1:26–27; 9:6), and that destruction was also contrary to the first commandment to populate the world (see Gen 1:28).

At the same time, the sages' understanding of fetal development was quite limited. Within the first forty days of pregnancy, they thought the mother to be carrying "merely water."[6] The fetus was not yet, to them, a "living being." In later stages of pregnancy, they viewed the fetus as a part of the mother ("*ubar yerach imo*"), like a limb or appendage of the mother.[7] In these opinions, they were maintaining the notion implicit in Exodus that an unborn fetus was not quite, not fully a person.

Applying these not entirely consistent viewpoints, at least from Mishnaic times, abortion was permitted, even required, to save the life of the mother. The mother's life was given more value than that of the unborn fetus, so much so that in the event of a danger to the mother's life, an abortion could be performed until the head of the baby emerged. Once the fetus's head emerged, however, the baby was considered alive, and its life, no less valuable than that of its mother, was protected.[8] At the beginning of the second millennium of the Common Era, Rashi accepted this principle, saying of the fetus, "*lav nefesh hu*," or "s/he is not a person."

The philosopher and physician Maimonides took a different view, though. When considering a threat to the mother's life from a fetus, Rambam analogized the fetus to a *rodef*, or pursuer, for whom one was not to have pity. Abortion was justified, even though the fetus was of high value, because the fetus was characterized as an active endangerment to the mother.[9]

MODERN APPROACHES

The position of contemporary American Jewish leaders is remarkably, although not entirely, uniform. In May, 2012, the Conservative Movement's Rabbinical Assembly ("RA") approved a resolution on "reproductive freedom." In that resolution, the RA briefly traced both the relevant

6. See b. Yebam. 69b, https://www.sefaria.org/Yevamot.69b?lang=bi.

7. See, e.g., b. Giṭ. 23b, https://www.sefaria.org/Gittin.23b?lang=bi.

8. See m. 'Ohal. 7:6; quoted in Rabbinical Assembly, "Resolution on Reproductive Freedom," para. 4.

9. See Maimonides, *Mishneh Torah, Hilchot Rotzeach u'Shemirat Nefesh* 1:9; see also b. Sanh. 72b, https://www.sefaria.org/Sanhedrin.72a?lang=bi.

sources and recent attempts at the state and federal level to define life as beginning at conception and to legislate restrictions, including criminal penalties, on abortion. Generally reaffirming prior resolutions on the topic, the RA expressed a reverence for the "sanctity of life," reaffirmed the traditional Jewish belief that personhood (and the rights attendant to it) begins at birth, not conception, and supported "the right of a woman to choose an abortion in cases where 'continuation of a pregnancy might cause the mother severe physical or psychological harm, or where the fetus is judged by competent medical opinion as severely defective.'"[10]

In his award-winning book *A Guide to Jewish Practice*, published by the Reconstructionist movement, rabbi and ethicist David Teutsch reaches similar conclusions. Teutsch recognizes that the interaction between physical and mental health is "complex," and that the "impact of the abortion itself requires careful consideration." According to Teutsch's analysis, abortion is permissible when the pregnancy results from rape or incest, and is therefore "psychologically devastating," or where the child, if born, would suffer a "very short" and "highly painful" existence due to a disease like Tay-Sachs. At the same time, because a condition like Down Syndrome does not necessarily preclude living a "happy" or "meaningful" life, Teutsch concludes that such a diagnosis "by itself" would not justify an abortion, but allows that some authorities consider factors such as the effect of an additional child on its family's economic or internal stability.[11]

Both the Conservative and Reconstructionist views echo a formal *responsum* (opinion) issued a generation ago by the CCAR, the umbrella organization of the Reform rabbinate. That *responsum* reviews the precedents in detail, and with references, noting the different and varied situations in which abortions have or have not been permitted. The *responsum* concludes with a call for caution in matters of abortion, with less hesitancy about abortion in the first forty days of fetal life and more certainty regarding the reasons for an abortion subsequently. While the *responsum* takes what it characterizes as the "liberal stance" of permitting abortion at any time when there is a serious danger to the life of the mother, it ends on a sobering note: "We do not encourage abortion, nor favor it for trivial reasons, or sanction it 'on demand.'"[12]

10. See Rabbinical Assembly, "Resolution on Reproductive Freedom in the United States," para. 15.

11. See Teutsch, *Guide to Jewish Practice*, 500–501.

12. See Central Conference of American Rabbis, "When Is Abortion Permitted," para. 18.

The issue of abortion in the Orthodox community is hotly debated. Indeed, one's approach to the topic is seen by some almost as a litmus test of one's commitment to Modern Orthodoxy or, alternatively, to a premodern, culturally conservative orthodoxy.[13] The latter tends to hold that abortion is permissible only where the danger to the mother's life is clear and direct. Otherwise, abortion is generally forbidden. Under this approach, abortion in the case of fetal abnormalities or deformities would not be allowed, nor indeed would be amniocentesis if the purpose of that procedure was to determine the presence of a birth defect. But there are exceptions. And one can even find rulings of respected Orthodox rabbis permitting abortions in cases of substantial emotional difficulty, such as when the expectant mother becomes suicidal or when pregnancy is the result of adultery.[14]

Needless to say, this limited review of the extensive Jewish literature on abortion is an insufficient basis for understanding the nuances of the challenges particular cases bring to Jewish values and principles. It is, however, hopefully helpful as part of an effort to understand both the relative uniqueness and the real limitations of the approach of the vast majority of American Jews to abortion.

While it is clear that, for over two thousand years, Judaism has understood that (1) personhood begins at birth and not conception and (2) that the life of a mother supersedes that of a fetus which threatens that mother, the notion reportedly expressed by roughly half of American Jews in the PRRI 2012 Jewish Values Survey that abortion should be permissible in *all* cases is absolutely unwarranted by Jewish tradition and values, whether filtered through an Orthodox, Conservative, Reconstructionist, or Reform lens. For instance, and without limitation, abortion for purposes of gender selection, convenience, or purely economic reasons, especially at any time in gestation, is not defensible in the Jewish faith.

THE REALITY OF HUMAN GESTATION

Today we have knowledge, tools, and insights that the ancient sages surely lacked. We know how a fertilized human egg develops from zygote to embryo to fetus and then, on birth, to a baby. We know, for instance, that by the fourth week of pregnancy, in an embryo barely one twenty-fifth

13. See Yuter. "Abortion Rhetoric within Orthodox Judaism."

14. See, e.g., Eisenberg, "Abortion in Jewish Law."

of an inch long, the embryo's brain and spinal cord and its heart have begun to form, and arm and leg buds have appeared. Within two weeks, the heart starts to beat, blood to flow, and the embryo is the size of a lentil, maybe a quarter of an inch long. Brain activity commences. By the eighth week, all essential organs and external body structures, including eyes and eyelids, have begun to form. At the end of the eighth week, the embryo is about an inch long, but still weighs less than one-eighth of an ounce.

Consequently, the idea that the developing embryo is not a life form, that it is mere water for forty days, is not only archaic, but it is without scientific foundation. The embryo is not at all viable at that stage, but to deny that it is alive, that it contains human DNA, that it is a genetically distinct and unique being, and might, without interference, develop someday into an independent human, is at best disingenuous.

The fetal stage begins after week eight. In an uneventful pregnancy, the fetus will grow to about three inches and almost an ounce at week twelve and to four to five inches and almost three ounces at week sixteen. A translucent skin begins to form, and the fetus can make sucking motions. If you want to read detailed descriptions or see images of fetal development week by week, they are readily available on websites sponsored by various private and public agencies.[15]

Estimates of the viability of a fetus (that is, the capacity to survive a premature birth) vary. Recent studies suggest that less than one-fifth of babies born at twenty-three weeks of gestation survive, with the rate increasing to about two-fifths by week twenty-four and 50 percent at week twenty-five. By week twenty-six, over 80 percent of babies born prematurely can survive, and that number increases to 90 percent by week twenty-seven.[16] However, survivability—especially before the twenty-sixth week—will normally not come without serious medical intervention, and preemies born prior to that time run a substantial risk of being disabled, often seriously so. Nevertheless, assuming viability, an abortion not related to saving a mother's life cannot fairly be analogized to ridding one's self of personal property or amputating one's limb.

Similarly, given what we know today, though a fetus may well be the direct or indirect cause of a mother's life-threatening condition, we

15. See, e.g., http://americanpregnancy.org/while-pregnant/first-trimester/, https://www.babycenter.com/fetal-development-week-by-week, and https://www.womenshealth.gov/pregnancy/youre-pregnant-now-what/stages-pregnancy.

16. See "Premature Birth Statistics"; "What Is a Viable Pregnancy."

cannot say that the fetus is a "pursuer." There is no evidence, and really never was in Rambam's day either, that the fetus possessed the capacity to form an intention to kill its mother or, indeed, do any harm. Rambam's purpose in drawing that analogy is unclear. Let us admit that his position was inventive, perhaps compassionate and perhaps restrictive, but in any event ultimately, deeply, and fundamentally flawed.

Science can and should inform this discussion way more than it does. Through sonograms in the first trimester, we can begin to evaluate a pregnancy. We can literally see the shape and specific features of a fetus. We can see its head, its limbs. We can monitor its heart beating. What we cannot do—and ought not do—is deny its essence.

WE COULD USE SOME MORE CRITICAL THINKING

To the extent that the Jewish position on abortion is premised on the notion that the unborn child is mere fluid, or is not alive, or is a malevolent pursuer, it rests not just on unsupported assumptions, it is grounded in ancient ideas that have been clearly refuted by empirical evidence. Consequently, when Judaism meets science in the womb, its conclusion, again to the extent based on these notions, is misconceived. To the extent that the position is premised on an unborn's lack of a soul, of not being a person, the position is based on a religious tenet which can neither be proved nor refuted. Absent extenuating circumstances, extinguishing life on such a premise is, at best, problematic.

Yet, to acknowledge that life is present is not to conclude the inquiry. Science cannot, for example, extinguish the rape or incest that may have caused a pregnancy. In short, medical science is informative, but not dispositive of the questions to be considered with respect to abortion.

In fact, modern medical science perhaps raises more questions than it answers. For instance, just as it can provide information that makes the fetus appear to look more like a baby, today medical science can also tell us if that potential child is afflicted with a serious defect or disease. Today the human genome has been mapped, and many of us can get tested for genetic anomalies at relatively nominal cost. What do we do with the information we learn? If we find that a female fetus has a mutation on either gene BRCA1 on chromosome seventeen or on BRCA2 on chromosome thirteen, and therefore has a statistically significantly greater likelihood of developing breast cancer than a mutation-free female, what then?

What about a finding of a mutation of the ApoE gene on chromosome nineteen, which suggests an increased chance of Alzheimer's after age sixty? What of the literally dozens of diseases that affect groups of Jews disproportionately, from ataxia telangiectasia to Werner syndrome?[17] We have only just begun to have a discussion about the need to have a discussion about these issues.[18]

Author and political commentator Dennis Prager has argued that the approach of Jewish Americans to abortion is both a matter of "moral disappointment" and also a part of the substitution of "leftism" for Judaism.[19] Prager's frustration with American Jewry on abortion is understandable, but his argument is not persuasive. Whatever he may mean by "leftism," it seems hard to sustain that assertion when the statistics indicate that 83 to 93 percent of the group is on one side of the issue. That is, if almost everyone is on the "left," then there is no "left" anymore, or "right" for that matter. Invoking the left/right dichotomy is generally not very helpful or productive on political matters. On issues as knotty as abortion, it is next to useless.

And Prager's suggestion that one can be pro-choice—i.e., anti-criminalization of abortion—and still recognize that "many abortions have no moral defense" is also problematic. He wants pro-choice Jews, "especially rabbis," to say that they regard "most abortions" as "immoral." But, to be polite, this approach lacks precision. How can "most abortions" be immoral if only "many abortions" have no moral defense? What precisely does he mean by "moral" and "immoral" in this context? And why "especially" rabbis, as if (a) they have an impeccable track record on moral issues and (b) the rest of us are too obtuse to understand what's at stake?

Rabbi Danya Ruttenberg, who was ordained in the Conservative movement, has suggested another reason that the vast majority of the Jewish community supports the legality of abortions. After first noting traditional Jewish views that a fetus is at first considered "mere fluid" and later simply a part of its mother, she adds that Jews "have historically been among the strongest supporters of the separation of church and state, and have long been wary of legal maneuvers that appear to be coming from

17. See Entine, *Abraham's Children*, 380–81. For more information on genetic disorders that affect Jews of Ashkenazi, Sephardi, and Mizrachi descent, and the role that genetic screening can play, see generally the Jewish Genetic Disease Consortium website at https://www.jewishgeneticdiseases.org/about-the-jgdc/.

18. See Shurkin, "Prenatal Whole Genome Sequencing Technology."

19. Prager, "Jews and Abortion."

a place of Christian religious conviction."[20] She may well be correct. At a minimum, history does warn us that caution is in order in such circumstances. Yet, while it may be prudent to be concerned, anyone who would modify Ruttenberg's description into a prescription and suggest that "we" should be against something simply because "they" are for it would only be encouraging unserious groupish-ness, or tribalism, at its worst. And, obviously, such an approach would neither be a persuasive stance to take in conversations with "them," a much larger group than "us," nor likely a productive one.

There are, of course, secular arguments that can be mustered in favor of precluding government interference in a decision concerning abortion, and even for abortion itself. These arguments have their roots in philosophies including but not limited to libertarianism, utilitarianism, individualism, and pragmatism. They may even find some support in the Ninth and Tenth Amendments to the US Constitution, which suggest that citizens have rights and powers not specifically stated elsewhere in the Constitution. For instance, one such argument may assert that an individual has an absolute right to control his or her own body. Yet, however appealing that argument may be in and to a secular society, it is not one rooted in traditional Jewish soil. Judaism has long held that one's body belongs to God and that we are but stewards of it.[21] In ancient days, this meant, among other things, that one could not mar one's body with tattoos (see Lev 19:28). In the present day, it means, among other things, prohibiting smoking or otherwise using tobacco products.[22]

Needless to say, Jewish Americans are not bound to apply Jewish principles to issues of secular policy. But if they want to act not simply as citizens, but as Jews, whether individually or collectively, if they seek to frame the abortion issue not as a secular matter, but as a moral matter grounded in Jewish principles and values, and further, if the approach is intended to be persuasive to modern Jews and to our neighbors of other faiths, as an independent argument and conceivably one under the Free Exercise Clause of the US Constitution, then it cannot, indeed must not, be presented based on the scientifically flawed arguments articulated in the Talmud and by medieval rabbis with which Rabbi Ruttenberg starts her opinion piece. Most Jewish Americans don't look to those sources for

20. See Ruttenberg, "Why Are Jews So Pro-Choice," para. 11.

21. See Artson, "Judaism and the Human Body."

22. See Telushkin, *Book of Jewish Values*, 51–54, 260.

guidance on much of anything these days. If we are not to dummy-down on matters as central as creation and Exodus, why look to those sources and do so on this particular issue, one not even discussed in the Torah? Naturally, non-Jews would care even less, if that is possible. Instead, Jews need to rethink the Jewish position on abortion, and develop one both consistent with traditional Jewish values and grounded in reality.

The data behind the dilemma

To begin that discussion, we should focus on what is really at stake. Some data may be helpful. We can begin in 1972, the year before the US Supreme Court ruled that a woman's right to privacy under the Fourteenth Amendment to the US Constitution encompassed a right to an abortion, subject to the state's interests in protecting the woman's health and the unborn.[23] The number of reported legal abortions in 1972, the year prior to that decision, was about 587,000.[24] The annual number of abortions increased until 1990 when it peaked at about 1,429,000 procedures and then began to decline. Similarly, the ratio of abortions to live births increased from 196 per one thousand in 1973 to 364 in 1984 when it, too, began to decline.[25] The reasons for these declines are not clear, but may be related to increased use of contraceptives, as well as fewer available provider locations, legal restrictions, or other factors.[26]

According to the CDC, by 2014, about 650,000 women (mostly in their twenties) had legally induced abortions. To put this number in some context, as we shall see in chapter 21, the number of legally induced abortions was more than eighteen times the number of incidents of death associated annually with firearms in the United States. Significantly, of these abortions, over 90 percent were performed at thirteen weeks or less of gestation. About 7 percent occurred between fourteen and twenty

23. See *Roe v. Wade*, 410 U.S. 113 (1973), at https://supreme.justia.com/cases/federal/us/410/113/case.html. *Roe* was modified subsequently in *Planned Parenthood of Southeastern Pa. v. Casey*, 505 U.S. 833 (1992), at https://supreme.justia.com/cases/federal/us/505/833/case.html, but its core holding regarding a woman's right to an abortion in the first trimester stands as of September 2018.

24. See Centers for Disease Control and Prevention, "Abortion Surveillance—United States, 1998," table 2.

25. See CNN Library, "Abortion Fast Facts."

26. See Wind, "U.S. Abortion Rate."

weeks of gestation. Just over 1 percent were performed at twenty-one weeks or greater.[27]

Parenthetically, how many of these procedures involve Jewish women is unknown, but it is also important if this discussion is to be something other than an academic discussion for the Jewish community.[28] Whatever else has and can be said about abortion, the consequence of the procedure is not consistent with the first commandment in the Torah, to reproduce the species (see Gen 1:28). This is, or at least should be, a matter of no small import when the population of the already-small Jewish community in North American is projected to decline, not just relative to other groups, but in absolute numbers.[29]

When we focus on the national numbers, what they tell us (among other things) is that the argument concerning abortion, while certainly invoking serious moral considerations and often heated, is also about a medical procedure that has been declining both in rate and absolute numbers for over a generation. This observation is not intended to suggest in any way that each decision is not anything but exceptionally consequential to the pregnant woman involved, and others. Nor does it assume that the present legal framework will remain intact. It is, though, to recognize that overwhelmingly, if not uniformly, the issue is one which is being addressed in the first trimester of pregnancy, long before viability becomes a consideration.

A matrix to consider

Perhaps Jewish scholars, medical ethicists, and others can profitably consider a matrix in which pregnancies are divided in two ways, one by time and one by other circumstances. First, there can be a vertical division

27. See Centers for Disease Control and Prevention, "Data and Statistics." This is the latest year for which (as of mid-2018) we have statistics from the Centers for Disease Control and Prevention. According to the Guttmacher Institute, about 926,000 abortions were performed in the same period. See Jerman et al., "Characteristics of U.S. Abortion Patients." Aside from possibly different source lists, the reason (if any) for the substantially different numbers recorded is not clear.

28. Of the abortions studied by the Guttmacher Institute for 2014, 30 percent identified as Protestant, 25 percent as Catholic, 8 percent as some other affiliation, and 38 percent claimed having no religious affiliation. By race, the abortion rate was about 2.7 times as high in the black community as in the white non-Hispanic community. See "Abortion Rates by Race and Ethnicity."

29. See Pew Research Center, "Future of World Religions: Jews."

between situations of pre-viability and those where the fetus is viable. Then there can be a horizontal division related to threat or concern about the mother's life or health. The resulting four quadrants create four different challenges. Two of those situations, though quite different, may be relatively easy to address: one where the fetus is not viable and the mother is seriously endangered, and one where the fetus is viable and the mother is not faced with any physical or emotional harm. The analysis of the other two possibilities may pose more difficult challenges. In one, where the fetus is viable and there is a serious threat to the mother, a horrendous choice may have to be made. Statistically, this would be a rare circumstance, but that does not diminish the gravity of the predicament. The final quadrant seems to be the area in which by far most abortions occur today, where the unborn is not yet viable and there is also no clear devastating physical or psychological danger facing the mother. And it is here, perhaps, that Jewish thinkers especially, but the rest of us also, need to focus.[30]

A TIME TO REFLECT

As we focus on new ways of thinking about abortion, let us try to apply reality-based religious ethics, rather than righteous enthusiasm. Let us, therefore, eschew labels like "pro-choice" and "pro-life." Those are false and incomplete options. The truth is that neither side has a monopoly on cherishing life, nor can either claim exclusivity over the metaphor of the image of God. Indeed, let us reject the temptation to produce any result that fits on a bumper sticker, and instead urge nuanced reflection, even if it takes a clause or two to articulate.

The biblical view on the status of a fetus is, fortunately, not one of those rules literally or figuratively written in stone. We remain free to struggle over how and where to draw the lines we inevitably must draw when faced with situations about which we would rather not know, much less contemplate and resolve. The true challenge, when considering matters of life and death, is to be cautious when others are certain, to be sensitive when others are strident, and to exercise humility when others

30. Obviously, one can imagine various modifications to this matrix and other kinds of analyses, as well. The point is not that this is the only way to analyze the abortion issue, but that the issue can and should be analyzed with some rigor rather than in a rote or reflexive fashion.

exhibit hubris. We all need to rise to this challenge, rabbis and laity, physicians and patients. When better to start than now?

Chapter 18

Jews, Judaism, and Genetically Modified Crops

GENETICALLY MODIFIED ("GM") CROPS are plant products which have been genetically altered for certain traits. Such traits include resistance to viruses, bacteria, fungi, nematodes, insects, herbicides, and drought, as well as aspects of product quality like improved yield, nutritional value, and longer shelf life. The characterization is somewhat of a misnomer. Modification of biological organisms is not a new process. It has been occurring in nature for billions of years. Indeed, the natural selection of some traits over others is the driving force of biological evolution, the process by which a species over time secures a competitive advantage in its environment.

Today, the label of GM foods is meant to identify those products that have been modified or engineered by human means. And yet, the intervention of humans in an otherwise natural process is not new either. For instance, an Assyrian relief, dated to 870 BCE, illustrates pollination of date palms by man almost three millennia ago.[1] But human activity in plant breeding can be traced farther back than that, even to ten thousand years ago.[2]

The Torah tells of Jacob manipulating his flocks of goats and lambs so that he would increase his herd with the fittest among them (see Gen 30:31—31:13). That the author ambiguously attributed Jacob's success to

1. See "History of Plant Breeding."
2. See Brown, "Plant Domestication," para. 1.

both magical sticks and God's miraculous power is irrelevant for present purposes. What is important is that the story is testament to the reality that at least since the text was written some twenty-five centuries ago, humans have recognized the desirability of and have sought to guide the alteration of existing species in ways thought to be beneficial. This guided intervention has produced a host of useful and now common food products, but it is (or was) slow, unpredictable, unreliable, costly, and inefficient.

CONTEMPORARY GM CROP PRODUCTION

Conscious and more cost-effective breeding activity accelerated in the twentieth century of the common era with the use of nuclear technologies, tissue cultures, haploid breeding and, most recently, transgenic technology.[3] The latter technology encompasses the placing of genetic material from one species into another, resulting in an organism described as transgenic. In this light, GM crops are really genetically engineered ("GE") crops or biotech crops. The terms will be used interchangeably here.

Genetically engineered crops were introduced on a commercial level in the United States in 1996. Since then, corn (field and sweet), soybean, and cotton farmers have adopted GM crops rapidly and robustly.[4] According to the USDA Economic Research Service, in 2018, the percentage of planted acreage using genetically engineered seed that was resistant to herbicides, insects, or both, was as much as 90% for corn, 94% for soybeans, and 91% for cotton.[5] Farmers in the United States like GM seeds primarily because they increase the crop yield, and also because their use reduces management time and decreases the cost of pesticides.[6] Other GM crops in the United States include apples, potatoes, canola, squash, papaya, and sugar beets.

The United States is far from the only country which has adopted GM seeds. According to the International Service for the Acquisition

3. See Brown, "Plant Domestication"; see also King et al., "Transgenic Breeding."

4. See USDA Economic Research Service, "Recent Trends in GE Adoption."

5. See USDA Economic Research Service, "Adoption of Genetically Engineered Crops in the U.S."

6. See USDA Economic Research Service, "Genetically Engineered Crops in the United States," 18, 47.

of Agri-biotech Applications ("ISAAA"), in 2017, GM crops were being produced in twenty-four countries, some industrialized, but most developing.[7] More than forty other countries import biotech crops.[8] For instance, a number of European countries currently prohibit production of GM crops, but Europe is a large consumer of such crops.[9]

The fulsome adoption of GM crops has not come without controversy. Two broad categories of claims have been advanced against GM crops. The first is that they are neither safe nor fit for consumption because they are unnatural and untested and will introduce toxins and allergens and otherwise harm consumers. The second is that the corporate purveyors of GM crop seed are improperly seeking to control the crop market to their financial advantage and the economic detriment of farmers and the general population.

According to the American Association for the Advancement of Science ("AAAS"), the charges related to food safety are unfounded. In the United States, the largest producer of GM crops, "each new GM crop must be subjected to rigorous analysis and testing in order to receive regulatory approval." The seed producer has the burden of demonstrating both the integrity of any new crop and that any proposed new protein trait is "neither toxic nor allergenic." Consequently, the overwhelming consensus in the reputable scientific community is that GM crops which have been subjected to national government analysis, testing, and approval are safe. More precisely, as the AAAS Board of Directors has put it: "consuming foods containing ingredients derived from GM crops is no riskier than consuming the same foods containing ingredients from crop plants modified by conventional plant improvement techniques."[10] The American Medical Association has concurred, as has the World Health Organization.[11]

The arguments concerning market control arise from political and economic philosophy, but seem equally dubious. In the United States, the seed market for soybeans and corn clearly has been dominated by two

7. See International Service for the Acquisition of Agri-biotech Applications, "Global Status," 3.

8. See International Service for the Acquisition of Agri-biotech Applications, "Global Status," 109.

9. See Genetic Literacy Project, "Where Are GMOs Grown."

10. See Pinholster, "AAAS Board of Directors."

11. See American Medical Association, "Reports Of The Council On Science And Public Health" 5; see also World Health Organzation, "Frequently Asked Questions."

companies, Monsanto and DuPont Pioneer, which split about 70 percent of each market.[12] Rather than demonstrating control by either, however, market analysis therefore shows that each company has a very strong competitor and numerous smaller competitors.

WHAT DO GM CROPS HAVE TO DO WITH JEWS AND JUDAISM?

At its crassest level, GM crops provide one more lightening rod for anti-Semites. On websites that reportedly are not secure, and will not be identified here, Monsanto is attacked not simply because it is a major producer of GM seeds, but also because of its alleged Jewish connections. These include Monsanto's allegedly original Jewish ownership and its more recent Jewish officers and investors. One site accuses the company of conspiring with Jews in the United States Food and Drug Adminstration, and (former) "Senator Jew [*sic*] Lieberman" to secure the "right to shut down farmers who refuse to purchase Jewsanto's [*sic*] GMO seeds" and to obtain a "global monopoly . . . forcing the populace to consume this poison." As we know, though, the haters will hate, and we must carry on.

More importantly, for our discussion, there are Jews who oppose GM crops and purport to do so based on Jewish beliefs and values. One such opponent is Rabbi Arthur Waskow, a leader of the Jewish Renewal movement, an advocate of Eco-Judaism, and a prolific writer. A few years ago, Waskow published a "Tu B'Shvat Seder to Heal a Wounded Earth."[13] Part of it is premised on the charge that Monsanto is "imposing" GM crops on increasing numbers of farmers, and that Monsanto "threatens the sustainability of agriculture." No factual support is offered for either of those or related accusations. The first is dubious, as market statistics demonstrate, and the second would be financially suicidal for Monsanto and therefore an extraordinarily improbable effort on its part. There may be a Jewish rationale for opposing GM crops, but what is offered here is, to be charitable, analytically weak, and more a negative, almost Pavlovian reflex to big business, in this case personified by an agribusiness giant, Monsanto.

12. See AgriMarketing, "Estimated U.S. Seed Market Shares." As of this writing, both of those companies are undergoing structural corporate changes.

13. See Waskow, "Tu B'Shvat Seder."

One can almost hear a contemporary Isaiah asking rhetorically, to those comfortable enough to celebrate: "Is this the *Seder* I desire, one that curses a producer of crop seed? Isn't the *Seder* I desire one that supports the expansion of agricultural resources, that increases the amount of grain available to the poor, that feeds more so that less are hungry?" (cf. Isa 58:3–7). The hard truth, however uncomfortable it may be for some who operate on a more mystical plane, is that in the real world the GM seeds produced by Monsanto and others, sown in hundreds of millions of acres around the globe from Spain to South Africa, from Paraguay to Pakistan, and from China to the Czech Republic put more nutritious food in more mouths than all the *Tu B'Shvat Sederim* ever held.

Another opponent of GM crops is Raphael Bratman, who has argued that such foods ought not be considered kosher. He also claims that the mixing of species violates a Jewish prohibition known as *kilayim*.[14]

Bratman's discussion begins appropriately enough, with reference to a statement by the Orthodox Union ("OU") concerning whether the introduction of nonkosher genetic material into an otherwise kosher product renders the altered product as nonkosher. Bratman reports that OU's position is that genetic engineering does not alter the kosher status of the recipient organism for two reasons. First, the amount of transferred genetic material is microscopic and insignificant. Second, the descendants of the altered item were not themselves recipients of nonkosher genetic material. Having found an answer he does not like, and from an authority he selected, instead of reconsidering his position, Bratman expresses his admitted frustration with "OU's ignorance of the issue" and rejects OU's statement as "miss(ing) the point completely." This is not surprising. He took essentially the same tack when his argument that injectable vaccines were not kosher, and that position was universally rejected by leading *kashrut* authorities around the world.

Oddly, the OU statement to which Bratman refers seems to have been deleted (as of this writing) from the OU website. But Chabad Rabbi Tzvi Freeman concurs with the basic idea. Addressing the issue of genetically modified foods, he says: "Although there are instances of genetic material of non-Kosher [*sic*] animals being used in kosher food,

14. See Bratman, "Why Genetically Modified Foods Should not be Considered Kosher." We have discussed Bratman's argument against vaccinations on the grounds that they are not kosher in chapter 16, above.

to date, no one has succeeded in demonstrating that this renders the food non-kosher[*sic*]."[15]

Bratman's second objection, based on the doctrine of *kilayim*, is more problematic, but he spends little time addressing the problems. At its biblical root, *kilayim* is based on two verses in the Torah, one in the Holiness Code in Leviticus and the other in Deuteronomy. The first bars the mating of two kinds of animals, the sowing of two kinds of seeds in a field, and the wearing of clothing made from two kinds of cloth (see Lev 19:19). The provision in Deuteronomy is similar, but not identical. It prohibits seeding a vineyard with two kinds of seeds, plowing with an ox and ass together, and the wearing of wool and linen together (see Deut 22:9–11). The prohibitions are consistent with a worldview that sees a certain order in the universe, believes that such order should be maintained, and emphasizes separation of like from unlike and the preservation of boundaries as defining features of holiness. At the same time, the intent of the authors expressing these rules is neither stated nor clear.

Not surprisingly, renowned rabbis across centuries have disagreed about the interpretation of these phrases. Long ago, Nachmanides held that these rules teach that humankind should not disturb the fundamental nature of God's creation. Centuries later, Rabbi Yehuda Lowe, the Maharal of Prague (c. 1525–1609), took a different tack. As summarized by Rabbi Freeman, the Maharal contended that "any change that human beings introduce into the world already existed in potential when the world was created. All that humans do is bring that potential into actuality."[16]

Within the last few years, committees of both the Reform and Conservative rabbinic associations have addressed the issue of genetically engineered foods. The Responsa Committee of the Reform CCAR focused on a narrow question: the permissibility of using a specific modified food known as Golden Rice, which is fortified with an injection of foreign genetic material designed to provide Vitamin A (the lack of which contributes to blindness). In November 2015, the Committee on Jewish Law and Standards of the Conservative RA approved a lengthy and detailed study by Rabbi Daniel Nevins regarding the interpretation and application of Jewish law with respect to a host of issues related to genetically modified organisms.[17]

15. See Freeman, "Genetically Modified Foods," para. 1.
16. Freeman, "Genetically Modified Foods," para. 3.
17. See Nevins, "Halakhic Perspectives."

For the CCAR Responsa Committee, the primary legal question was whether the manufacture of Golden Rice explicitly or implicitly violates a provision in the Torah, specifically the principle of *kilayim*. The Committee concluded that Golden Rice does not do so for three reasons: 1) the prohibition only applied to a natural mixing, not to synthetic engineering; 2) having been transformed in a laboratory, the transferred gene segment was not from a different species of plant, but was a "different substance altogether (*davar chadash*)"; and 3) the resulting product was not a new species, but "a member of the same species bearing with new characteristics."[18]

While finding that "the prohibition of *kilayim* does not apply to contemporary techniques of genetic modification,"[19] the Committee could not reach a consensus on the effects of GM crops generally or Golden Rice specifically on the ecosystem, another and historic matter of Jewish concern (see, e.g., Deut 20:19–20). It then suggested that those who could make educated judgments about such matters, whether for or against Golden Rice, would stand "on good Jewish grounds."[20]

For the record, Vitamin A deficiency adversely affects the health of tens of millions around the globe, especially pregnant women and children. By some accounts, it is responsible for five hundred thousand cases of juvenile blindness and up to two million deaths annually. In 2015, after the CCAR Responsa was issued, President Obama's White House Office of Science and Technology Policy announced that Golden Rice was awarded a Patents for Humanity Award, given for its contribution to improving global health and raising the living standards of underserved populations.[21] Apparently, Pope Francis has also blessed Golden Rice.[22]

The RA study addressed considerably more issues than did the CCAR Committee. Like the latter group, the RA spent no meaningful time discussing *kashrut*. It began with a brief review of evolution and a more detailed one concerning recent developments in genetic engineering. In the course of the review, the Committee made several crucial observations.

18. See Central Conference of American Rabbis, "CCAR Responsa 5774.5," 1.
19. Central Conference of American Rabbis, "CCAR Responsa 5774.5," 2.
20. Central Conference of American Rabbis, "CCAR Responsa 5774.5," 3.
21. See "Golden Rice Project Wins."
22. See Pundit Planet, "Pope Francis Blesses Golden Rice."

Initially, the RA recognized that the traditional notion of an initial creation of fully formed species and "the stability of these species across time" has become "untenable in the past two centuries."[23] The RA also noted that current life forms not only share common parents, but that sometimes one species acquires genes from another. The Committee accepted data showing that "as many as 145 genes (from among 20,000 in the human genome) have been picked up from other species."[24] Further, while acknowledging that continued study is warranted, the RA held that "[g]enetic engineering is a field of great promise in combating hunger and disease," and "we are obligated to feed the hungry, heal the ill and to preserve human health."[25]

The RA then proceeded to discuss two methods of *halakhik* analysis, one based on legal formalism and the other on values-informed interpretation, as they may relate to the field of transgenics. Noting with approval the Talmudic principle that "stringent positions in *halakha* bear the burden of proof," the RA recognized that one could limit forbidden activities to those precisely prohibited in Torah.[26] By contrast, a values-informed analysis looks to the purpose of a law.[27] Here the RA noted that the sages were concerned about blending species out of "respect for the creation."[28] At the same time, while those sages would clearly forbid the act of forming a new hybrid species, it was not clear to the Committee what the sages would prohibit other than full blending, that is, "the transfer of (limited) sequences of DNA from one organism to another."[29] In any event, as the Committee observed, the sages "were also clear in permitting the produce of [any] such forbidden efforts."[30]

The RA concluded that the Torah's ban on *kilayim* "does not extend formally to the modification of gene sequences via the introduction of foreign DNA in order to convey a specific capability in the new organism."[31] Cautioning that the "health implications of genetically modified foods

23. See Nevins, "Halakhic Perspectives," 3–4; see also 21–24.

24. Nevins, "Halakhic Perspectives," 8.

25. Nevins, "Halakhic Perspectives," 5, 12.

26. Nevins, "Halakhic Perspectives," 30–31.

27. Nevins, "Halakhic Perspectives," 33.

28. Nevins, "Halakhic Perspectives," 35.

29. Nevins, "Halakhic Perspectives," 37.

30. Nevins, "Halakhic Perspectives," 36.

31. Nevins, "Halakhic Perspectives," 44.

must be examined on an individual basis," it further recognized that "Jews may benefit from the fruits of hybridized plants."[32]

WHERE JUDAISM AND SCIENCE CAN MEET ON GM CROPS

There seems to be as much of a consensus as there might ever be when it comes to an understanding of Jewish law as applied to new technology. GM crops, to date, do not raise any serious *kashrut* issues, nor does the principle of *kilayim* necessarily preclude either the production or the consumption of GM crops. The only serious issue is whether such foods are safe and beneficial or not, matters best left to scientists than rabbis.

Do new technologies of genetic engineering raise concerns? Sure. Does the application of any new technology to the production and consumption of food products warrant heightened scrutiny? Of course. But after twenty years of increased commercialization, subject to government protocols and reviews, with hundreds of millions of acres of genetically modified crops being produced and consumed, with all the data that has been accumulated and dissected, with all the studies that have been generated, it would seem that the initial reasonable concerns of the past have been addressed and the originally feared scenarios have not materialized.

This conclusion is buttressed by a report published online in 2013 in the *Critical Review of Biotechnology*.[33] The report concerned a survey of 1,783 research papers and other documents published in the previous decade regarding various aspects of GM crop safety. Such a meta-survey is important because it avoids problems inherent in cherry-picked data and puts an anomalous result from any single study in proper perspective. The main findings were quite instructive: 1) no significant hazard was detected in connection with the use of GM crops; 2) not a single credible example of a detrimental effect from the consumption of such crops was identified; 3) there was no evidence that GM crops were uniquely allergenic, much less toxic; 4) genetic segments of DNA from GM crops have not been and cannot be integrated into our cells; 5) there was little to no evidence of damage to the environment from biotech crops; and 6) usage of GM crops was less likely to reduce biodiversity than non-GM

32. Nevins, "Halakhic Perspectives," 44.

33. Nicolia et al., "Overview of the Last 10 Years."

crops.[34] Similarly, and more specifically, another review in 2016 found that Golden Rice was both effective in providing Vitamin A and did so without any negative health or safety effects.[35]

Just as the haters will hate, the science deniers will deny. Whether the rejection of science-based studies comes from one end of the political spectrum or the other is irrelevant. It is, by definition, non-rational, and often irrational. It is also counterproductive to the work our tradition teaches we need to do. With the important caveats that new data might warrant different conclusions,[36] and that vigilance is always warranted when our bodies, our food, and our environment are involved, reality-based Judaism supports the introduction and usage of tested and approved GM crops that help people. It does so because reality-based Judaism seeks to address the world as it is, not as it might have been in some ancient mythical garden, or as it might be in some medieval mystical construct of ten *sefirot*,[37] or even as it is experienced in the comfortable confines of academia or similar social bubble; or, for that matter, as we might like it to be. And it does so, not with the exclusion of organic or non-GM crops, but because tested and approved GM crops can be a useful means to achieve a desired end.

Given the overwhelming consensus on what Judaism permits, the data should drive us—that is, good data from reputable, independent sources, rigorously applying the scientific method we have discussed above. And evidence should trump anecdote and ideology every time, whether the issue is abortion, vaccinations, or something else.

The writers of the Torah and the ancient sages can be forgiven for some of their assumptions of the nature of our world. They did not have the benefit of our scientific methods or tools. But today there is no excuse for ignorance, or for either romanticizing or dummying down the truth. When real people are hungry or seek relief from disease, there ought not be much room for those who would close the door to the possibility of better health by bending or ignoring the facts in order to conform to some preexisting political or economic bias. We can do better. We can heed the lesson from another, more traditional *Seder*. Does not the

34. See generally "Review of 10 Years of GMO Research," and Pomeroy. "Massive Review Reveals Consensus."

35. See Albaugh, "Golden Rice: Effectiveness and Safety," 12.

36. See, e.g., Wilson and Latham, "GMO Golden Rice."

37. A mystical notion concerning ten forces that some argue operate between God and humans. See Segal, "Kabbalah."

Passover *Haggadah* call on us to open the door? Does it not read: "Let all who are hungry, come and eat!"?

Chapter 19

To Frack or Not to Frack, Is that a Jewish Question?

DID YOU KNOW THAT fracking, an industrial process of extracting natural gas from shale rock is *treif*, or that it violates Jewish values? Who knew? You could read the entire Torah, study the Talmud, go through the commentaries of the rationalist Maimonides and the more mystical Nachmanides, and (for good measure) review Joseph Caro's legal compendium, the *Shulchan Aruch*, and you will never once see the technology discussed, or the word even uttered. And yet, there are Jewish individuals and organizations that insist that fracking is so contrary to Jewish values that it must be banned.

WHAT IS FRACKING ANYHOW?

Before we drill down a bit into the Jewish aspects of these arguments, let's take a moment to review some basic facts about fracking. Fracking, or fracturing, is a method of drilling for gas or oil trapped in rock formations deep in the earth as a result of the cooking of organic material tens of millions of years ago.[1] The fracking process has been utilized in the United States for many years, but recent advances in technology have allowed drillers to recover hydrocarbons in the form of gas or oil from relatively tight, impermeable shale rock. It has also led to the production of fuel from huge reservoirs or basins in a number of areas across

1. See Gold, *Boom*, 14–15.

the continental United States, from the Marcellus, which runs from Pennsylvania into Kentucky, to the Barnett and Permian in Texas, and up to the Bakken in North Dakota, among others.

Fracking is messy. It requires the injection of highly pressurized water, combined with 1 percent of other materials such as sand, plastic balls, and chemicals, through a pipe thousands of feet underground and into a shale rock formation. The infusion fractures, or cracks, the dark grey, dense rock or expands existing cracks, releasing natural gas or oil, and perhaps other items, which had developed and accumulated naturally in the shale. The natural gas or oil then can flow back up the pipe to the wellhead. Today, fracking is often accomplished in conjunction with horizontal drilling, which allows gas drillers greater access than previously possible to a vertical shale formation, resulting in a more extensive and economical frack.[2]

The earliest efforts to fracture rock to stimulate the release of hydrocarbons occurred over a century and a half ago, and involved placing small bombs, or torpedoes, down wells filled with water. Subsequently, hydrochloric acid and napalm were the media of choice to free the desired fuel.[3] The results were uneven. In July 1998, however, a slickwater fracturing technique, applying more water and enhanced pump pressures than previously attempted, proved commercially successful.[4] That achievement in the Barnett Shale of north Texas initiated an energy and economic boom that has not ceased yet. Fracking technology not only transformed energy production in the United States, but it has had a dramatic effect on our environment and disrupted geopolitics as well.

Determining the number of fracked wells in operation today is not easy, as records are kept (or not) in different states in different ways. According to one study based on data available in 2016, thirty-four states hosted oil and gas drilling activity involving almost 1.2 million wells.[5] Not all of these wells use fracking technology, but the trend toward fracking is undeniable. Between 2010 and 2015, over 130,000 wells were drilled.[6] By 2011, the majority of new wells drilled used fracking technology, and

2. See Gold, *Boom*, 28–30.
3. See Gold, *Boom*, 64–73.
4. See Gold, *Boom*, 115–23; see also Gold, "Texas Well."
5. See Rubright, "34 States Have Active Oil & Gas Activity."
6. See Meko and Karklis, "United States of Oil and Gas."

by 2016 over two-thirds did so.[7] Not surprisingly, then, according to the US Energy Information Administration ("EIA"), by 2015, about two-thirds of all natural gas produced in the United States came from fracked wells.[8] Moreover, models studied by the EIA suggest that natural gas, now largely produced by fracking of shale, will be the fuel on which we rely increasingly over the next several decades, even as non-hydroelectric renewables increase their share of energy consumption.[9]

Before fracking, energy companies were building terminals to import natural gas. Now the United States can export that fuel. As a result of the dramatically increased supply of natural gas, a fuel which cost more than ten dollars per British Thermal Unit ("BTU") in 2008 cost less than four dollars per BTU in December, 2016. Similarly, before fracking, oil production had fallen to below five billion barrels a day. In 2008, however, fracking technology advanced to the stage where it became applicable to the longer molecules that comprise oil. In June 2018, oil was produced at a record 10.9 billion barrels a day.[10] Naturally, with all that activity comes jobs, both directly involved in fracking and in supporting roles as well. According to one analysis, "[t]he gas boom in Pennsylvania at the beginning of 2009 accounted for more than 23,000 new jobs and added $1.9 billion to the state economy."[11] At the same time, this energy abundance, in turn, contributed to the closure of more than two hundred coal mines and some nuclear plants, and the jobs associated with those facilities.[12]

The effect of fracking on the environment is the subject of considerable controversy. According to the EIA, because natural gas is mostly methane, it has a relatively high energy content when compared to other fuel sources and emits a relatively low amount of carbon dioxide per energy content.[13] Consequently, burning gas to heat homes or fuel electric generators is better for the environment than using heating oil or coal. While other factors are certainly involved, data from the Environmental

7. See Cook, "Hydraulically Fractured Horizontal Wells," https://www.eia.gov/todayinenergy/detail.php?id=34732.

8. See Perrin and Cook, "Hydraulically Fractured Wells," https://www.eia.gov/todayinenergy/detail.php?id=26112.

9. See "Annual Energy Outlook 2018," 12, 18.

10. See Gold, "Texas Well."

11. See Meko and Karklis, "United States of Oil and Gas," para. 16.

12. See Gold, "Texas Well."

13. See "How Much Carbon Dioxide," https://www.eia.gov/tools/faqs/faq.php?id=73&t=11.

Protection Agency ("EPA") shows that greenhouse gas emissions in the United States started to decline around 2008, when domestic production of shale gas obtained by fracking accelerated.[14]

At the same time, the production process is a dirty, noisy, and smelly intrusion into often previously bucolic settings. In 2013, the Union of Concerned Scientists published an extensive paper that raised questions about the contribution of fracking to local environmental issues including, but not limited to, water contamination, air pollution, seismic activity, and land use, as well as the global matter of climate change.[15] The same year, the Environment America Research and Policy Center published its own report, "Fracking by the Numbers." The subtitle provides a strong clue about the contents: "Key Impacts of Dirty Drilling at the State and National Level." This report, too, alleges "grave threats" arising from contaminated drinking water, air pollution, the exacerbation of global warming, and other adverse consequences of fracking. Its policy recommendations included the following: "In states where fracking is already underway, an immediate moratorium is in order. In all other states, banning fracking is the prudent and necessary course to protect the environment and public health."[16] Taking a more balanced approach, the president of the Environmental Defense Fund has been quoted as praising fracking for lowering energy costs, greenhouse gas emissions, and local air pollution, while also pressing for the process to be as clean as possible.[17]

While the effect of fracking on the domestic environment is in dispute, there is little doubt that fracking has disrupted the Organization of the Petroleum Exporting Countries ("OPEC"), a cartel which sought to leverage its production dominance to fix prices and command compliance with certain foreign policy positions. As a result of shale fracking, the United States passed Russia in November 2012 as the world's premier producer of gas.[18] It is currently approaching parity with Russia and Saudi Arabia, the world's dominant producers of oil, and may pass them

14. See "Greenhouse Gas Emissions," https://www.epa.gov/ghgemissions/sources-greenhouse-gas-emissions.

15. See Goldman, et al., "Toward an Evidence-Based Fracking Debate."

16. See Ridlington and Rumpler. "Fracking by the Numbers."

17. See Gold, "The Texas Well."

18. See EIA, "United States remains largest producer of petroleum and natural gas hydrocarbons," https://www.eia.gov/todayinenergy/detail.php?id=26352.

in 2019, if not sooner.[19] One stunning result of all this production is that, as a result of fracking, the United States became a net exporter of liquefied natural gas in 2017,[20] and it is projected to become a net exporter of energy by 2020.[21] Domestic fracking therefore insulates the United States from the efforts of those who would, as in the 1970s, use oil as an economic and political weapon against it. Fracking assures previously unobtainable levels of energy independence and price stability.

In sum, as the EPA has concluded, "[r]esponsible development of America's shale gas resources offers important economic, energy security, and environmental benefits."[22]

WHY DO SOME JEWS OBJECT TO FRACKING?

Let's consider a few of the objections.

Rabbi David Seidenberg has argued that fracking is "a danger to the well-being of the planet" and, moreover, "[c]onflicts with Kabbalah," a mystical approach to Judaism. Seidenberg notes that in Kabbalah, water is "the very symbol of blessing and life," and asserts that "water that stays in that fracked rock is deprived of fulfilling its deepest purpose."[23]

The appropriately named Jews Against Hydrofracking ("JAH"), founded by Dr. Mirele Goldsmith, sees fracking as an extreme, risky, poisonous, and polluting activity. In one of her essays, Dr. Goldsmith, who is not a medical doctor but an environmental psychologist, characterizes fracking as a "destructive practice." Her case begins with the proposition that "[w]e [Jews] value life above all else," and argues that Jews should be sensitive to the "ethical implications of hydrofracking."[24]

In another essay on the JAH website, Rabbi Lawrence Troster invokes the prophet Ezekiel's vision of a deep river flowing eastward out of a rebuilt temple 2,500 years ago, creating a new garden of Eden (see Ezek 47:1–12). From this, he argues that "[f]racking is a transgression against God's Creation, delaying the redemption of the world." He understands access to clean water to be "a basic human right," one that is impaired

19. See di Cristopher, "American Oil Drillers' Output."
20. See Mobilia, "United States Exported More Natural Gas."
21. See "Annual Energy Outlook 2018," 11–12.
22. See "Unconventional Oil and Natural Gas Development."
23. Seidenberg, "How Fracking Conflicts," paras. 6, 14.
24. Goldsmith, "Why Jews Should be Against Hydrofracking," para. 7.

by fracking, among "other things," which uses a large amount of fresh water.[25]

The more established Religious Action Center ("RAC"), an arm of the Reform movement, once acknowledged that natural gas is often seen as a cleaner source of energy than other fossil fuels "because it produces 43 percent less carbon emissions than coal for each unit of energy delivered, and 30 percent less emissions than oil." Nevertheless, it also urged that fracking not be permitted until "we can confirm that (the process) poses no danger to our communities." In its discussion of Jewish values and fracking, RAC referred to the calling in Genesis to "till and tend God's Earth" and the Midrash that notes that no one will be present to repair the damage if we fail to do so (see Gen 2:15; Eccl. Rab. 7:13). RAC added that "[t]he Talmudic law of Bal Tashchit, do not destroy, derived from Deuteronomy 20:19 paired with our 'till and tend' mandate, emphasizes the need to act as guardians" of the environment.[26]

That four opponents of fracking assert their opposition is premised on Jewish values, but do not agree as to which principles are applicable, suggests that something is amiss in the argument that fracking is clearly *treif*. Whether taken separately or collectively, however, the Jewish aspect of these arguments is neither conclusive nor, really, even persuasive, except perhaps to those predisposed to their conclusions.

For example, the contribution of the Kabbalah to a serious discussion of fracking issues seems quite limited. The *Zohar* is a major work of Kabbalah first disclosed in the thirteenth century, but purporting to be of much older origins. It understands the universe to consist of four basic elements: earth, air, fire, and water, each of which is seen as imbued with a spiritual dimension.[27] Whether inadvertent or not, Seidenberg acknowledges that this approach is how "the natural world is *imagined* in Kabbalah" (emphasis added).[28] But imaginative musings, however inventive and instructive they may be in other circumstances or for other purposes, are not the firmest of foundations for an ethical stance to be applied in the world as it truly exists.

25. Troster, "Redemption and Hydraulic Fracturing."

26. These statements were published on RAC's website in 2013. At some subsequent point, they were removed from the website but, through the miracle of the Wayback Machine, the identical relevant language can still accessed. See Reform Jewish Voice, "Hydrofracturing." As of July 2018, RAC's website appears to be silent on fracking.

27. See "Four Elements in Kabbalah."

28. Seidenberg, "How Fracking Conflicts," para. 5.

JAH's assertion that Jews value life above all else taps into an older and deeper Jewish value system, but as stated goes too far and undermines JAH's credibility. JAH refers to no source in the Jewish tradition for its sweeping statement that Jews value life "above all else," and, as written, it is not accurate as a matter of theory or in practice. For instance, as discussed in chapter 15, rabbinic tradition has long taken the position that death is preferable to committing idolatry, murder, and adultery and incest.[29] And today, as we have seen, studies show that most Jews in America are willing to permit taking the life of an unborn fetus in all or most circumstances. Jewish tradition places a high value on life, to be sure. But it is not absolute. Neither is Jewish practice.

JAH's reference to Eden does not help its argument either. Just as Dorothy in *The Wizard of Oz* came to understand that she was not in Kansas anymore, so too must those who seek to restrain the exploitation of natural resources understand that we do not now live in Eden and, in fact, never did. Nor, indeed, should we want to do so. To the contrary, we should embrace knowledge whether from a tree, the ground, or elsewhere. Similarly, a vision of a mythical river extending from a once-desired temple (which it is doubtful the author wants rebuilt today) is hardly a reasonable metaphor for the present. The worthwhile effort to protect limited fresh-water supplies is too serious a matter to be subject to such irrelevant references. To be plausible, reasonable water protection and conservation policies must address the actual and complex social and economic situations they seek to affect.

RAC's position was based on an even worse source. Of the three resources RAC identified, the first was the 2010 movie *Gasland*, which RAC described as a "[d]ocumentary showing the effects of hydraulic fracturing on local communities." The central moving, and literally most explosive, feature of the film is a scene in which a Colorado resident appears to ignite tap water that is flowing out of a kitchen faucet. What burns is not water, of course, but methane, and the implication is that the methane was present due to fracking. There is a fatal flaw in the presentation, however. The methane that burned did not enter the kitchen faucet as a result of any fracking activity. It was present naturally in the water that entered the subject house. Indeed, there were reports of lighting water in the area for at least eighty years, long before any fracking activity.[30]

29. See b. Sanh. 74a, http://halakhah.com/sanhedrin/sanhedrin_74.html.

30. See McAleer, "Gasland Movie"; see also Economides, "Don't Be Swayed."

The writer and director of the film knew these material facts, but failed to disclose them. *Gasland*, it turns out, was not just pseudoscience, but a fraud. And RAC relied on it, as did others.

On the other hand, RAC's reminder of a call to till and tend the land and be guardians of the environment are well grounded in the tradition. The problem here is whether those principles are applicable, and, if so, how they are to be applied. RAC has expressed concern about "the safety of our clean water supply, radioactive chemicals present in the Marcellus Shale, and the chemicals used in natural gas extraction."[31] The concerns are fair, but let's have a reality check.

DRILLING DEEPER INTO THE ALLEGED RISKS OF FRACKING

RAC has stated that fracking "threatens to release" naturally occurring radioactive chemicals, creates a "risk" of exposure to the chemicals used in the fracking process, and implies that fracking might contaminate clean water. To drive home the absolutist nature of its position, RAC sought to require that communities be assured "that radioactivity present in the rock and the chemicals used in gas extraction have *absolutely no possibility* of contaminating our clean water supply" (emphasis added). Radioactivity is pretty scary stuff in general, and there is no doubt that radioactive isotopes such as radium 226 and 228 are found in association with shale. Indeed, exploitable shale deposits are often discovered by the presence of such isotopes. But the mere presence of such isotopes does not necessarily indicate a health hazard, and RAC did not cite a single incident of the release of radioactive chemicals due to fracking in all of the years and all of the places fracking has occurred. To the contrary, tests of water used in fracking in the Marcellus Basin in Pennsylvania found no evidence of levels of radioactivity that would adversely affect the public.[32]

Moreover, if RAC's requirement of complete assurance of risk-free activity were applied consistently to manufacturing processes and the provision of services, there would be precious little, if any, innovation or even conventional manufacturing or delivery of goods and services. Risk-free living is neither realistic nor even reasonable. Farming and preparation of food products sometimes cause illness. We still eat. Ingestion of

31. Reform Jewish Voice, "Hydrofracturing," para. 8.

32. See McGraw, "Is Fracking Safe," 13–14.

pharmaceuticals can cause adverse reactions. We still take pills. Traveling on cruise ships can expose us to viruses. We still sail. The important question is not whether there is a risk of harm or contamination (though that is a fair question), but how serious the risk is and whether it can be managed. Experience teaches that radioactivity due to fracking is not a real problem.

Anti-frackers also raise two arguments with respect to water. One concerns the chemicals involved, and the other the amount of fresh water used in the process.

Rabbi Seidenberg, for instance, asserts that the water used in the fracking process contains "poison," and that the water which returns to the surface after fracking is "lethal and extremely difficult to treat." The assertion, however, is not supported by any scientific study or details whatsoever. Let's assume, though, that some of the chemicals used by some of the drillers can, in certain concentrations, be considered toxic. That circumstance still would not compel a conclusion that, as actually used in the field, those chemicals are dangerous. Again, given the number of fracked wells in operation and the duration of fracking in the United States, if fracking were as poisonous and lethal as Seidenberg contends, one would expect to have heard by now of substantial numbers of deaths, or at least poisonings, directly attributable to the process. We have not.

While fearmongering is unwarranted, caution certainly is appropriate. The final report of the US Environmental Protection Agency ("EPA") in 2016 regarding the impact of fracking on drinking water states that EPA found "scientific evidence that activities in the hydraulic fracturing water cycle can impact drinking water resources under some circumstances." At the same time, the report also identified "uncertainties and data gaps that limited EPA's ability to fully assess impacts to drinking water resources both locally and nationally."[33]

The fair conclusion, based on the evidence to date, is that whatever known risks exist are generally being managed appropriately. Rather than banning fracking, it seems more reasonable to require the disclosure of chemicals used, implementation of appropriate safeguards, and a monitoring of the process.

JAH contends that fracking squanders limited water resources. There is no doubt that fracking uses large quantities of water, perhaps millions of gallons of water for each frack. But fracking is hardly the

33. See "Questions and Answers," paras. 1–2.

largest user of fresh water. Again, data and context are important. A Penn State hydrogeologist reportedly noted that natural gas exploitation in Pennsylvania utilizes 1.9 million gallons of water a day ("MGD"). That's a lot, of course, but the usage pales in comparison to the 62 MGD used by livestock, the 96 MGD used in other mining activity, and the 770 MGA used in other industrial processes.[34] As a percentage of the 9.5 billion gallons of water used daily in Pennsylvania, fracking's share is less than two one-hundredths of one percent.

Moreover, even one of JAH's leading contributors, Rabbi Troster, acknowledges that there are multiple causes of potential fresh water shortages—"the rapid increase in world population," "an increase in contaminated water from human effluents," "an increase in the rate of water consumption per capita," and "[m]odern agricultural methods, power generation and industrial use."[35] But JAH is not calling for population limits, a return to premodern agriculture, or a reduction in power generation and industry. Consequently, its singular focus on fracking reflects limited science, and even undermines its purported seriousness about preserving fresh water.

In short, arguments against fracking based on water usage and contamination have not yet been supported by any objective study and have been refuted by others. Hyperbolic claims cannot overcome another set of stubborn facts.

WHAT ABOUT FRACKING AND ECONOMICS?

While the eco-warriors rail against fracking, they tend to avoid one "eco" issue—that of economics. As Gary Becker, a University of Chicago professor and Nobel Laureate in economics, pointed out back in 2012, because the cost of fracking became so competitive (1), most domestic natural gas production came from fracking and (2) the price of natural gas per BTU fell dramatically.[36] Rabbi Seidenberg does not tell us how the Kabbalah would address either micro- or macro-economic issues, whether domestic or international.

Dr. Goldsmith at least has an economic argument, and it is straightforward. "[I]t makes no sense," she says, "to invest in new

34. See McGraw, "Is Fracking Safe," 3–4.

35. Troster, "Return of the Healing Waters," para. 5.

36. See Becker, "Fracking and Self-Sufficiency."

infrastructure for outdated fossil fuels when we could put the same money into renewables."[37] But who are "we," what "same money" is she talking about, and why does she think "we could"? Apparently Dr. Goldsmith believes that energy sources are fungible in the production of consumable fuel, and that the demand for various fuels is also quite insensitive to price, i.e., inelastic. The bases for any such beliefs are not evident. Perhaps some would be willing to bet on and pay more for expensive solar panels and windmills that do not generate energy at all times and in all weather conditions instead of continuing with proven technology that delivers energy in an economical and thoroughly reliable fashion. Most Americans don't have that luxury, though, and just want cheap, dependable energy.

For its part, when it published its statement on fracking, RAC conceded that "ninety percent of new natural gas wells utilize" fracking, but it would still have placed an unprecedented burden on fracking drillers. Had it been adopted, the logical consequence of RAC's position would have been to diminish the domestic production of natural gas, which in turn would not only have condemned America to use even more coal and oil than it does presently, but to do so at a greater financial cost than would be incurred were the supply of gas not artificially restricted. RAC also never explained how a goal of guarding the environment is served by greater use of more expensive and dirtier sources of energy than natural gas.

DO ALL JEWS OPPOSE FRACKING?

Not all Jews or Jewish organizations oppose fracking. Support comes from multiple directions, and is subject to a variety of conditions.

Referring to the significant reduction in imported oil due to fracking, the American Jewish Committee in 2012 asserted its belief that the extraction of natural gas from shale could be effected "safely."[38] Similarly, the Jewish Council for Public Affairs has recognized that fracking could reduce energy costs and create jobs.[39] It called for a number of reasonable safeguards to protect the environment and public health. Nationally prominent Jews like Malcolm Hoenlein, Sandy Eisenstat, and Neil Gold-

37. See Goldsmith, "Fracking Not the Answer to Energy Security."

38. See Harris, "Jewish Groups Take On Fracking,"

39. See Jewish Council for Public Affairs, "Resolution on Hydrofracking," para. 3.

stein, through the Council for a Secure America (as well as lesser known individuals like the Libertarian Jew), also have advocated for fracking.[40] Some Jewish camps have even leased some of their property to drillers.[41]

NEITHER FEAR NOR QUOTE MINING SHOULD OVERCOME FACTS

The problem with opposing fracking on Jewish principles is not that fracking is not mentioned in the Torah or Talmud. Jewish tradition is organic and can adapt to new circumstances. The problem is that the opponents to date have not made a compelling case based on those principles. They assumed a burden they then failed to meet. Instead, they have relied on quote mining for selected Jewish adages, aphorisms, and irrelevant fables and metaphors, which is not persuasive advocacy in complex matters of science, technology, economics, and societal well-being, other than to those already inclined to accept such a narrative. Worse, inappropriately waving the *tikkun olam* (repair the world) flag in a situation that does not clearly warrant it diminishes the importance of the idea and the impact of the approach.

The reality is that fracking is, as University of Michigan chemical engineering professor Johannes Schwank characterizes it, a "remarkable feat of engineering."[42] It is also, as he recognizes, an imperfect process, and there are abundant reasons for concern and caution about fracking, just as there are for many industrial activities in which we currently or might engage. Though it seems reasonably safe when performed properly, further experience and studies may reveal facts that call for a reevaluation and increased regulation. At present, though, we have the experience of hundreds of thousands of wells that have been drilled with hydraulic fracturing. The pervasive doom that was predicted in previous years has not materialized. Until such time as the data changes, however, fear should not trump facts.

Wall Street Journal investigative reporter Russell Gold has observed that "[e]nergy systems change when something better and cheaper comes along."[43] That better system may turn out to be one based on renewable

40. See "Is There a Link."

41. See Nathan-Kazis, "Fracking Comes to Jewish Summer Camp."

42. See Moore, "Fracktopia," para. 5.

43. See Gold, *Boom*, 35.

energy sources, such as wind and solar. Paradoxically, according to Gold, fracking both challenges and paves the way for those alternative systems to emerge. Yet until those technologies improve, especially with respect to batteries needed to store energy for times when the wind does not blow and the Sun does not shine, Gold concludes that "natural gas is the best available option for reducing carbon emissions, without grinding the wheels of modern economies to a halt."[44]

Fracking is a multidimensional issue which requires rigorous analysis. Let's all act accordingly.

44. Gold, *Boom*, 35–36, 306.

Chapter 20

Judaism, Neuroscience, and the Free-Will Hypothesis

FORGET MOSES'S IMPASSIONED PLEA to the Israelites concerning their choices among the many blessings and curses that God reportedly set before them as they were about to cross the Jordan river into their promised land (see Deut 11:26–28; 30:15, 19). Evolutionary biologist Jerry Coyne claims we have no ability to choose freely among alternatives. According to Coyne, "we couldn't have had that V8, and Robert Frost couldn't have taken the other road."[1] Presumably, the Israelites in the story did not have much choice either.

Coyne argues that the free will we sense when we make a decision, the feeling that we are choosing among available alternatives, does not exist. In reality, he contends, our conduct is predetermined by physics. This result follows, he says, because our brains and bodies, the "vehicles that make 'choices,' are composed of molecules, and the arrangement of those molecules is entirely determined by [our] genes and [our] environment." The decisions we think we make are, in his opinion, merely "[the] result from molecular-based electrical impulses and chemical substances transmitted from one brain cell to another."[2]

Coyne's view is shared by his friend, author and philosopher Sam Harris, who also has training as a neuroscientist, and with whom he also shares a strong antipathy toward religion. In his exceedingly short

1. Coyne, "You Don't Have Free Will," para. 7.
2. Coyne, "You Don't Have Free Will," para. 2.

booklet *Free Will*, and elsewhere, Harris has written that free will is an "illusion." He, like Coyne, is of the camp that argues that human behavior, including decision-making, is completely determined by "a chain of causes that precede conscious awareness and over which we exert no ultimate control." While Harris concedes that our choices "matter," and that their emergence is the result of a "fundamentally mysterious process," he nevertheless maintains with absolute certitude that they are the result of "causal states of the brain." Though human beings can "imagine and plan for the future, (and) weigh competing desires," he claims that, ultimately, a human being is nothing more than a "biomechanical puppet."[3]

Such determinism (call it hard determinism) is but one of four classic understandings of free will, each with variations. Another approach, known as (hard) indeterminism, agrees that there is no such thing as free will, but not because our decisions are predetermined. Rather, it views the universe as dominated by randomness over which there is no control. Two other views, compatibilism and (nonpolitical) libertarianism, hold that free will exists, but are split on whether reality is determined.[4]

The absolute determinism of Coyne and Harris is not surprising. Both are known for holding firm convictions, rather than espousing nuance. But their viewpoint comes with a healthy dose of irony. Both men are scholars, teachers, and authors. They spend a good deal of their energy attempting to marshal facts and arguments in efforts to persuade people of various positions, especially in areas where they believe science or logic or both demonstrate the invalidity of religious texts or principles. Yet, if our conduct is hard-wired and predetermined, if we really could not have chosen that V-8 or taken that other road, what is the point of listening to them or reading their books? After all, if their thoughts and words were predetermined by their genes and environment, why should they be assigned any greater value than someone else's predetermined thoughts and words? And anyhow, would what they say make any difference with respect to a listener's or a reader's future conduct? Perhaps they would say that their words can somehow affect the brain's synapses so that the next time a decision needs to be made, the wiring will be different (i.e., predetermined) but with a better outcome. But if that is so, shouldn't it be so for those who hear Moses's words, too? Some might call that learning how to think, reason, and make choices.

3. See Harris, *Free Will*, 5, 36–37, 42, 47.

4. See Caouette, "Free Will Problem."

In any case, it is one thing to argue, as Coyne and Harris do, that the biblical account of creation or evolution is not supported by science. (Of course it isn't. The Torah was never intended to be a science text, and its various authors lacked requisite knowledge anyhow.) It is quite another thing to argue that humans have no free will. That's getting personal.

WHAT IS THE JEWISH UNDERSTANDING OF FREE WILL?

There is no question that Judaism, from the first millennia before the common era, through the Talmudic period, into the Middle Ages, and in the present has assumed that individuals have considerable free will, and it did so initially for good reason. If you were the authors and editors of the Torah text, and were trying to establish rules and regulations for what you envisioned to be a sacred society, a religious civilization, you would want to encourage the members of your community to be responsible for their conduct, not just for their own sake, but for the wellbeing of the polity. One way to do so would be to create a foundational text which is premised on the idea that individuals are free to act and that those actions have consequences for the actors and the greater social order. In theory, the resulting social pressure would then reinforce the desired behavior.

The biblical understanding of free will is not asserted philosophically or analytically. Initially, it is expressed conceptually and through fables. At the very outset of Genesis, the Torah asserts a profound position: humankind, the final creation, is fashioned *b'tzelem Elohim*, or "in the image of God" (see Gen 1:26–27). As we have discussed, the meaning of that phrase is debatable, but the human capacity to exercise moral freedom is clear from the beginning. The text contains a unique story. After the first two humans are created, they are told not to eat fruit from the tree of knowledge of good and evil. Though they know the rule, they disobey anyway, and there are consequences (see Gen 2:16–17; 3:2–3, 6, 16–24). Free will, it seems, even includes "the power to disobey the deity."[5]

Shortly thereafter, we are told, Adam and Eve's son Cain was upset when God did not pay the same attention to him and his offering as God did to Abel and his offering. God advises Cain that he has the power to overcome an evil inclination, but Cain does not do so, and again there are

5. See Hayes, *Introduction to the Bible*, 41.

consequences (see Gen 4:3–16). That these stories are pure fiction does not detract from the fact that they reflect a fundamental understanding of human nature, one that includes the power to make decisions.

When we reach the book of Deuteronomy, we have a new and revised set of governing principles presented in a new literary style and with a more personal and urgent tone, but the importance of free will is, if anything, heightened. It is here that we find Moses's moving speech to the assembled Israelites, the speech that Coyne and Harris would, presumably, consider unfounded, misleading, and futile. Just before he dies, Moses tells the people that the instruction being given to them is neither too difficult to understand nor beyond their reach (see Deut 30:11). Rather, they have choices to make. He has set before them two paths: one to life and prosperity, and another to adversity and death. Moses urges the people to follow the rules that have been outlined—that is, to choose life, so that they and their offspring might live (see Deut 11:26–28; 30:15, 19). Again, the historicity of the speech (or lack of it) is not important now. What is critical is the demonstration of the underlying strength of the idea that humans have the capacity to choose among available courses of conduct and that there will be consequences to their choices, and not just personal consequences, but societal ones as well.

As Professor Hayes has noted, in the Deuteronomist's view of history, "Israel's fate is totally conditioned on her obedience to the covenant."[6] In the writings that follow in the books we know as Joshua, Judges, Samuel, and Kings, what we have is historiosophy, an argument based on a selected historical (some undoubtedly fictional) touchstones, all of which show a pattern of reward and punishment tied to Israel's (dis)obedience.[7] The anarchy described in Judges led to stories of a monarchy. The failure of the kings, and of the people, then led to national ruin.

With the fall of the kingdoms of Israel and Judah, community leaders needed to rationalize how their omnipotent deity failed to preserve their homeland and, indeed, his own home, the temple in Jerusalem. As previously discussed, attributing the catastrophe to some absence or defect in the deity was not an option. What was clear, therefore, was that the people had failed to observe the laws provided to them, and as Moses and subsequent prophets like Isaiah, Amos, and Jeremiah had warned, their bad choices had led to social disorder and collapse.

6. Hayes, *Introduction to the Bible*, 169.

7. See Hayes, *Introduction to the Bible*, 182–84, 186.

Centuries later, Jewish sects like the Pharisees, Sadducees, and Essenes anticipated modern diversity of thought about free will. According to the account of the historian Josephus, the rural and righteous Essenes were fatalists, seeing everything as divinely predestined, while the Sadducees, associated by Josephus with the priesthood and financial power, favored absolute free will unencumbered by divine providence.[8] Both groups essentially disappeared after the fall of the Second Temple in 70 CE. The Pharisees, both more populist and less Hellenized than the Sadducees, accepted a mix of determinism and free will, and became the intellectual ancestors of modern Judaism.

So strong was the view that each person was responsible for his or her actions, that we find in the Talmud the following statement: A person is always responsible for his actions, whether awake or asleep.[9] Subsequently, Jewish philosophers, whether rationalists or mystics, treated free will almost as an axiom. They differed greatly about the apparent paradox that free will could coexist with God's omniscience, and whether, for instance, God even knew about particular human activities in advance of their occurrence, but, in general, they did not dispute that humans had free will.

For instance, Maimonides claimed both that God knew everything that would happen before it did but he also expressly rejected the idea that God decreed a man's character, whether righteous or wicked, from his birth, and asserted that free will was "granted to every man." If it were otherwise, he reasoned, there was no purpose to the prophetic urgings to improve our behavior.[10] On the other hand, Rabbi Levi ben Gerson (1288–1344), known as Gersonides (or Ralbag), agreed that man had free will, but asserted that God did not know in advance how any person would exercise it.[11] On the third hand, the mystic Kabbalists believed that, through a process of contraction or self-limitation known as *tsimtsum*, God constricted his essence in order to allow the world to exist and humanity to have free will. The late Rabbi Eliezer Berkowitz incorporated this concept while trying to synthesize rationalism and post-Holocaust theology.[12]

8. See Kohler and Broydé, "Predestination."

9. See b. B. Qam. 3b, https://www.sefaria.org/Bava_Kamma.3b?lang=bi.

10. See m. Torah 5:2, 4.

11. See "Gersonides," sec. 4, paras. 7–9.

12. See Mohl, "Tsimtsum."

Obviously, the Jewish acceptance of free will is deeply rooted and enduring. Rabbi Jonathan Sacks, among many others, continues to link personal freedom to moral (and social) responsibility.[13] But we have seen elsewhere that Jewish thought is often a product of its time and, as such, potentially deeply flawed when viewed from a more modern and informed perspective. The authors of the creation and flood texts were neither scientists nor historians, and the later sages who opined on the process of gestation were typically not physicians, and, in any event, lacked the knowledge we have today about fetal development. Their stories and views may have value, but taking their statements as factually true is unwarranted. Similarly, Jewish thinking on free will should be subject to the same sort of critical analysis, and possible revision, that we have seen in other contexts. In short, are Coyne and Harris correct? Is the free-will hypothesis false? And, if so, what are the implications for Judaism? What does modern neuroscience have to say?

WHAT DOES NEUROSCIENCE TEACH TODAY ABOUT FREE WILL?

The history of neuroscientists' efforts to explore the free will phenomenon was reviewed in 2016 by philosopher and neuroethicist Andrea Lavazza in the journal *Frontiers in Human Neuroscience*.[14] The setting for our current understanding was drawn a half century ago with the discovery by Hans Kornhuber and Luder Deeke of the Readiness Potential ("RP"), a measurement of increased bioelectric activity in the brain. The RP was measured by an electroencephalogram ("EEG"), a procedure in which electrodes were placed on a subject's scalp to allow for the recording of bioelectric activity. This activity was seen as an indication of preparation for a volitional act.

One question raised by the discovery of RP was whether an individual was conscious of an intention to act before RP appeared. In the early 1980s, Benjamin Libet, a son of Jewish immigrants from the Ukraine who became a neuroscientist at the University of California, Davis, sought to answer that question. Libet and his team designed a relatively simple test. First, subjects were wired for an EEG. To record muscle contraction, electrodes were also placed on subjects' fingers. Then the subjects were asked

13. See, e.g., Sacks, "Freewill (Vaera 5775)."

14. See generally Lavazza, "Free Will and Neuroscience."

to do two things: spontaneously move their right finger or wrist, and, with the aid of a clock in front of them, report to researchers the time they thought they decided to do so.

What Libet found was that conscious awareness of the decision to move a finger preceded the actual movement of the finger by two hundred milliseconds, but also that RP was evident 350 milliseconds before such consciousness.[15] While Libet recognized that his observations had "profound implications for the nature of free will, for individual responsibility and guilt,"[16] his report appropriately contained several caveats. First, it noted that the "present evidence for the unconscious initiation of a voluntary act of course applies to one very limited form of such acts." Second, it allowed for the possibility that there could be a "conscious 'veto' that aborts the performance . . . [of] the self-initiated act under study here." Finally, it acknowledged that the possibility of "conscious initiation and control" in those situations that were not spontaneous or quickly performed.[17]

Not surprisingly, and despite the caveats, some interpreted Libet's experimental results as proof that one's actions are not freely made, but rather predetermined by unconscious neural activity. In the years following the publication of Libet's report, other experimenters have not only replicated his work, with more sophisticated measuring devices, but they have extended it. For instance, early experiments involved index fingers on both hands, and calculations of both RP and lateralized RP over the motor cortex contralateral to the responding hand. Subsequently, scientists at the Max Plank Institute for Human Cognitive and Brain Sciences also utilized right and left index fingers, this time to press a button, and the subjects reported awareness of action not by observing a clock, but by identifying one of many letters streaming by. Brain activity was detected by using functional magnetic resonance imaging ("fMRI") signals. In a 2008 publication in *Nature* authored by Chun Siong Soon and others, it was claimed that brain activity encoding a decision could be detected in the prefrontal and parietal cortex for up to ten seconds before the subject became aware.[18]

15. See generally Libet et al., "Time of Conscious Intention," 641, 636.
16. See Maugh, "Benjamin Libet," para. 10.
17. Libet et al., "Time of Conscious Intention," 640–41.
18. See Soon et al., "Unconscious Determinants of Free Decisions."

Also not surprisingly, the assumptions in and the interpretations of results from these experiments drew criticism, beyond the obvious concern about mistaking correlation (of recorded brain activity) with causation (of a decision to act). And they continue to do so. After all, the average human brain contains billions and billions of nerve cells, called neurons. We have recently learned that the number of neurons is approximately eighty-six billion.[19] Each neuron is connected to other neurons by perhaps thousands of synapses, junctions through which neuroactive molecules or electrical impulses travel. The total number of these synaptic connections exceeds a hundred trillion.[20] Moreover, while we once believed the brain to be fixed, now we know that it is more plastic and changes constantly.[21]

Even if we were thoroughly familiar with all of these connections, and all of the electrical and chemical processes which operate (or not), and when and why they do (or do not), and also had a complete grasp of neuroplasticity (which understandings we do not currently possess), we clearly do not understand what has been called the Hard Problem, the nature of consciousness.[22] If we do not understand that, then obviously we also do not understand the nature of subconsciousness. So what exactly, if anything, were Libet and Soon observing other than some sort of recordable activity? The answer is not apparent.

More narrow objections could be and were raised, as well, regarding the early tests. Florida State philosophy professor Alfred Mele suggested that because subjects might have different understandings of the "awareness of the intention to move" they were to report, the term was too ambiguous to measure to any degree of scientific value. Moreover, even if some readiness potential could be measured, isn't it possible that RP itself is indicative of nothing more than the result of various stimuli, including being placed in a control room, hearing instructions, and focusing on a specific task? In this view, it would be akin to heightened anxiety that a patient might feel prior to or during a conventional physical examination.

In addition, the tests performed were narrow in scope and duration. They generally involved very simple motor functions being undertaken, or not, within seconds of some signal. But Princeton psychologist and

19. See Herculano-Houzel, "The human brain in numbers."

20. See Sukel, "The Synapses-A Primer"

21. See Wesson, "A Primer on Neuroplasticity."

22. See Weisberg, "The Hard Problem of Consciousness."

Nobel-Prize-winning economist Daniel Kahneman teaches that we think fast and slow.[23] His core observation is that humans operate with two different thought modes. In the first, known as System 1, the brain "operates automatically and quickly, with little or no effort and no sense of voluntary control." In the other, known as System 2, the brain "allocates attention to the effortful mental activities that demand it, including complex computations." Kahneman associates the operation of System 2 with what we feel as agency and choice.

Is it possible that a person's brain activity, as recorded by an EEG, an fMRI, or some other mode of neuroimaging, would display different results in circumstances where more complex actions are involved, especially over an extended period? Is it conceivable that brain activity would be different if the subjects were in a kitchen and asked to choose how many, if any, eggs they wanted for breakfast, and how they preferred them cooked, and with what bread, what spread, and what fruit and drink? And might that kind of brain activity be different still than the kind involved in deciding over the course of a presidential campaign which candidate to support, or deciding during courtship whether to select a certain someone for a life partner?

We have no EEGs or other scans that address breakfast, political, or marital choices, but some recent experiments suggest that the death of free will, as announced by Jerry Coyne and Sam Harris, may have been not just premature, but unwarranted. In 2012, French neuroscientists published a report in the *Proceedings of the National Academy of Sciences* concerning a study about RP which included a variation on Libet's experiments, specifically an audible cue to the participants to make a movement in response to an unpredictable noise.[24] Rather than reflecting the final causal stages of planning and preparation for movement, Aaron Schurger et al. found that neural activity in the brain fluctuated normally. At such times, "when the imperative to produce a movement is weak, the precise moment at which the decision threshold is crossed leading to movement is largely determined by spontaneous subthreshold fluctuations in neuronal activity."[25] Consequently, decisions about self-initiated movement were "partly determined by spontaneous fluctuations

23. See Kahneman, "Of 2 Minds."

24. See Shurger et al., "Accumulator Model."

25. Shurger et al., "Accumulator Model," 1.

that are temporally autocorrelated" with a neural decision to move.[26] In other words, movement might not be determined subconsciously, but may simply occur when the brain is in a sufficient state of arousal.

Similarly, a study undertaken by graduate student Prescott Alexander and his team attempted to isolate motor and non-motor contributions to RP. As reported in 2016 in the journal *Consciousness and Cognition*, they found that "robust RPs occurred in the absence of movement and that motor-related processes did not significantly modulate the RP."[27] This suggested to Alexander et al. that the RP measured was "unlikely to reflect preconscious motor planning or preparation of an ensuing movement, and instead may reflect decision-related or anticipatory processes that are non-motoric in nature."[28] They concluded, in part, that "RP does not primarily reflect processes unique to motor execution or preparation, and may not even be primarily generated by the neural activity involved in making a free choice."[29]

VOLITION AND RESPONSIBILITY

What do all these studies show? At a minimum, Schurger and Alexander, and their teams, have interrupted what seemed to be developing scientific support for hard determinism and against free will. They have provided scientific grounding for an alternative understanding of previously accumulative data. In the words of cognitive neuroscientist Anil Seth (speaking of Schuger et al.), they have opened "the door towards a richer understanding of the neural basis of the conscious experience of volition."[30]

Consequently, when Alfred Mele argues that science has not disproved free will, he is correct. Science has not falsified the free-will hypothesis even once, let alone in the kind of replicable experiment that is the hallmark of the scientific method. At the same time, science has not confirmed the free-will hypothesis either. The unsettled state of affairs is not necessarily bad, though, for at least two reasons.

26. See Shurger et al., "Accumulator Model," 11.

27. See Prescott et al., "Readiness Potentials," 38.

28. Prescott et al., "Readiness Potentials," 38.

29. Prescott et al., "Readiness Potentials," 45

30. See Caouette, "Neuroscience and Free Will," para. 15.

First, the reality is that we are at the early stages of our understanding of both the human brain and levels of consciousness, and we undoubtedly do not even know fully what we don't know. For instance, in 2015, neuroscientists were acknowledging that no one knew how the human brain was wired and bemoaning the fact that they could not even map a mouse's brain, let alone a human one.[31] About a year later, scientists were able to produce a map of the brain's cerebral cortex with a "new mapping paradigm," but even so, a participating researcher conceded the limitations of the new map.[32]

Similarly, in early March of 2017, an interdisciplinary group of researchers at UCLA published a report in the journal *Science* in which they claim that the brain is much more active than previously believed.[33] Moreover, according to senior author Mayank Mehta, the results of the new research demonstrate that neurons are not purely digital devices, as scientists have held for sixty years, but also "show large analog fluctuations."[34] If so, according to Mehta, this changes the way we understand how the brain computes information.

The idea of a more powerful, dynamic brain may trigger yet more revisionism concerning free will, as well. Indeed, it is at least conceivable that the reductionists are looking at the picture in the wrong way, zooming in to try to locate and record each signal the brain emits, rather than stepping back for a broader perspective.[35] That is, for all its amazing discoveries and insights, perhaps neuroscience, as commonly practiced today, is too narrow a science. Perhaps there must be some consideration for the possibility that the vast number of neurons and synapses, and their intricate interconnectedness, in conjunction with neural plasticity, yields something greater than the individual cells themselves, even as water is more than its component molecules made of hydrogen and oxygen. Perhaps consciousness is an emergent phenomenon. In this view, at a certain level of collective complexity, consciousness emerges.[36] And with it, free will.

31. See Lee, "Race to Map the Brain."
32. See Turk, "Scientists Made a New Map," paras. 15–16.
33. See Moore et al., "Dynamics of Cortical Dendritic Membrane."
34. See Gordon, "Brain Is 10 Times More Active," para. 9.
35. See Yong, "How Brain Scientists Forgot."
36. See, e.g., Nelson, *Emergence of God*, 32–35.

From the history of science and technology, we can assume that the pace of our progress will be uneven and the results surprising. Perhaps we will move faster than our ancestors did on the centuries-long path from Ptolemy to Copernicus to Hubble (both the man and the telescope). But how much time we will need is not clear. Consider the journey from Wilbur Wright's first step onto a biplane at Kitty Hawk to Neil Armstrong's first step off the lunar module Eagle on the Moon, and whether neuroscience is arguably more complex than rocket science.

Second, another reality is that the stakes in the multidisciplinary debate between free-will advocates and determinists go far beyond the musings of philosophers and the reputations of neuroscientists seeking grants and fame. Should science somehow disprove free will, should it show that we are not just influenced by our genes and our physical and social environment, but that our response to each option available to us is truly compelled rather than chosen, it is not too hard to imagine at least two dystopian results.

In the first case, should it be generally known that humans have no free will, and that conduct is in fact predetermined, significant numbers of individuals might well feel released from whatever tenuous social bonds are now attached to them, free to engage in disruptive behavior. We already have some experimental evidence from psychologists Kathleen Vos and Roy Baumeister that supports the idea that weakening a belief in free will leads to "cheating, stealing, aggression, and reduced helping."[37]

A second worrisome situation that might arise concerns potential screening of individuals for genetic or environmental or other predispositions to anti-social behavior. Might individuals found to possess an antisocial gene be incarcerated or subjected to gene therapy to alter or remove the problematic genetic material? If so, it is not too difficult a leap to rounding up groups of people who, by virtue of their color, ethnicity, geographic origin, socioeconomic status, or other trait likely all have the offending gene. The infamous Nazi medical experiments on Jews, Roma, and others provide a chilling example of the depraved capacity of some humans to mistreat the Other, and to do so ostensibly in the "interest of science" or some asserted "greater good." Social historian Yuval Harari

37. See Vos and Baumeister, "Addiction and free will," 1.

has warned recently about the merger of Big Data with Big Brother.[38] It is a warning worth heeding.

In many ways, then, the free-will hypothesis is more important than the understanding laid out in Genesis with respect to creation and evolution. We have learned a great deal about how our universe came into existence and how life forms have evolved. And we have learned that we can survive quite well with such knowledge. But if the free-will hypothesis is incorrect, if we are only products of our genes and our environment and of the purely mechanical interplay of chemistry and physics, if we do not have any meaningful capacity to make choices, then how can we proceed responsibly? Well, we could still act *as if* we were free and our decisions mattered (a path advocated by some determinists like Israeli philosopher Saul Smilansky),[39] but there would be a cloud hanging over us, one difficult to dissipate. Many could not overcome.

Yet, even in the most dire circumstances, some do overcome. Recounting the horrors of the concentration camps, psychiatrist and neurologist Viktor Frankl noted that despite the conditions, the actions of some showed that "[e]verything can be taken away from a man but one thing: the last of the human freedoms—to choose one's attitude in any given set of circumstances, to choose one's own way."[40]

Why some react one way under pressure (or without it) and others do not remains a mystery, as even Sam Harris has acknowledged. Maybe science will solve that mystery someday, but maybe not. So, perhaps Descartes (1596–1650) was not quite right when he declared "*Cogito, ergo sum*," that is, "I think, therefore I am." Perhaps thinking is a necessary but not sufficient element of being. Perhaps we need to be able to choose to be fully alive and vital. Consequently, until (if ever) scientists prove otherwise, *eligo, ergo sum*—I choose, therefore I am. Or at least, I think I do. And at least once every year I am grateful to the Deuteronomist for reminding me of the extensive menu of blessings and curses that is set out before me, and for his emphatic call to choose life.

38. See Harari, "Yuval Noah Harari."

39. See "Saul Smilansky."

40. See "Viktor E. Frankl Quotes."

Chapter 21

The Torah and Tachlis of Violence with Firearms

THESE DAYS IN THE United States we see and hear much violence associated with firearms. This violence does not discriminate. It cruelly takes its toll on people of all colors and creeds, all ages, all genders, all nationalities. It knows no boundaries. It invades spaces both secular and sacred. Sometimes it erupts in a mass shooting at a college or an elementary school, a nightclub or an outdoor concert, a church or a synagogue. Sometimes it comes with the steady staccato of a semiautomatic weapon fired by a sociopath at a group of individuals s/he may not even know. Sometimes it comes by way of a single bullet, the shooter being the same person who is shot.

However it manifests itself, the sadness that follows is palpable. Our hearts are broken at the loss of life, of what might have been, of possibilities foreclosed permanently. And we are angry, too—angry at the perpetrator and angry about the conditions that permitted (if not caused) a person to become so hateful, self-righteous, desirous of notoriety, callous, or full of despair that s/he acted to take a human life.

When such violence strikes, to the extent that its senses and sensibilities have not been numbed, the Jewish community has not been shy. With sermons and articles and resolutions and more, it has spoken—loudly, passionately, and repeatedly. But it has not spoken uniformly, much less always wisely. There is in the Jewish community, as there is in the nation as a whole, a variety of viewpoints. The question before us is whether our tradition can offer both Torah and *tachlis*—that is, instruction grounded in Jewish values and ideas that are also practical and productive.

To answer that question, we first need to understand the nature and extent of violence involving firearms in America today. That is, before moving forward, we need to take a step back. We need some perspective. We need context. We need to look at the grim statistics and break them down. And in the course of the inquiry, we need to be mindful that there are statistics and then there are statistics. We will try to keep cherry-picking to a minimum. After that, we can consider the spectrum of Jewish ethical values and see how, if at all, they could inform a productive approach to the challenges presented by violence with firearms.

THE NATURE AND EXTENT OF VIOLENCE INVOLVING FIREARMS IN THE UNITED STATES

We begin with some basic numbers. In recent years in the United States, the annual number of deaths associated with firearms—whether guns, rifles, or other such devices—has approximated thirty-four thousand. That averages to more than ninety a day. This number more than doubles if one includes the physically wounded,[1] and related adverse psychological effects further exacerbate the problem.[2]

But there is always a danger when one looks at large amounts of data like these numbers. They both reveal and conceal important information. How can we understand these numbers?

According to the Centers for Disease Control and Prevention, in 2015 just over 2,710,000 resident individuals died in the United States. By far, the leading natural causes of death were heart disease and malignant neoplasms (cancerous tumors), followed by chronic lower respiratory diseases. Those three natural causes together accounted for just over half of the reported deaths in 2015. Of the top ten causes of death, two were not diseases. Unintentional injuries (accidents) and self-inflicted harm (suicide) ranked respectively as the fourth and tenth leading cause of death that year.[3] Typically, about half of all suicides involved firearms,

1. See Fowler, "Firearm Injuries in the United States."

2. See generally Cornell and Guerra, "Gun Violence," https://www.apa.org/pubs/info/reports/gun-violence-report.pdf.

3. See Murphy et al., "Deaths: Final Data for 2015," Table B.

meaning about twenty-three thousand.[4] In addition, in 2015, there were almost thirteen thousand homicides related to firearms.

Still, even including suicides, deaths involving firearms would not be ranked in the top ten, nor even the top fifteen causes of death in America. For instance, in 2015, there were fewer deaths involving firearms than deaths attributed to each of kidney disease, septicemia, and nephritis and nephrosis,[5] but deaths by those causes rarely get the headlines that firearm deaths do. Deaths involving motor vehicles, averaging 103 per day in 2015,[6] are almost three times more common than firearm-related deaths, yet do not generate similar national or even local pressure for additional regulation.

There is a plausible explanation, of course. In the normal course, one expects to die of something, which, if not an accident, most likely will be one of the dozen or so diseases which are the leading causes of death. These diseases, in turn, can and often are diagnosed and managed, so preparations can be made. In the normal course of life, one does not expect to be shot, and we cannot really prepare for such an event. It tends to come suddenly, literally explosively. We are saddened by Alzheimer's disease, the sixth leading cause of death in 2016, but we are shocked by violence coming out of a barrel of a gun or rifle.

When we focus solely on deaths associated with firearms, now looking at final numbers for 2015, we see that about three out of five such deaths (22,018) were due to suicide—that is, they were intentional and self-inflicted.[7] Just over one-third of firearm-related deaths that year (12,979) were the result of homicides, either murder or manslaughter.[8] As bad as that was, it was also much lower than in 1993, when firearm-related homicides peaked at 17,075.[9]

Some types of activities are more likely than others to involve firearm-related deaths: gang activity,[10] commission of a felony, and domestic

4. See, e.g., Centers for Disease Control and Prevention, "Suicide and Self-Inflicted Injury," https://www.cdc.gov/nchs/fastats/suicide.htm.

5. See Murphy et al., "Deaths: Final Data for 2015," Table B.

6. See Centers for Disease Control and Prevention, "*QuickStats*: Average Number of Deaths," para. 1.

7. See Murphy et al., "Deaths: Final Data for 2015," Table 11.

8. Murphy et al., "Deaths: Final Data for 2015," Table 11.

9. See National Institute of Justice, "Gun Violence," para. 3.

10. See National Institute of Justice, "Gun Violence," para. 4.

disputes,[11] including arguments and romantic triangles.[12] Aside from these categories, some homicides involving firearms in 2015 were the result of legal intervention (484), others were unintentional (489), and a smaller number (282) were undetermined.[13] Over the course of the last thirty-five years, the number of individuals killed in mass shooting incidents has averaged twenty-three per year,[14] with several notable exceptions. Of course, that could change, and some argue that it has begun to do so.[15] Whether relatively recent events in Las Vegas, Orlando, Parkland, and Pittsburgh indicate a new trend in frequency, or intensity, remains to be seen.[16]

Further, while mass shootings and high-powered rifles garner the attention of the press, the public, and the politicians, in those homicides in 2015 and 2016 for which the Federal Bureau of Investigation ("FBI") has received weapons data, and where the weapon involved was a firearm, over two-thirds of the time the weapon used was a handgun.[17] Moreover, in half of the 2015 cases where the relationship of the offender and the victim was known, about 60 percent of the time the victim was killed by a friend or acquaintance and 26 percent of the time the killer was a family member.[18]

The nature of the incidents involving firearm deaths and those involved may help us focus on possible approaches to reducing them, but, again, we still need to be careful about the numbers we have. For instance, restricting the analysis to homicides involving firearms, the numbers translated to about thirty-six per day, every day, in the United States in 2015. This average daily number of deaths is horrific. It is also quite misleading. The frequency and distribution of such incidents varies widely depending on a few key factors like location, gender, age, and race. Even the time of year or the day of the week can be important.

11. See Sugarmann, "For Women."
12. See Federal Bureau of Investigation, "Expanded Homicide."
13. See Murphy et al., "Deaths: Final Data for 2015," Table 11.
14. See Reynolds, "Mass Shootings," para. 11.
15. See Thompson, "Mass Shootings in America."
16. See Duwe, "Mass Shootings Are Getting Deadlier."
17. See Federal Bureau of Investigation, "Murder Victims by Weapon, 2012–2016,"
18. See Federal Bureau of Investigation, "Expanded Homicide Data Table 10."

Where does violence with firearms occur?

No geographic area is immune from the scourge of homicides, but the death rate in 2016 attributable to firearms was highest in the states of Alaska and Alabama and the lowest in Massachusetts and Rhode Island.[19] It was about average in Illinois and Maryland, but tell that to the citizens of Chicago and Baltimore.

We can narrow our focus to large urban areas, like Chicago, but the complexity of the problem does not diminish. In the last few years, Chicago has sustained more firearm-related deaths than any other community in the country: 415 in 2014, 473 in 2015, and 762 in 2016.[20] The last number was greater than the number of homicides in New York City and Los Angeles combined.[21] On average, this means that the evening news in Chicago reported no less than a homicide a day, every day, but summers are typically worse than winters, and weekends and holidays are generally worse than mid-week. During the July 4, 2017 extended holiday weekend, 102 individuals were shot and fifteen died.[22]

These numbers, though, mask the potentially crucial fact that the homicide rate in some neighborhoods in Chicago is drastically different than that in other neighborhoods. Many areas are free of firearm-related violent crime, but others approach and may even exceed the homicide rate in third-world countries.[23] And Chicago is not unique. Mass shootings aside, homicides with guns tend to be spatially clustered to a limited number of "hot spots."[24]

Similarly, while the body count is high, Chicago does not have the worst homicide rate in the United States when deaths per one hundred thousand citizens are calculated. In fact, in 2016, it ranked no higher than eighth on one dishonorable list of community dysfunction. Taking a slightly longer term view over a five-year period, Chicago ranks twelfth, behind Detroit, St. Louis, New Orleans, Baltimore, and others.

19. See Centers for Disease Control and Prevention, "Firearm Mortality by State."

20. See Marx, "Chicago Homicide Data," and Rosenberg-Douglas and Briscoe, "2016 Ends with 762 Homicides."

21. See Crimesider Staff, "Chicago Saw More 2016 Murders."

22. See D'Onofrio et al., "102 Shot, 15 Fatally."

23. See Lucido, "Chicago's Safest And Most Dangerous," paras. 3, 7.

24. See Verbruggen, "Reducing Gun Violence."

The homicide rates in those four named cities are twice that of Chicago over the five-year period.[25]

By contrast, Newtown, Connecticut is a small, picturesque, financially comfortable New England town. But for one incident, neither its firearm homicide numbers nor its rate or ranking would be noticeable. Yet, on December 14, 2012, in less than five minutes, one individual armed with a Bushmaster .223 caliber model XM15 semiautomatic rifle loaded with exploding hollow point rounds shot 154 bullets into Sandy Hook Elementary School, killing six adults and twenty children and physically wounding and psychologically scarring untold others.[26]

When considering who resorts to homicide with firearms, geography matters, but so does gender, race, and age. According to FBI statistics for 2016, where the gender of the offender was known, almost nine out of ten times the killer was a male.[27] Where race was known, about 54 percent of the offenders were black or African American and 44 percent white.[28] Where age was known, the greatest number of offenders was found in the twenty-to-twenty-four-year-old cohort, followed by the twenty-five-to-twenty-nine-year-old bracket.[29]

Who are the victims of the use of firearms?

The composition of the victims of homicidal violence tends to parallel that of the perpetrators. That is, according to the FBI, the victims are overwhelmingly male and they are more likely to be black than white.[30] Where the relationship between offender and victim is known, as noted above, the data is clear: a victim is more likely to be killed by someone s/he knew, such as a friend or acquaintance, or even by a family member, than by a stranger.[31] In other words, perpetrators tend to kill victims who resemble themselves.

A report in the July 2017 issue of *Pediatrics*, the journal of the American Academy of Pediatrics confirms that the pattern observed by

25. See Mirabile, "Chicago Still Isn't the Murder Capital," fig. 3.
26. See Jameson, "Sandy Hook Atrocity."
27. See Federal Bureau of Investigation, "Expanded Homicide Data Table 2."
28. See Federal Bureau of Investigation, "Expanded Homicide Data Table 2."
29. See Federal Bureau of Investigation, "Expanded Homicide Data Table 2."
30. See Federal Bureau of Investigation, "Expanded Homicide Table 1."
31. See Federal Bureau of Investigation, "Expanded Homicide."

the FBI reaches deep down to affect our nation's children. There is an enormous gap between boys and girls as victims. Whether the issue is death or injury, boys are involved in just over four of five incidents and girls in just under one in five. Similarly, there are significant differences in the rates of mortality between racial or ethnic groups. With respect to homicides involving firearms, the annual mortality rate for African American children is twice that of American Indian children, four times the rate for Hispanic children, and about ten times the rate for White and Asian American children. Interestingly enough, the situation is different with respect to suicides. White and American Indian children have the highest rates of suicide involving firearms, rates four times higher than that of African American children and Hispanic children, and five times the rate for Asian American children.[32]

The economic cost of violence with firearms

Researchers at John Hopkins University have estimated the cost of emergency department care for such violence at 2.8 billion dollars annually.[33] If you think that number sounds high, consider another estimate from an investigator at the Pacific Institute for Research and Evaluation: eight billion dollars in direct costs and another $221 billion in indirect costs.[34] Putting issues of definitions and methodology aside, the economic cost of firearm violence is substantial.

How many guns are there in the United States?

While violence involving firearms has a disparate impact in different neighborhoods and among different groups of people, the fact remains that a common denominator in all of this carnage is a firearm. Given the clear and devastating consequences of violence associated with firearms, some would like to ban or restrict their use, either by regulating the weapons themselves or the ammunition used. On the other hand, some gun advocates oppose virtually all such regulation, with the succinct slogan of "Guns don't kill people, people kill people."[35]

32. See Fowler et al., "Childhood Firearm Injuries."
33. See Gani, "Price Of Gun Violence," para. 4.
34. See Follman et al., "True Cost of Gun Violence," paras. 22–23.
35. Ironically, the slogan was apparently initially intended to promote gun safety,

Though gun-control advocates don't like to acknowledge it, the gun advocates' argument is largely true, but it is also incomplete and somewhat irrelevant (except in a not-to-be-underestimated political context). That is, the argument is accurate to the extent that firearms, by themselves and unloaded, are inert objects which cannot cause any more harm than any other solid object like a hammer. If you want to be serious about addressing violence with firearms, you have to recognize the reality that relatively few firearms are ever used to commit violence against humanity.

The point is underscored by looking at the number of firearms and firearm owners in the United States. We don't have precise figures, but we have what appear to be good estimates of both. According to the United States Census Bureau, on July 4, 2018, the population of the United States was around 328 million people.[36] NORC, a nonpartisan research organization at the University of Chicago, reported in 2015 that, while household gun ownership has declined in recent decades, almost one-third of all households report owning one or more weapons.[37] Numbers provided by the also nonpartisan Pew Research Center in its 2017 study, "America's Complex Relationship with Guns," are consistent with that report.[38] If there are individuals who would not self-report possession of a firearm, the fraction of homes with weapons may approach or even exceed two-fifths.

Whatever the actual number of households with firearms, and wherever they are located, on average each such household appears to possess more than one firearm. After all, there are firearms and then there are firearms. Someone may use a small pistol for target practice, a rifle for small game, and another weapon for home defense. Unfortunately, there is no precise record of the number of firearms held by civilians in the United States, but a recent survey suggests that the number of civilian guns in the United States passed the population of the country back in 2008.[39] And the number of firearms manufactured in the United States, including guns, rifles, shotguns, and other weapons, has exceeded eight

in line with a manufacturer's concern that only people who knew how to use guns should have them. See Popik, "Guns Don't Kill People."

36. See United States Census Bureau, "U.S. and World Population Clock."

37. See Smith and Son, "Trends in Gun Ownership," Table 1.

38. See Parker et al., "America's Complex Relationship With Guns," para. 3.

39. See Ingraham, "There Are More Guns Than People," Table 1.

million in recent years.[40] Even eliminating the highest estimates, there are probably in excess of 350 million firearms in civilian hands in the United States today.

Who owns and possesses all of these firearms?

As one might expect, there are demographic differences in household firearms ownership. NORC found household firearms ownership was greatest in the East South Central region and smallest in the Pacific and Northeast regions. Ownership was concentrated in rural areas and highest in counties with no town over ten thousand people. It was also more than twice as likely among households with income over ninety thousand dollars than with income below twenty-five thousand dollars. More than twice as many white respondents acknowledged owning firearms than did African-American respondents or Hispanics.[41]

To a degree, then, the data we have on firearms possessed by residents of the United States can be read to support a key contention of the gun advocates. If we compare the number of deaths associated with firearms (say, thirty-five thousand) with the number of available firearms (say, 350 million), the result is an exceedingly small percentage, about one-hundreth of one percent. This will be of absolutely no comfort to anyone who has lost a family member or friend due to violence, but it demonstrates the difficulty of dealing with the violence problem as a gun problem because it suggests strongly that 99.99 percent of all firearms in the United States are not being used irresponsibly.

The gun advocates' argument is incomplete and somewhat irrelevant, though, because, when loaded, and when used intentionally for one of the purposes for which they were manufactured and sold, or used recklessly (or simply used negligently), firearms can pack not just a powerful punch, but a lethal one. And that punch literally can extend well beyond the natural reach of the individual trying to harm another person. So, a firearm is different qualitatively than a knife or some other kind of weapon. And the difference in the nature of the device is why we don't see drive-by knife attacks or mass murders committed with a bow

40. See United States Department of Justice Bureau, "Firearms Commerce in the United States," 1.

41. Smith and Son, "Trends in Gun Ownership," 1–2, Table 4.

and arrows. Even with respect to suicides, while suffocation or poison is often used, the method of choice seems to be a small firearm.[42]

Why have a firearm?

The reason people want to have guns and rifles is, according to another survey by Pew, multifaceted.[43] Many, especially those who live in comfortable urban and suburban surroundings, do not understand a perceived need for, much less the attraction of, guns and rifles. Yet others do. The primary reason offered is usually for protection. For example, from 2014 to 2017, the greatest increase in applications for concealed-carry gun permits in Chicago came from black women, often living in the more dangerous parts of the city.[44] Whether their decision is wise may well be disputed,[45] but none of the studies that contend that guns offer no protection or do not deter crime are, for obvious reasons, classic double-blind or repeatable experiments, and the evidence that is presented is more suggestive than definitive. Academicians can correlate all they want, but does anyone not understand what motivates these women? Similarly, rural residents, living in isolated areas, may also look to firearms for protection.

Then there are those who just like to hunt and eat game animals or shoot at targets or just collect firearms or have a keepsake inherited from an ancestor. Still others want to have firearms because they can. That is, they have a legal right and want to exercise it. For them, firearms are a matter of freedom and independence. One man explained his accumulation of about thirty guns by analogizing the collection to buying shoes, while a woman with a similar number of firearms, claimed they were like tattoos, you could not have just one.[46] Who knew?

42. See Centers for Disease Control and Prevention, "Suicide and Self-Inflicted Injury."

43. See Igielnik and Brown, "Key Takeaways on Americans' Views."

44. See Hilbring, "Number Of Black Women."

45. See Moyer, "More Guns Do Not Stop."

46. See Beckett, "Meet America's Gun Super-Owners," paras. 16, 33.

JEWISH ETHICAL CONSIDERATIONS

In response to the carnage associated with firearms, the Jewish community has done what it always has done when faced with a social issue—it has reached back for instruction, primarily to its foundational texts. The concern is clear, but can lessons forged in ancient hills of Judah and refined in living rooms in Babylon and, later, small communities in Europe have any applicability to the vastly different American society of the present? Let's look at some of approaches that have been advanced.

The sanctity of life argument

As we have discussed in several other contexts, respect for life has many and deep roots in the Jewish tradition. To review briefly, the argument begins with two propositions asserted in the Torah and a third from the Talmud. The first, part of the Eden story found in Genesis, is that each person is made *b'tzelem Elohim*, or in God's image (Gen 1:27). The second, from the Ten Commandments, is the injunction "You are not to murder" (Exod 20:13). The sages recorded in the Talmud add the third, saying, with a bit of poetic hyperbole, that if anyone were to take a life, it would be as though he has destroyed the universe, but that if anyone saved a life, it would be as though he has saved the universe.[47]

Based largely on these propositions, as far back as 1975, the Reform movement, through its Religious Action Center, called for the elimination of "the manufacture, importation, advertising, sale, transfer and possession of handguns, except in limited instances."[48] And, over the intervening years, the movement has been a consistent supporter of a wide variety of gun control measures.

As its name suggests, though, RAC is biased in favor of action and somewhat less focused on considered analysis or persuasion. For example, in a gun-control statement which (as of this writing) appears on its website, RAC notes that Jewish tradition favors life and adds references to the prophet Isaiah's dream that we should beat our swords into plowshares and our spears into pruning hooks (see Isa 2:4) and reflections in

47. See m. Sanh. 4:5, https://www.sefaria.org/Mishnah_Sanhedrin.4.5?lang=bi; see also b. Sanh. 37a, https://www.sefaria.org/Sanhedrin.37a?lang=bi.

48. See Religious Action Center of Reform Judaism, "Gun Violence Prevention," para. 3.

the Talmud about a flaming sword held by *cherubim* (winged creatures) at the east of the garden of Eden, where gehenna was located.[49]

Unfortunately, these sentiments do not seriously address the problem. In fact, RAC simultaneously fails to present an honest and comprehensive view of the Jewish approach to violence and weapons, thereby leaving its credibility open to challenge, and resorts to language that is tone deaf and quite unlikely to have any effect on people who actually own, possess, and utilize firearms.

For instance, RAC fails to address the numerous places in the Torah, and later in the *Tanakh*, where life was disrespected and the murder of individuals, groups, and nations was either required or rewarded. Cain, having committed the first murder, was physically marked, but founded a city and begat many descendants (see Gen 4:8, 15, 17–22). Pinchas slew an Israelite man and his Midianite woman, but later was elevated to the position of high priest (see Num 25:8; Judg 2:28). Though hitting a rock twice was enough to keep Moses himself out of the Promised Land, striking an Egyptian overlord until he died resulted in no punishment at all (see Exod 2:11; Num 20:8–12).

Moreover, RAC apparently fails to recognize that we are neither in a mythical garden of Eden nor at the end of days of which Isaiah was speaking. Does RAC seriously think that members of the notorious MS-13, Bloods, or Mongols gangs, or even peaceful rural citizens in the Bible Belt, many of whom possess and use firearms, care one whit about a prophet's messianic musings or what some rabbi said two thousand years ago about winged beings near a valley where children were sacrificed?

Another Jewish anti-firearm group takes a similar tack. It calls itself Rabbis Against Gun Violence, as if there are any rabbis *for* gun violence.[50] This organization begins and essentially ends the Jewish underpinnings of its position with the famous phrase which is found at the end of Deuteronomy, read on the holiest of holy days, Yom Kippur, and generally translated as "choose life" (see Deut 30:19). The problem here is that the phrase is taken out of context and is an incomplete, simplistic, and, therefore, misleading invocation.

The authors of Deuteronomy were not talking about life in its physical sense, that is, the beating of a heart, the inhaling and exhaling of breath, or the firing of synapses. They were talking about a large

49. See Religious Action Center of Reform Judaism, "Gun Violence Prevention," para. 2.

50. See Rabbis Against Gun Violence at http://www.rabbisagainstgunviolence.org/.

collection of rules and regulations which if followed, they said, would bring the blessing of a worthwhile life to each and all on the land promised by their God (see Deut 26:16—28:69). They were making a political plea, not a medical or even an ethical one.

The shame of weapons argument

Orthodox Rabbi Ari Hart has offered a more creative approach. He notes that there is a view in the Talmud that sees weapons carried on Shabbat as shameful.[51] He uses this concept as a basis for building a case for gun control. Rabbi Hart is correct in his reference, but looking at the discussion as a whole, the rabbis involved seem more concerned with the sanctity of the Sabbath than any particular device used in violating that sanctity.

In any event, the noted thirteenth-century Spanish commentator Nachmanides had a sharply different view. He wrote about Lamech, who is mentioned early in the Torah as a great-great-great grandson of the murderer Cain (Gen 4:18). In Nachmanides's telling, Lamech taught his son Tubal-cain the art of metalworking. Nachmanides then imagines Lamech's wives being worried that he, Lamech, would be punished by God for helping produce swords and, thus, facilitating murder. Anticipating a now familiar argument, Nachmanides says that Lamech comforted his wives by observing that the sword would not be the agent of death, rather the person who chose to wield the weapon would be.[52] Apparently Nachmanides was a card-carrying member of the Local Sword Association, the motto of which was "Swords don't kill people, people kill people."

The right of self-defense argument

There is yet one more argument that RAC and others who seek to ban or heavily regulate firearms tend to ignore. The Torah expressly exonerates a person who kills a pursuer who had an intent to kill him (Exod 22:1). In twelfth-century Spain, Maimonides (perhaps the greatest premodern Jewish philosopher) went further and argued that if a pursuer is warned not to proceed and continues his pursuit, there is an *obligation* to kill

51. See Hart, "Weapon's Shame."

52. See Yaffe, "What Does Judaism Say."

the pursuer.[53] That is, he acknowledged, a right of self-defense and self-preservation. And, to his credit, so does Rabbi Hart.

There is even a passage in the Talmud that goes further still. Threatened by a whistleblower who was about to disclose that a rabbi had slandered a local official, the rabbi characterized the man as a pursuer and killed him first.[54]

The danger of disarmament argument

Where might the thinking of Maimonides and Nachmanides lead? You can find out on a website called Jews for the Preservation of Firearms Ownership (JPFO). JPFO was founded almost 30 years ago with the stated purpose of educating Jewish Americans about what it says are the "historical evils that Jews have suffered when they have been disarmed."[55]

JPFO is not the most coherent website online. But if you can manage to wade through it, you can find at least two rationales for its mission. First, highlighting biblical episodes RAC and others fail to address, JPFO notes incidents, including one set at the time of the Judge and Prophet Deborah and one at the time of King Saul. In both cases, the people of Israel were facing disaster at least in part because they were unarmed (see Judg 5:7–8; 1 Sam 13:19–22). In both cases, and with identical language, the Hebrew Bible stresses that there was neither a sword nor a spear among the entire population. And JPFO argues that today, in contrast to biblical times, the Jewish people cannot rely on a miracle to defend themselves.

Second, JPFO offers a list of situations, mostly in the twentieth century, in which a variety of nations have enacted strict gun control laws of various kinds and then targeted for elimination disfavored groups such as political opponents and ethnic minorities. Think Ottoman Turkey and Armenians, Nazi Germany and Jews and gypsies, Uganda and Christians, Rwanda and Tutsis.

Now, you may believe that such consequences could not happen here and that JPFO is paranoid, and you may be right. The narrative

53. See Hilchos Rotzeach uShmirat Nefesh 1:6–7, https://www.chabad.org/library/article_cdo/aid/1088917/jewish/Rotzeach-uShmirat-Nefesh-Chapter-One.htm.

54. See b. Ber. 58a, https://www.sefaria.org/Berakhot.57b?lang=bi; see also b. Sanh. 72a–b, https://www.sefaria.org/Sanhedrin.72a?lang=bi.

55. See Jews for the Preservation of Firearms Ownership at http://jpfo.org/filegen-a-m/about.htm.

that Jews with firearms could have stopped the Third Reich, when better armed people from France to Russia could not, has been debunked by those with at least as strong a sense of history (and armaments) as JPFO, including Jewish defense organizations.[56] The Anti-Defamation League ("ADL"), for instance, even argues that the Holocaust has no place in the domestic gun debate.[57] Still, based on the ADL's own analysis and recent FBI statistics, anyone who in the last couple of years has not observed an increase in anti-Jewish incidents in the United States, and a darker, more brazen tone to them (including deadly consequences), has either not been paying attention or is in denial.[58]

None of this is intended to suggest an equivalency between RAC and JPFO as organizations, much less between a maximalist and minimalist orientation toward the control of firearms. Rather, it is to recognize that, with respect to weapons of violence, and over more than 2,500 years, Jewish communities have generated a considerable variety of views and precedents regarding the sanctity of life and the propriety or obligation of self-defense. Denying that truth is neither intellectually honest, nor likely to be productive in resolving the similar tensions in the broader American discussion.

The concern for the safety of persons, places, and things argument

The Jewish ethical tradition has even more to say, especially about safety. Here are just two examples.

First, the Torah contains three closely related commandments. One, found in the Holiness Code, prohibits placing a "stumbling-block" in front of the blind (Lev 19:14). Another affirmatively requires building a parapet around the roof of a house, lest someone fall and spill blood (Deut 22:8). A third says, "take you care, take exceeding care for your self" (Deut 4:9). Each maxim seeks to keep a person from harm either because he or she cannot see or anticipate a dangerous condition or because the circumstances are inherently dangerous.

Note, though, how these principles can support both opponents as well as proponents of gun control laws. For instance, some would argue

56. See Seitz-Wald, "Hitler Gun Control Lie."

57. See Greenblatt, "Why the Holocaust Has No Place."

58. See "U.S. Anti-Semitic Incidents Spike"; see also Federal Bureau of Investigation, "Victims."

that readily accessible guns constitute a stumbling block, while others would suggest that restricting access to weapons impedes their ability to protect themselves. Some would argue that guarding yourself requires possession of a weapon, while others would contend that it means eliminating or securing them.

Similarly, the Talmud reports an argument about a dangerous dog. The majority view sought to prohibit the ownership of a mean or dangerous dog, or at least require that anyone who had such a dog remove the risk of danger.[59] Jewish gun-control advocates naturally view this as a precedent for banning possession of a firearm or, minimally, requiring that weapons be securely locked and stored. As you may have guessed by now, the rabbis involved in this discussion also recognized exceptions to the general rule. For example, they allowed those living in dangerous towns to unchain their dogs at night for protection. In sum, and continuing with a weapons metaphor, general safety principles can be used both as a sword and a shield.

APPLYING JEWISH ETHICAL GUIDELINES TO TODAY'S REALITY

Applying the Jewish ethical principles honestly and productively to the dilemma of violence associated with firearms in the United States is both difficult and frustrating. First, as we have seen, an authentic Jewish ethical approach to the ownership, possession, and use of weapons is neither simple nor uniform. Instead, it is complex and nuanced. As Chabad Rabbi Shlomo Yaffe has noted, correctly and wisely, the Torah merely required the placement of parapets around the roof of a house.[60] It did not prohibit flat roofs. Second, the violence that is sought to be quelled manifests itself in so many ways that a plausible solution for one aspect would do nothing to alleviate the harm that arises in another setting. There will have to be many solutions for the varieties of violence we face. For both these reasons, those who insist on applying biblical bumper stickers instead of urging reason and responsibility are not being as helpful as they could be. We need fewer references to myths and messianism, and more emphasis on reality-based ethics and evidence-based solutions.

59. See b. B. Qam. 15b, 46a, https://www.sefaria.org/Bava_Kamma.15b.19?lang=bi&with=all&lang2=bi and https://www.sefaria.org/Bava_Kamma.46a?lang=bi.

60. See Yaffe, "What Does Judaism Say."

The discussion is also complicated by two other factors, one legal and the other practical. First, any effort to address violence associated with firearms in America must do so within the context of the Second Amendment to the United States Constitution. In *District of Columbia v. Heller*, the United States Supreme Court ruled that individuals have a right under the Second Amendment to possess guns in their home for traditionally lawful purposes, such as self-defense. Consequently, it held Washington DC's ban on handguns and certain restrictions on the possession of rifles to be unconstitutional.[61] Subsequently, the ruling in *Heller* was made applicable to state and local governments in *McDonald v. City of Chicago*.[62]

At the same time, writing for the majority in *Heller*, the late Justice Antonin Scalia said that the Second Amendment right, like other constitutional rights, was "not unlimited." More precisely, he acknowledged that the right was historically limited to those weapons "in common use at the time," and consistent with "prohibiting the carrying of 'dangerous and unusual weapons.'" It was not, therefore, "a right to keep and carry any weapon whatsoever in any manner whatsoever and for whatever purpose."[63] Justice Scalia then identified, without limitation, individuals such as felons and the mentally ill and "sensitive places" such as schools and government building as being proper subjects of restrictions. Similarly, he recognized that there might be appropriate laws "imposing conditions and qualifications on the commercial sale of arms" and on "the storage of firearms to prevent accidents."[64]

Invoking some of the language discussed above, RAC (among other organizations) predictably condemned *Heller* as "misguided,"[65] but taken as a whole, *Heller* was not (and is not) inconsistent with the traditional Jewish view, also seen in its entirety. The principle of self-defense was affirmed, and reasonable restrictions on one or more points of the trajectory of violence associated with firearms were declared to be presumptively valid. Moreover, recent history suggests that concerns over *Heller* were misplaced. According to the Giffords Law Center to Prevent Gun Violence, over a thousand cases involving restrictions on firearms have

61. See *District of Columbia v. Heller*, 554 U.S. 570 (2008).

62. See *McDonald v. City of Chicago*, 561 U.S. 742 (2010).

63. See *District of Columbia v. Heller*, 554 U.S. at 626.

64. See generally *District of Columbia v. Heller*, 554 U.S. at 626–32.

65. Religious Action Center of Reform Judaism, "Reform Jewish Leader Disappointed."

been decided since *Heller*, and the restrictions have been upheld in over 90 percent of them.[66] The permissibility of placing reasonable parapets is, therefore, established.

The second complicating factor is more daunting. It requires determining what kind of regulation might be effective in reducing the likelihood or incidence of violence involving firearms. From the design and manufacture of the weapon itself, to its sale and distribution, to its ownership, possession and use, what works? And does what works in one setting work in others?

A recent case illustrates the problem, and gives reason for concern. Not long before the October 2017 mass shooting emanating from the Mandalay Bay Resort and Casino in Las Vegas, which generated deserved outrage and predictable cries for gun control, US District Court Judge Roger Benitez (Southern District, California) considered the constitutionality of a new California law which barred the possession by gun owners of high-capacity ammunition magazines—i.e., those holding more than ten rounds. According to the law's proponents, these magazines were not needed by civilians for defense or hunters for sport, but were used in mass shootings. Nevertheless, in *Duncan v. Becerra*, Judge Benitez issued an injunction against enforcement of the new law. The standards applied by Judge Benitez, and his legal reasoning, may or may not make sense to higher courts or legal scholars.[67] But his findings of fact offer instruction for everyone.

In the course of his sixty-six-page opinion, Judge Benitez reviewed the facts of ninety-two mass shootings in the United States and the testimony of numerous proffered experts to try to determine if the new law would have made any real difference in the outcomes of those situations.[68] He found that it would not have made any substantial difference because magazines with over ten rounds were used in only six of the ninety-two cases considered, and half of those six involved already-illegal acquisitions. Moreover, none of the state's four experts provided persuasive, science-based studies that showed the effectiveness of any ban on such

66. See "Post-*Heller* Litigation Summary," 2.

67. On July 17, 2018, by a two-to-one vote, a three-judge panel of the Ninth Circuit Court of Appeals affirmed the trial court's issuance of an injunction. Whether there will be further appellate proceedings is not known at this writing. Nor is the final outcome of any trial in the trial court. Cf. *Kolbe v. Hogan*, 849 F.3d 114 (4th Cir. 2017) (En banc, ten to four; petition for certiorari denied by the US Supreme Court).

68. See "Order Granting Preliminary Injunction," 27.

magazines. One conceded that "robust supporting data is missing" and that "'available data and statistical models are unable to discern [any] effect.'"[69]

Writing before they occurred, Judge Benitez did not consider either the massacre in Las Vegas, which left fifty-eight dead victims and over eight hundred with injuries, or the shooting at Stoneman Douglas High School in Parkland, Florida, which followed the Mandalay Bay shooting by about four months and resulted in seventeen deaths and a similar number of injuries. Would the facts of those cases have made any difference in his analysis?

Obviously, both incidents took place outside of California—that is, beyond the California court's jurisdiction. Aside from that, though, in both situations, the shooter passed then current background checks when purchasing his weapons. The Florida shooter used a single weapon with ten-round magazines, so the law Judge Benitez was reviewing would not have applied. On the other hand, the killer in Las Vegas brought numerous weapons to his war room. Twelve of his semiautomatic rifles (just a portion of his arsenal) were equipped with hundred-round magazines, and those rifles, which in normal use can fire as fast as a user can pull the trigger, had been modified with bump stocks to allow for even more rapid fire, akin to automatic fire. Still, neither bump stocks nor hundred-round magazines were illegal in Nevada. And the additional fact that he purchased many weapons in a relatively short time before he acted was not illegal, either. So, at most, one could argue that a proposed ban on magazines in excess of ten rounds would have slowed down the carnage, and that is conceivable, but it is persuasive only if you also believe that someone intent on wreaking a horrific slaughter would have cared about such a law and also been unable to secure the ammunition he wanted.[70]

For those concerned with evidence as well as ethics, the absence of reliable research is a real problem, and it extends beyond Judge Benitez's courtroom. For instance, in 2017, Leah Libresco, a former statistician and writer for FiveThirtyEight, created a social media stir when she wrote a short opinion piece for the Washington Post to the effect that research persuaded her that gun control was not the answer to the problem of violence with firearms. More precisely, she said that certain popular proposals, including banning assault weapons, restricting silencers, and

69. "Order Granting Preliminary Injunction," at 43.

70. See Jansen, "Florida Shooting Suspect"; Horton, "Las Vegas Shooter."

reducing the size of magazines, are not likely to reduce violence where it frequently appears: with young gang members, abused partners, and suicide victims.[71]

The reaction to Libresco's essay was swift. One headline on Vox blared, "The research is clear: gun control saves lives." Yet, proving once again that text does not always follow where a headline leads, the author concludes limply saying that "gun control does, at least to some extent, reduce gun deaths."[72] And that underwhelming conclusion was based on the results following a gun confiscation program in Australia, not the United States. He then concedes that gun control cannot stop all violence, that other factors like poverty, urbanization, and alcohol consumption play a role, and that "we could always use more research into gun policy."[73] Indeed.

Consider the individuals who are the source of much of the violence with firearms. Given the enormous gender disparity among homicide offenders who use firearms, "[a]ny account of gun violence in the United States, must," as the American Psychological Association recognizes, "be able to explain both why males are the perpetrators of the vast majority of gun violence and why the vast majority of males never perpetrate gun violence."[74] The APA calls for the development of programs and settings "that aim to change gendered expectations for males that emphasize self-sufficiency, toughness, and violence, including gun violence."[75] That is, it wants more research. And if we are to be serious about addressing violence with firearms used by street gangs or directed toward female partners and friends, such research seems vital.

Similarly, more research concerning the intersection of mental illness and weapons with respect to both homicides and suicides seems necessary. Today, many assume that mental illness, however defined, is causally connected to or at least correlated with violence involving firearms, especially mass shootings. But, according to the APA, "most people suffering from a mental illness are not dangerous."[76] Studies suggest that even people with diagnosable serious psychiatric disorders do not, absent

71. See Libresco, "I Used to Think."

72. See Lopez, "Research Is Clear."

73. Lopez, "Research Is Clear," para. 30.

74. See Cornell and Guerra, "Gun Violence," para. 4.

75. Cornell and Guerra, "Gun Violence," para. 4.

76. Cornell and Guerra, "Gun Violence," para. 5.

a substance abuse disorder, present a likely risk of violent disorder.[77] And, perhaps more importantly, predicting who might engage in a violent act is "a very inexact science."[78]

At the same time, federal and state laws respecting individuals who have been reported or adjudicated to have a mental illness are not consistent, comprehensive, coordinated, or even well-enforced. Consequently, more research is necessary to fill the gaps in our knowledge about the effectiveness of regulations that can reduce risks of harm to self and others, but neither reinforce stigmas about mental illness nor deter those who need it from seeking help.[79]

Because Jewish tradition tends to be pragmatic, it does not require us to research the best parapet when people are at risk of falling off a roof. Nor does it require us to save every life that is at risk. The oft-quoted teaching of Rabbi Tarfon may well be applicable here: While you are not required to complete the task, neither may you desist from it.[80] So, let's recognize both the legal, but not absolute, right of citizens to keep and bear certain arms as established in *Heller* and its progeny, and also the inherent and inalienable right of individuals and the public to life and the pursuit of happiness, marked at least by safety in their homes and in their normal movement in an open society, if not a risk-free environment. Let's also accept (1) that guns don't kill, people do; (2) that when a person has access to a firearm the damage s/he can do with it can be lethal; and (3) when people kill with firearms, they do so sometimes maliciously, sometimes recklessly, sometimes negligently, sometimes under the influence of chemicals, and sometimes due to internal disorders we may not fully understand. Let's focus on human behavior and some potential means for increasing safety and reducing risk in those more common situations where acts of violence cluster. And, finally, let's draw on the wisdom of a variety of social scientists, including (among others) criminologists, behavioral economists, psychologists, psychiatrists, sociologists, and epidemiologists, published in evidenced-based studies at respectable institutions like the Johns Hopkins Center for Gun Research and Policy, the Joyce Foundation Gun Violence Prevention Program, the RAND Corporation, and elsewhere.

77. See generally Swanson et al., "Mental Illness and Reduction," para. 13.

78. Swanson et al., "Mental Illness and Reduction," para. 27.

79. See Swanson et al., "Mental Illness and Reduction," para. 47.

80. Pirqe 'Abot R. Nat. 2:16, https://www.sefaria.org/Pirkei_Avot.2.16?lang=bi.

When we do this, we might be able to develop a non-exhaustive list of reasonable parapets which individually and collectively can lead to fewer adverse consequences in a variety of situations involving firearms. Here are eighteen such programs:[81]

- Each state should establish a firearms safety program designed to teach individuals how to use and store firearms in a safe and secure manner. Licenses for those who successfully pass their safety exam should be granted for a limited period of time, subject to renewal.
- No person should be able to purchase a firearm, or possess or use one, without having passed the established safety standard in his or her state.
- No person under the age of eighteen should possess or use a firearm, except in the presence of an adult who would be responsible for the conduct of the minor.
- States should enact safe lock and storage laws to keep minors, thieves, and other unauthorized persons from accessing a functional firearm. Without limitation, allowing an unsupervised minor access to an unsecured but functional firearm should be penalized as a felony.
- Firearm owners should be required to report a lost or stolen firearm within a reasonable, but short, period of time. If any such owner fails to do so and that firearm is used in crime, s/he should be held responsible to the same extent as would be the user of the firearm.
- Police and community intervention initiatives that address root causes of violence, provide modalities foe conflict resolution, encourage impulse control and personal responsibility, and offer positive and mediate alternatives to high-risk cultures should be instituted and expanded.[82]
- No person who commits a felony, while in possession of a firearm, should be entitled to own, possess, or use any firearm for a limited period of time subsequent to the completion of his/her sentence for such felony, and should be required to turn over temporarily to local authorities all firearms in his/her possession or control.

81. Recall that eighteen is the numerical value of the Hebrew word *chai*, meaning life.

82. See, e.g., the Boston TenPoint Coalition at http://btpcoalition.org/.

- No person convicted of more than one offense within a five-year period of using a controlled substance or driving under the influence of alcohol or drugs should, for a limited period of time, be entitled to own, possess, or use a firearm, and should be required to turn over temporarily to local authorities all firearms in his/her possession or control.
- No person convicted of abuse or stalking, whether with respect to a spouse, dating partner, friend, or otherwise, nor any person subject to a restraining order prohibiting harassment, threats, or abuse, should, for a limited period of time, be entitled to own, possess, or use a firearm, and should be required to turn over temporarily to local authorities all firearms in his/her possession or control.
- Family members, intimate partners, and specified law enforcement officers should be entitled to seek civil relief in the form of a gun violence restraining order, which would authorize the temporary removal of firearms from a household based on a credible risk of harm to any person in that household. This safety measure is sometimes called a domestic violence protection or "red-flag" law.
- Courts, agencies, and other governmental bodies that may have information relevant to whether a person may be prohibited under federal law from purchasing a firearm should be provided such funds and personnel as are necessary and appropriate to transmit relevant records to the FBI's National Instant Criminal Background Check System ("NICS").
- All states should require the reporting to NICS by mental health-care providers of the identification of any person who has been "adjudicated as a mental defective" or "committed to any mental institution" as defined in the federal Gun Control Act of 1968, as amended.
- Background checks should be required of every potential purchaser of a firearm prior to the sale of any firearm to that person, whether the sale is by a licensed dealer or private seller, and that information should be reported to a database in the state where the sale is made and then transmitted to NICS or such other database as may be established.

- States should prohibit the delivery of any newly sold firearm until a background check is completed, but in no event in less than two weeks after the sale is made.
- States should limit the number of firearms that can be purchased by an individual at any one time and in any month.
- States should continually evaluate how, if at all, laws that permit the carrying of concealed weapons and laws that allow a person to stand his/her ground affect the frequency or rate of homicides or crime in the jurisdiction.
- Government agencies and organizations such as CDC, the National Institute of Health, the National Institute of Justice, and the Bureau of Alcohol, Tobacco, Firearms, and Explosives should be directed and suitably funded to proceed in a coordinated and efficient manner to collect data on the frequency, causes, and effects of violence involving firearms.
- Private foundations should accelerate research regarding appropriate policies to reduce deaths and injuries associated with the possession and use of firearms.

A FEW FINAL WORDS

First, language drives a discussion in complex ways. That is, as discussed elsewhere, words matter. When there is a multi-vehicle incident that results in deaths and injuries, we do not characterize the event as "car violence" and we do not talk about "car control." So why do we call violence with firearms "gun violence" and instinctively seek "gun control"? And, more importantly, do those characterizations cause us to focus attention on an object instead of the behavior that activates the object or, further, the root causes of that behavior? Do they also and unnecessarily antagonize many individuals who might otherwise be willing to work to reduce violence associated with firearms?

Second, the Jewish tradition, developed over an extended period of time in many settings, is fundamentally rooted in the reality of human nature and experience. It recognizes the tensions inherent in communal life and, consequently, is sensitive to core desires for both safety and security. With insights developed over millennia, Judaism ought to be able to absorb what sociologists, psychologists, criminologists, and others have to say about the agonizing and complicated problem of violence associated with firearms and act constructively and productively.

What is crucial as we move ahead is that we tone down unhelpful rhetoric and focus on facts, however disturbing they may be. Fortunately, when addressing relationships between individuals or between one individual and the larger society, Judaism does not call on us to act self-righteously or make a show of utopian virtue. Rather, it seeks practical solutions to often-complex problems. We can do with less preaching and less posturing, and substitute instead more listening and more learning. With less demonizing and more dialogue about guns, who knows—maybe together Judaism and science can help us make the world a little safer tomorrow than it is today.

Chapter 22

Some Lessons from the Bible Code Controversy

In Part Two of this book, we looked at what happens when the ancient Torah is challenged by modern science. In Part Three, we have looked at whether Torah principles remain relevant for us. What if those two themes were joined together? What if we could use the tools of modernity to discover previously hidden secrets within the text of the Torah? And what if those secrets could reveal important information about the events of today and, even, tomorrow?

Let's return to the beginning of our journey, the biblical book of Genesis—or, more specifically, to the Hebrew text of it, known as *B'reishit*—and look at the first four verses. We will start with the first word in the first verse, *b'reishit*. The last letter of *b'reishit* is the Hebrew letter *tav*. Now, let's look for the letter which is fifty letters away from that *tav*. Let's repeat that process two more times, each time skipping forty-nine letters and seeking the next letter that is fifty letters away from the one we just found. If you count carefully, when you reach the third letter in the second word of the fifth verse in the book *B'reishit*, the four Hebrew letters you find in this sequence are *tav*, *vav*, *resh*, and *hey*. Together, in that order, they spell "Torah," the first five books of the *Tanakh*.

Congratulations! You have just uncovered a hidden Bible code, one formed by an equidistant letter sequence, or ELS. Skeptics can repeat the exercise and get the same result in the beginning of the next book in the Hebrew Bible, the book of Exodus, known in Hebrew as *Sh'mot*. Find the first *tav* in the first verse of *Sh'mot* (it's at the end of the second word) and the next three letters, each fifty letters apart. Again, if you are careful, you should find the sequence *tav-vav-resh-hey*, or Torah.

Too simple? A mere coincidence, you say? Wait, there's more.

This time, start at Exodus 11:9. Find the first letter in the name Moshe (Moses), a *mem*. Now apply the fifty-letter ELS process and you should find the letters *shin, nun,* and *hey*. If so, you have uncovered the word *Mishneh*. Next, go back to the *mem* in Moshe. Skip 613 letters to reach the *tav* in the third word in Exodus 12:11 and begin the fifty-letter ELS process once more. Ending at the second *hey* in the first word of Exodus 12:11, you should, once again, find the *tav-vav-resh-hey* sequence, which spells "Torah." And there you have the great work of the incomparable sage Maimonides. His legal code, the *Mishneh Torah*, is not only referenced in the Torah itself, but the two title words are linked by the exact number of commandments in the Torah: 613.

For any remaining doubters, the first letters in the last four words in Exodus 11:9 are *resh-mem-bet-mem*, which spell Rambam, the acronym for Maimonides's full Hebrew name, Rabbi Moshe ben Maimon. So, in addition to the ELS method, we also have related words in proximity to each other, a phenomenon known as a word cluster. There is even more in Exodus 11:9—12:13, but you get the point.[1]

Some argue that these sorts of sequences and clusters and even more complex word associations in the Torah contain hidden messages placed there when God gave the Torah to Moses over three millennia ago. In this view, the placement of the words *Mishneh* and Torah proves not only that the *Mishneh Torah* encompasses all of Jewish law, but also that the Author of the Torah anticipated and approved of Rambam's commentary.

The general contention is not new. Consider the very first word of the Ten Commandments, found in Exodus 20:2. In Hebrew, the word *anochi* ("I am") is formed by the letters *aleph, nun, khaf,* and *yud*. Centuries ago, the Talmud reported that Rabbi Johanan suggested that these letters were an acronym for the phrase "I Myself have written the Script."[2]

1. See Tigay, "Bible 'Codes,'" para. 4.
2. See b. Šabb.105a, http://halakhah.com/shabbath/shabbath_105.html.

THE MODERN DISCOVERY OF ALLEGED CODES AND MESSAGES

Until recently, however, there was little effort to discover meanings hidden in the Torah text and decode encrypted information. At the end of the thirteenth century, Spanish Rabbi Bahya ben Asher is said to have mentioned an ELS when commenting on a verse in *B'reishit*/Genesis. About two hundred and fifty years later, in Tzfat, Rabbi Moshe Cordovero wrote of secrets revealed through the "skipping of letters."[3] The first reported reference in modern times dates to the middle of the nineteenth century, when Prague Rabbi Haim Michael Dov Weissmandl is said to have first noticed the fifty-letter ELS's, discussed earlier, that disclose the word "Torah" in *B'reishit* (Genesis) and *Sh'mot* (Exodus).

Two major assertions regarding Bible codes were made in the 1990s. First, Hebrew University associate professor of mathematics Eliyahu Rips and two researchers, Doron Witztum and Yoav Rosenberg (collectively, "WRR"), published an article in *Statistical Science*.[4] The article described what has become known as the Famous Rabbis Experiment.

B'resishit has 78,064 Hebrew letters.[5] Focusing on those letters in the Koren edition of the Hebrew Bible, a currently accepted text, WRR looked for the presence of the names of over thirty rabbis and the dates of their birth or death in Genesis. After comparing their results to a control test consisting of a similarly sized Hebrew version of Tolstoy's *War and Peace*, WRR claimed to have found in Genesis a large number of names formed by ELS's in proximity to related words. The WRR article was significant because the journal in which the findings were published was considered respectable, and the authors concluded that "this proximity . . . is *not due to chance*."[6]

A few years later, Michael Drosnin authored *The Bible Code*. Drosnin's pedigree did not include the formal mathematics expertise of Rips. Rather, he was a reporter and bestselling author. But his claims were much more sensational than those of the WRR group. Drosnin purported to disclose previously hidden references, not just about past events like World War II, the Moon landing, and Watergate, among others, but also a prediction and reported warning to then Israeli Premier Yitzhak

3. See "Torah Codes Explained."
4. See Witztum et al., "Equidistant Letter Sequences."
5. See "Pamphlet 9."
6. Witztum et al., "Equidistant Letter Sequences," 434 (emphasis added).

Rabin about an assassination attempt one year before Rabin was gunned down. The Bible, Drosnin argued, was more than an ancient story. It was a computer program![7] And it needed to be studied and decoded.

Needless to say, *The Bible Code* became an international bestseller. And it begat *Bible Code II: The Countdown*, which claimed that the air assault on the Twin Towers in New York in 2001 was anticipated in the Torah, and predicted an atomic holocaust in 2006, specifically a nuclear world war, precipitated by terrorism in the Middle East. A third book, *The Bible Code III*, followed as well, and promoted the notion that Osama bin Laden possessed a nuclear bomb.[8]

The primary lure of the Bible Code books was, of course, the implication that God encrypted messages in the text of the ancient Torah, messages that Drosnin, building on the work of WRR, was able to tease out of the text. The messages might also be taken as scientific proof that God exists.

Drosnin asserts, repeatedly, that he is a nonbeliever.[9] But he also argues that "no human" could have anticipated the events which he claims are referenced in the text.[10] Rather, "[s]ome intelligence," some "nonhuman intelligence" designed the code, proving that "we are not alone."[11]

As to the source of the nonhuman intelligence, Drosnin quotes Rips as confirming that his (Drosnin's) findings are "non-random," and that they were placed in the text by a "higher intelligence," by which Rips, an Orthodox Jew, means God, a deity who acted intentionally in encoding the Torah text with hidden messages.[12] If the reader has missed the point, Drosnin quotes Israeli Prime Minister Benjamin Netanyahu's father, the late secular historian Benzion Netanyahu, as saying of the code: "If it's real, then I will believe in God, not only God, but the God of Israel, and I will have to become religious."[13] Is it real? Drosnin also quotes Hebrew University game theorist and Nobel Prize winner Robert J. Aumann as concluding that the code has passed the highest, most stringent standard

7. See, e.g., Drosnin, *Bible Code*, 25, 45, 98, 179.
8. See Drosnin, *Bible Code III*.
9. See Drosnin, *Bible Code*, 61, 79, 181.
10. Drosnin, *Bible Code*, 50–51.
11. Drosnin, *Bible Code*, 90, 97, 179.
12. See Drosnin, *Bible Code*, 20, 40, 103, 174.
13. Drosnin, *Bible Code*, 79.

of propriety and acceptability. Said Aumann, the code is "not just Kosher, it's Glatt Kosher."[14]

In *Bible Code II*, Drosnin dances a similar two-step. He continues to make references to the "Lord of the Code" and the "Code of God," as well as to Rips's belief that the Bible Code did not arise on Earth, but came from God.[15] And Drosnin continues to assert his lack of belief in the Deity.[16] Here, though, Drosnin also and more explicitly suggests that the code was the product of alien—that is, extraterrestrial—life. Referencing the claim of Nobel Prize winner Francis Crick, codiscoverer of the spiral structure of DNA, to the effect that the DNA molecule did not spontaneously develop on Earth, but was, essentially, seeded here by an advanced civilization, Drosnin contends that the Bible code, too, was brought to Earth in a vehicle by an alien, indeed, the same one who brought the DNA molecule.[17]

Today, the interest in and search for Bible codes extends far beyond academics and bestselling authors. Not surprisingly, Bible codes are an active topic on the Internet. A quick Google search for Bible codes recently returned about thirty-six million results for potential further exploration. In fact, the search for and commentary on Bible codes has become something of a cottage industry. In addition to discussing various aspects of the codes, many websites offer computer software for sale that is advertised as a tool to assist in locating encoded messages. These programs purportedly can rapidly (1) comb through the entire biblical text to search for letter sequences at various static and progressive spacings, as well as in different directions and angles, (2) identify related word clusters, and (3) create grids of letters to display the results.

THE ASSUMPTIONS BEHIND THE CODE CLAIMS

The idea that there is a code hidden in the Torah, and the further idea that this code reveals divine messages placed there by God when God dictated the text to Moses, depends, of course, on two independent sets of assumptions. One is mostly (but not entirely) grounded on religious

14. Drosnin, *Bible Code*, 43.

15. See, e.g., Drosnin, *Bible Code II*, 33–34, 148–49, 209–10.

16. See, e.g., Drosnin, *Bible Code II*, 5, 28, 93, 181.

17. See Drosnin, *Bible Code II*, 144–45, 155, 181.

beliefs and history, while the other is secular and primarily based on mathematics.

The first assumption is that the text of the Torah we have today was originally dictated by God to Moses about 3,300 years ago, and that the words we see today—and specifically the way they are spelled—are identical to what was written back then. If true, it would mean that the original tablets or scrolls containing these verses, words, and letters have been transcribed time and again over thirty-three centuries and survived intact despite constant movement, vagaries of weather and security measures, wars, exile, possible purposeful editing, and plain human negligence. We have discussed this assumption in chapter 11 and found it wanting. The second assumption is that hidden messages are exceedingly improbable, not of human origin, and difficult to encode. Let's examine the remaining pillar of the Bible code argument.

The common occurrence of improbable events

Naturally, the publication of the Famous Rabbi's Experiment drew responses from academicians and skeptics alike. At first, the responses generally challenged the methodology used by WRR, suggesting that the data was manipulated. Some tried to replicate the experiment, and failed to achieve the reported results. But while the secular criticism of Bible codes was initially directed at the Famous Rabbi's Experiment, it extends well beyond that episode.

The WRR paper and Drosnin's *Bible Code* books repeatedly assert that their findings are highly improbable, that the results cannot be attributed to mere chance. Yet, virtually all respected mathematicians and statisticians who have reviewed their claims have disputed them. Many of the reasons for this professional rejection are technical, based on the idea that code proponents misconceive or misrepresent the nature of mathematical probabilities generally or erred in their calculations specifically. For instance, IBM professor of mathematics and theoretical physics at Cal Tech Barry Simon contends that the WRR calculations assume an independence of events when no such assumption is warranted.[18]

Similarly, Jordan Ellenberg, a professor of mathematics at the University of Wisconsin, likens the results to the classic stockbroker's con, in which a broker sends out unsolicited stock tips to a wide swath of

18. See Simon, "Case against the Codes," paras. 27–29.

potential clients and then narrows the number of recipients over time to those who happen to be that tiny minority of originally solicited investors who receive a string of successful calls. Not knowing of his transmissions of unsuccessful tips sent to others, the members of this smaller group reasonably think that his record proves the broker to be a genius and worthy of hiring. There is an underlying flaw in the con and the codes, however, according to Ellenberg. It is that "[i]mprobable things happen a lot."[19] The late Cal State Fullerton professor of mathematics and statistical mechanics Mark Perakh provided another illustration. He noted that, while the chance for any one lottery ticket to win may be extremely small, the likelihood that some ticket will win is much larger. And, in fact, winners of lotteries are announced frequently. Likewise, the odds of finding a specific ELS in the Torah may be small, but the odds of finding some ELS in association with other words are, again, much larger.[20]

Putting aside the technical difficulties of the Famous Rabbis Experiment, one of the core attractive features of the code claims, the finding of an impressive number of seemingly significant letter and word associations in an ancient text (and, for some, a holy one, at that) was negated early on. University of Toronto mathematics professor Dror Bar-Natan and Australian National University computer science professor Brendan McKay showed, contrary to WRR, that results similar to those reported by WRR could be achieved by a search of the Hebrew version of Tolstoy's *War and Peace*.[21] Then, responding to a specific challenge from Drosnin to find an encrypted assassination message of a political leader in *Moby Dick*, McKay found several.[22]

McKay, Bar-Natan, and others subsequently drafted a comprehensive rebuttal to the Famous Rabbis Experiment called "Solving the Bible Code Puzzle" and published it, fittingly enough, in *Statistical Science*.[23] Subsequently, Perakh, not knowing of the work of Bar-Natan and McKay, reviewed several of his own writings, both in Russian and in English, and found numerous ELS's.[24] Presented with the criticism of some of the opponents of the codes, and based on his own new research, Nobel laureate

19. See Ellenberg, *How Not to be Wrong*, 98.

20. See Perakh, "Rise and Fall," paras. 19–25.

21. See Bar-Natan and McKay, "Equidistant Letter Sequences."

22. See McKay, "Assassinations Foretold in Moby Dick!"

23. See Bar-Hillel et al., "Solving the Bible Code Puzzle"; see also McKay et al., "Scientific Refutation."

24. See Perakh, "Rise and Fall," paras. 19–25.

Aumann changed his mind. He concluded that "the thesis of the Codes research seems wildly improbable," that research he supervised "failed to confirm the existence of the codes," and, so, "the Codes phenomenon is improbable."[25]

The ease of human creation

The respective assumptions of Rips and Drosnin that God or extraterrestrial aliens must have created all of the ELS's in the Bible have also been deflated by Perakh and others, who have shown that ELS's are not so complex as to require a supernatural deity or an extraterrestrial alien to create them. Perakh proved his point in two ways. First, he wrote a short poem, consisting of just 558 letters, in which he quickly found thirty-seven ELS's. Then he purposely encoded ELS's into another writing. So, whether an ELS is created inadvertently or intentionally, the fact remains, as Perakh has written, that a "human mind is quite capable of creating arrays of ELS not unlike those found in the Bible."[26] Note, he is not arguing that the author(s) of the biblical text in fact encoded messages, only that one need not be otherworldly to do so. Moreover, he has shown that humans can create texts embedded with ELS's with modest efforts.

Theological gooses and ganders

If sometimes there is too little proof for Bible codes, at other times there may be too much, at least for certain tastes. Once the idea of previously hidden messages gained currency, the search was open to all. As could be expected, Christians also combed through the Hebrew Bible. They were looking for hidden messages that Jesus, known in Hebrew as Yehoshua (*yud-hay-vav-shin-ayin*) or, in shortened form, Yeshua (*yud-shin-vav-ayin*), is the Messiah. Lo and behold, they found what they sought. The shortened name Yeshua appears frequently as an ELS in the Torah.

Did Professor Rips, an Orthodox Jew, inadvertently uncover a hidden encryption system dictated by God in Hebrew to the Jewish prophet Moses that turns out to confirm that Yeshua/Jesus is the long awaited *Moshaich*/Messiah? Not quite. First, some of the Hebrew letters involved are among the most frequently found in the Torah. In particular, *yud*, the

25. See Aumann, "Personal Perspective," 3.

26. See Perakh, "Rise and Fall," para. 33.

first letter in the name, is the letter most frequently found in the Torah. Moreover, finding Yeshua in the Torah is no more probative than finding the names in the Famous Rabbis Experiment. Perakh searched for Yeshua in a contemporary, thoroughly secular Israeli work, and found ELS's and word clusters similar to those found in the Torah for "Jesus is my name," "Jesus is my teacher," "Jesus is able," and "Blood of Jesus."

THE FAILURE OF THE CASE FOR BIBLE CODES

The validity of Bible codes should not rise or fall depending on whose theological ox is being gored, and it does not. It fails for more fundamental, more objective reasons, and it fails decisively. In 2004, over fifty mathematicians and statisticians endorsed the Mathematicians' Statement on the Bible Codes. The signatories to the statement, having examined the evidence regarding Bible codes, found it "entirely unconvincing." Specifically referencing the WRR paper and Drosnin's books, the scholars stated that word clusters identified by code promoters are an "uncontrolled phenomenon" and expected in texts of "similar length." Moreover, "[a]ll claims of incredible probabilities for such clusters are *bogus*, since they are computed contrary to standard rules of probability and statistics."[27]

"Bogus" is a strong word, but the statement is noteworthy for more than its obvious strength and clarity. The signatories are highly credentialed scholars from around the world, including ten from Israel who could be expected to have an understanding of the Hebrew text at issue in addition to the mathematics.

Further, a number of the signatories are reportedly *frum*—that is, religiously observant. But instead of seeing Bible codes as proof of God, and a scientific one to boot, at least one appears to be offended by the whole process. Professor Barry Simon describes himself as a "halachic layperson," meaning one who observes Jewish law. He plainly and forcefully objects to the use by some individuals of Bible codes (or Torah codes, as he calls them) to enhance the religiosity of others, specifically "to get some non-religious Jews to start thinking seriously about yiddishkeit." To him, it is simply impermissible to "lie to non-religious Jews to get them to keep Shabbos."[28]

27. See "Mathematicians' Statement on the Bible Codes," para. 3 (emphasis added).
28. See Simon, "Case against the Codes," paras. 1–2.

THE REAL LESSONS OF THE BIBLE CODES STORY

In the end, the real value of the debate over Bible codes might be not the scholarly shedding of light on yet more pseudoscience, though that effort is exceedingly significant. Rather, two other important points emerge. The first is that faith and commitment need not and should not depend on pseudoscience, even and maybe especially a pious fraud like the purported Bible codes. The second is that one can be dedicated both to Judaism and to reason. These are not bad results at all for an exercise that began on such shaky ground.

Part Four

ANTICIPATING THE FUTURE

Chapter 23

New Planets, a God for the Cosmos, and Exotheology

We are blessed to live in an age of great discoveries. Before about twenty-five years ago, astronomers had not been able to identify planets in orbit around stars beyond our solar system. These planets are known as extrasolar planets, or exoplanets.

The existence of not one, but two, exoplanets was discovered in 1992. They were orbiting a pulsar in the constellation Virgo.[1] A type of neutron star, an isolated pulsar is a small, compact object that emits beams of radiation as it rotates.[2] Consequently, these exoplanets could not sustain organic life. Just three years later, though, 51 Pegasi b, a planet about half the size of Jupiter, was confirmed as the first single exoplanet orbiting a main-sequence star, one that, like our Sun, fuses hydrogen atoms into helium atoms.[3] That planet, though, was a "roaster," one whose orbit almost skirts the surface of its host star, another situation not likely to allow life to evolve.[4]

Of course, there are exoplanets and there are exoplanets. After finding that exoplanets did in fact exists, the next challenge was to find

1. See National Aeronautics and Space Administration, "First Exoplanets Discovered."

2. National Aeronautics and Space Administration, "Neutron Stars, Pulsars."

3. See National Aeronautics and Space Administration, "Exoplanets 101," para. 8.

4. See National Aeronautics and Space Administration, "First Exoplanets Discovered."

an exoplanet within a habitable zone of a star, i.e., an area in which the surface of a planet, under appropriate atmospheric pressure, can possess liquid water, presumed to be necessary to support biological life. The first such exoplanet, HD 25185 b, was discovered in 2001.[5] It was, however, six times large than Jupiter, the largest planet in our solar system. Therefore, it was not likely to have a rocky surface, as does Earth.

In 2009, NASA launched the Kepler space observatory.[6] The purpose of the Kepler mission was to search for exoplanets in the habitable zones of stars within a relatively small star field in the constellations Cygnus and Lyra, perhaps the extent of the sky obscured by an average extended fist. With this advanced technology, astronomers began to locate several Earth-sized exoplanets. Among the first to be found, Kepler-20e is somewhat smaller than Earth, and Kepler-20f is somewhat larger. Again, though, neither seems suitable for life. The smaller of the two exoplanets orbits so close to its parent star that its surface temperature approaches 1,400 degrees Fahrenheit. The other, by comparison, is relatively cooler, but still registers around eight hundred degrees Fahrenheit.[7] Subsequently, the Kepler team also found an exoplanet which orbits in the middle of the habitable zone of its star. Dubbed Kepler-22b, it is about 2.4 times as wide as Earth and circles its host star in around 289.9 days. Unfortunately, it, too, is probably too large to have a rocky surface.

The first rocky exoplanet was discovered in 2011. It was similar in size to Earth, but exceptionally dense and extraordinary hot.[8] Finally, in 2014, the mission found Kepler-186f, an exoplanet both Earth-sized and in the habitable zone of its host star.[9]

Perhaps better still, in the summer of 2016, astronomers at the European Space Observatory in Chile announced the discovery of a potentially habitable Earth-sized and probably rocky exoplanet in the star system closest to Earth, that of Proxima Centauri in the Alpha Centauri

5. See National Aeronautics and Space Administration, "First Planet Found within the 'Habitable Zone.'"

6. See National Aeronautics and Space Administration, "Kepler Planet-Finding Mission Launches."

7. See National Aeronautics and Space Administration, "NASA Discovers First Earth-Size Planets," paras. 1–3.

8. See National Aeronautics and Space Administration, "Kepler's First Rocky Exoplanet."

9. See National Aeronautics and Space Administration, "First Earth-Sized Planet"; see also National Aeronautics and Space Administration, "ESO Discovers Earth-Size Planet."

system.[10] At a distance of just over 4.2 light years from Earth, though, Proxima Centauri is still almost twenty-five trillion miles away.[11] NASA's New Horizons spacecraft, traveling over thirty-six thousand miles per hour, would still need over seventy-eight thousand years to reach it.[12] Obviously Earthbound readers of this book will not be alive when the first probe to Proxima Centauri, if ever launched, would report its findings. That may not matter, as Proxima Centauri is a red dwarf star, and such stars generally have characteristics that interfere with the habitability of exoplanets that surround them. Happily, in 2017, a team using the European Southern Observatory found four exoplanets, two in the habitable zone, in orbit around Tau Ceti, the nearest sunlike star to our own and only about twelve light years away.[13]

As of mid-2018, scientists had confirmed the existence of over 3,700 exoplanets in diverse areas of the known universe.[14] About the same time, the Planetary Habitability Laboratory at the University of Puerto Rico listed forty potentially habitable exoplanets, thirteen of which seem to be the right size in the right location and with other characteristics that make them good candidates to be a home for biological life as we presently understand it.[15] Even before these discoveries were made, science writer Timothy Ferris put it well: "We live in a changing universe, and few things are changing faster than our conception of it."[16]

Of course, we may not be looking for the right clues. We are carbon-based, and our search has focused on the main elements critical to life on Earth: carbon, nitrogen, oxygen, phosphorus, sulfur, and hydrogen, especially in combinations with which we are familiar. There are other natural elements and many ways to combine the common and uncommon ones. Our search needs to be creative and expansive, as well as focused—not an easy task.[17]

For sure, it is much too soon to claim that we are close to discovering life on other planets, especially intelligent life with which (or whom)

10. See National Aeronautics and Space Administration, "ESO Discovers Earth-Size Planet."

11. See Sharp, "Alpha Centauri: Nearest Star," para. 1.

12. See Byrd, "How Long to Travel," para. 5.

13. See Stephens, "Four Earth-Sized Planets Detected."

14. National Aeronautics and Space Administration, "Exoplanet Exploration."

15. See Planetary Habitability Laboratory, "Habitable Exoplanets Catalog."

16. Ferris, *Whole Shebang*, 11.

17. See National Aeronautics and Space Administration, "Looking for Life."

we could communicate. As we have seen, primitive life forms emerged relatively soon after our own planet coalesced and water formed, but well over four billion years had to pass before our species evolved. And even if there were life on an exoplanet, and intelligent life at that, communication with it would be problematic. Our own language skills have been developed only recently, by the cosmic clock, and our ability to utilize electromagnetic waves for communication is barely more than a century old.

It is not too soon, though, to contemplate the implications of a discovery of life on other planets. Indeed, people have speculated about other worlds for centuries, even millennia. The Jewish commentary is rather sparse, though, yet still provocative. For instance, speaking of a kingdom of God that extends to "*kol olamim*," Psalm 145, according to one translation, refers to "all worlds" (see Ps 145:13).[18] Then, too, the Talmud includes a discussion about what God has done since the destruction of the Second Temple. It contains a suggestion that "he rides a light cherub and floats in 18,000 worlds."[19] The number eighteen thousand may be derived from a perceived allusion in Ezekiel (Ezek 48:35) to a circumference of eighteen thousand. In any event, the *Tikunei Zohar*, an important text in Jewish mystical literature, composed early in the second millennium of the Common Era, continues the theme, contending that the eighteen thousand worlds are to be presided over by eighteen thousand *Tzaddikim* (righteous men).[20] We do not know, though, whether these references are to physical worlds or spiritual worlds.

Subsequently, different rabbis considered the issue of extraterrestrial life and produced, naturally, different results. In the fourteenth century, the Spanish Rabbi Chadai Crescas wrote in *Or HaAdonai* (*HaShem*) that nothing in Torah precluded the existence of life on other worlds (4:2). His student, Rabbi Yosef Albo (d. 1444?), on the other hand, held a different view. He reasoned in *Sefer HaIkarrim* that such creatures would have no free will, and therefore there would be no reason for them to exist. For him, as a theological matter, they could not and, therefore, did not exist.

Some four hundred years later, the Vilna kaballist Rabbi Pinchas Eliyahu Horowitz took a position between Crescas and Albo. In *Sefer HaBris*, he agreed that extraterrestrial beings would have no free will and no

18. See Ps 145:13; see also Kaplan, "Understanding God," para. 10.

19. See 'Abod. Zar. 3b; see http://halakhah.com/pdf/nezikim/Avodah_Zarah.pdf.

20. See Kaplan, *Aryeh Kaplan Reader*, 173.

moral responsibility, but thought that they might still exist.[21] Concluding his review of the literature, Rabbi Aryeh Kaplan said: "We therefore have a most fascinating reason why the stars were created, and why they contain intelligent life. Since an overcrowded Earth will not give the *Tzaddikim* the breadth they require, each one will be given his own planet, with its entire population to enhance his spiritual growth."[22]

Writing shortly after Neil Armstrong placed the first human foot on the Moon, Rabbi Norman Lamm, who was to become Chancellor of Yeshiva University in New York, considered at length man's place in the universe and the religious implications of extraterrestrial life. He feared neither technological advances nor mankind's changing role in the universe. He saw "no need to exaggerate man's importance" or "to exercise a kind of racial or global arrogance, in order to discover the sources of man's significance and uniqueness."[23] Moreover, while recognizing the difference between conjecture and proof, Lamm acknowledged that "[n]o religious position is loyally served by refusing to consider annoying theories which may well turn out to be facts."[24] Consequently, while Judaism has seen mankind as the purpose of creation, and man as made in the image of God, Lamm asserts that "there is nothing in . . . the Biblical doctrine per se . . . that insists upon man's singularity . . . Judaism . . . can very well accept a scientific finding that man is not the only intelligent and bio-spiritual resident in God's world."[25]

Half a century has passed since Lamm wrote those comments. And today, exoplanets are more than a theoretical possibility to be considered by philosophers. If the astrobiologists actually found life elsewhere (a second genesis event, if you will), the discovery would be stunning, maybe literally so. Whether others who learn of it will be as sensitive and humble as Lamm is uncertain.

Surely there will be those who will welcome the development with open arms. For instance, Father Jose Funes, a Jesuit astronomer at the Vatican Observatory, believes there would be no conflict with his faith, because the creative freedom of God cannot be limited: "As a multiplicity

21. See Kaplan, *Aryeh Kaplan Reader*, 170–73.

22. Kaplan, *Aryeh Kaplan Reader*, 173.

23. Lamm, *Faith and Doubt*, 99.

24. Lamm, *Faith and Doubt*, 124.

25. Lamm, *Faith and Doubt*, 128, 133.

of creatures exists on Earth, so there could be other beings, also intelligent, created by God."[26]

Others are not so sure. For Christians who hold that humanity was initially subjected to original sin and that a Savior, in the form of God incarnate, came to save it, what does extraterrestrial life say about sin, about saving, and about the Savior? At the 100 Year Starship Symposium, held in 2011 and sponsored by the US Defense Advanced Research Projects Agency, Christian Weidemann, a philosophy professor, asked, presumably in all seriousness: "Did Jesus die for Klingons too?"[27]

Jews don't have to answer that question. But they will have some of their own dilemmas to confront. That there really are—not just theoretically, but really are—actual planets out there that may serve as the hosts for extraterrestrial life is a fact that colors a question the late social critic Christopher Hitchens asked some years ago in an essay for the Templeton Foundation. The essay was in response to a general question that Templeton posed to over a dozen scientists and nonscientists: Does science make belief in God obsolete? Hitchens's answer was "No, but it should." In his fuller response, he was unsubtle, impolitic, acerbic, possibly anti-Semitic, and exquisitely sharp. He asked what planner would design a doomed galaxy like our own, subject our species to near extinction, and then just three thousand years ago disclose a saving revelation "to gaping peasants in remote and violent and illiterate areas of the Middle East?"[28]

It is easy to dismiss Hitchens because his tone is so off-putting, but Rabbi Arthur Green, founding dean and currently rector of the Rabbinical School at Hebrew College in Boston, a person as devoted to Judaism and the Jewish people as anyone, has asked essentially the same question. Using similar words, but in a different context and no doubt with a different purpose, Green has asked, "Can we imagine a God so arbitrary as to choose one nation, one place, and one moment in human history in which the eternal divine will was to be manifest for all time? Why should the ongoing traditions, institutions, and prejudices of the Western Semitic tribes of that era be visited on humanity as the basis for fulfilling the will of God?"[29]

26. See Chivers, "Vatican Joins the Search," para. 6.

27. See Schapiro, "Professor Asks."

28. See "Does Science Make Belief," 15–16.

29. Green, *Seek My Face*, 105.

To be sure, Hitchens and Green proceeded from different backgrounds, sensibilities, and assumptions to reach their independent, but convergent, recognition of the origin and nature of biblical stories. One saw myth, the other Myth. One found at best nothing special, while the other sees the basis for a morality applicable to all humanity. But, if the underlying question being raised by both Green and Hitchens is a good one, why isn't it a better one when raised to the cosmic level? Can the God which once spoke sparingly to selected individuals, and then became the God of a family, of a tribe, and ultimately a people and a nation, now expand its reach not just around our globe and to everything that lives in this biosphere, but also beyond, to other star systems, even other galaxies? Can we on Earth accommodate such a God? Put another way: new discoveries have shown that the previously accepted scientific consensus of a geocentric universe was flawed. Can we now also agree that geocentric theology is similarly wanting?

Some eighty years ago, Mordecai Kaplan observed that "[r]eligion conceived in terms of supernatural origin is the astrology and alchemy stage of religion. The religion which is about to emerge is the astronomy and chemistry stage."[30] Were he still with us, perhaps Kaplan would have helped us move to the quantum and cosmic stage. Fortunately, there are others who are trying. Even without the benefit of the discoveries of the Kepler mission, this is how Rabbi Jeremy Kalmanofsky has framed the issue: "Theology that cannot face brute facts about cosmology and evolutionary biology is hopeless," he has written. "Contemporary Judaism needs a faith befitting a cosmos."[31]

Thinking about the existence of extraterrestrial life, the science fiction writer Arthur C. Clarke is credited with saying, "Sometimes I think we're alone in the universe, and sometimes I think we're not. In either case, the idea is quite staggering."[32] True enough. Now, though, we have some good evidence that there are exoplanets in a position to host life forms. Shouldn't we be responding to Rabbi Kalmanofsky, and engage in developing a theology for the cosmos, an exotheology for this new reality? And won't we need a new liturgy as well, one that is universal in the fullest sense of that word?

30. Kaplan, *Judaism as a Civilization*, 399.

31. Kalmanofsky, "Cosmic Theology and Earthly Religion," 24–25.

32. See https://www.goodreads.com/quotes/14901-sometimes-i-think-we-re-alone-in-the-universe-and-sometimes.

Chapter 24

When a Jewdroid Walks Into *Shul*

In a short story written expressly for inclusion in *Wandering Stars,* a groundbreaking anthology of Jewish science fiction and fantasy, the British writer William Tenn imagined a future galaxy populated with Jews who, consistent with their ancestors' history, traveled far and wide in search of a better life.[1] Among these Jews, or at least creatures who claimed to be Jews, was a certain group of small, brown pillow-shaped beings covered with grey spots out of which protruded tentacles. Residents of the fourth planet in the Rigel star system (Rigel being a star in the Orion constellation, as seen from Earth), they claimed to be Jewish by descent from a community of Orthodox Jews who lived in and around Paramus, New Jersey. Their nonhuman appearance was the result, they said, of natural relationships, over time, with the native inhabitants of their new planet. In Tenn's tale, the Bulbas, as they were known, traveled to Venus in the year 2859 CE in order to participate in the First Interstellar Neo-Zionist Convention, which was convened for the purpose of discussing a renewed claim to Israel, an area on Earth then without any Jewish population. The question presented was whether the Bulbas could be accredited as Jews.

While set some eight centuries in the future, Tenn's story asked age-old questions about the nature of Jewishness. And if the context of the story seems far ahead of our times, as we just saw in the preceding chapter, the reality is that the pace of discovery regarding potential life on other planets continues to accelerate. Indeed, dramatic advances in

1. See Tenn, "On Venus, Have We Got a Rabbi," 7–40.

science and technology are raising the issue of Jewishness in yet another context. If the claim of the temporally and geographically distant Bulbas, who did not resemble our species in the slightest, was challenging, how will we consider the Jewishness of an android, a robot designed to look like us, and programmed with considerable intelligence, artificial though it may be?

THE RISE OF THE ROBOTS

The idea of artificial beings has been with us for millennia, originally in the form of ancient myths and musings from numerous disparate communities, including what are now India, Greece, Scandinavia, and China. Subsequently, efforts were made to fashion the imagined characters. Leonardo da Vinci (1452–1519), for instance, produced sketches of mechanical knights in armor.[2]

In order to protect his sixteenth century community, Rabbi Judah Loew ben Bezalel (1525–1609), the Maharal of Prague, reportedly created an artificial being from clay and brought it to life by invoking God's name.[3] But the Maharal was not the first Jew alleged to have formed a humanlike being. A passage in the Talmud states that Rava created a "man" and sent him to Rav Zeira, who spoke to him but received no answer. Talmud being Talmud, the passage is somewhat obscure and may have been intended as metaphor, though Rashi seems to have treated it as true and attributed the creation to a mystical invocation of God's name. Either way, clearly the writer envisioned the creation by a human of a "man."[4]

Modern humanoid robots—indeed, the word robot itself—trace to the Czech author Karel Capek's 1921 drama *R.U.R*, which stands for Rossum's Universal Robots. The word robot is, in fact, an English variant on the Czech *robota*, meaning forced labor. Since *R.U.R.*, robots have become a fixture in popular art forms like short stories, novels, movies, radio, and television. In the 1960s, for instance, Rosie served as a maid in *The Jetsons*, a popular animated television show, and the robot from *Lost in Space*, another television series, was an important part of the spaceship crew. Subsequently, such portrayals have been more varied and textured.

2. See "Robotic Knight."
3. See Mindel, "Rabbi Judah Loew," para. 4.
4. See b. Sanh. 65b, https://www.sefaria.org/Sanhedrin.65b?lang=bi.

As seen in the Star Wars movie saga, at a time long ago, in a distant galaxy, far, far away, an autonomous droid identified as C3PO, with the size and the general shape of an adult human, could (among other attributes) translate over seven million forms of communication and assist in matters of protocol and etiquette. Set a thousand years into the future, another clearly mechanical character with humanoid features, Bender Bending Rodriguez, Sr., aka Bender, stars in the animated television sitcom *Futurama* as one who was trained as a metal bender, but works for a cargo delivery service, and is prone to drinking, smoking cigars, and womanizing with both female robots and humans. In the computer animated movie, *WALL-E,* the last robot on unpopulated Earth, is a Waste Allocation Load Lifter-Earth class entity, who after centuries of being alone, notices an attractive probe-bot named EVE, and adventure ensues.

For all their charm, though, with their mechanical heads and other appendages, C3PO, Bender, and Wall-E, are each distinctly not human. By stark contrast, Lt. Cmdr. Data, from the television series *Star Trek: The Next Generation*, and David, from the movie *A.I.: Artificial Intelligence*, look like us. They have their quirks (don't we all), but if you did not know they were not natural-born and carbon-based, and that their insides consisted of chips and circuits with silicon everywhere, you would not have guessed.

Both Data and David would have passed the Turing Test, an evaluation of machine intelligence in which a subject is asked a series of questions by a judge who is seeking to determine whether the subject is human.[5] If the subject has been programmed sufficiently, then the judge will not be able to distinguish the machine with artificial intelligence from a real human. In other words, the subject would pass the Turing Test.

The movie *Ex Machina* provides a twist to the Turing test. As the Turing Test was originally designed, the interrogator would pose questions in writing to two unseen subjects, one being human and one not.[6] Both would then respond in writing. In this way, the interrogator's decision about the subject's ability to think like a human would not be biased by either visible or aural clues as to the true nature of the subject. In *Ex Machina*, however, Ava's artificial composition is not hidden. Through her transparent neck, torso, and limbs, many of her internal components

5. The Turing Test was developed to determine whether, and to what degree, an artificially intelligent machine could think independently. See generally "Turing Test, 1950."

6. See Marzano, "Turing Test," 65.

are plainly visible. Nor is her ability to think and move like a human in any doubt from the first time that a computer programmer named Caleb, playing the role of interrogator, sees her through a glass partition. Rather, a primary question posed by the meetings of Caleb and Ava, a physically attractive android, is whether one or both would be as emotionally drawn to the other as humans might be.

As it often does, life is imitating art, and the age of the robots is, in fact, rapidly approaching. In 2000, Honda introduced ASIMO, a mobile android. The current version has facial, voice, and posture recognition capabilities, can communicate with humans, and not only walks, but also runs.[7] Not to be outdone, Toyota produced a violin-playing robot which displayed such sufficient arm strength and hand dexterity that it could even achieve, Toyota claimed, vibrato similar to that accomplished by human violinists.[8] Pepper, created by SoftBank Robotics, reportedly can recognize human emotions and respond accordingly.[9]

Robots have, in fact, proven capable of engaging successfully in a wide range of activities. NASA has sent a robot, or, more accurately, a robonaut into space on a shuttle mission, and claims that its Robonaut 2 series has dexterous hand movement "approaching" that of a human.[10] NASA also plans to send a Robonaut 5, aka Valkyrie, to help explore Mars.[11] Meanwhile, back on Earth, nurse robots have been deployed to assist elderly patients, and greater usage has been planned.[12] Knightscope robots have been placed in service as security guards in parking lots, corporate headquarters, and other facilities, with the ability to record, store, and send video; read license plates; and, through thermal imaging, locate humans in restricted areas.[13] In June, 2018, an IBM robot, trained in the fine art of debate, and able to draw on hundreds of millions of articles to support its case, performed admirably while debating an experienced human debater, even attempting to inject humor into its argument, albeit somewhat awkwardly.[14]

7. See "History of ASIMO."
8. See Rotkop, "Toyota's Violin Playing Robot," para. 1.
9. See SoftBank Robotics, "Pepper."
10. See National Aeronautics and Space Administration, "R2 Robonaut."
11. See Renstrom, "Humanoid Robot Space Explorer," paras. 5–6.
12. See "Pearl"; see also Renstrom, "Robot Nurses."
13. See Shan, "Robots Are Becoming Security Guards," para. 7.
14. See Newcomb, "New IBM Robot."

Tasks previously restricted to Jewish humans are not immune from this trend. At the Jewish Museum Berlin in 2014, a robot demonstrated that it could write a Torah scroll, all 260 feet of it, flawlessly.[15]

Physical agility is impressive, of course, as are recognition technologies, but can a robot think both strategically and tactically? In the last two decades or so, two exercises began to demonstrate the power of artificial intelligence. In 1996 and 1997, the reigning world chess champion Gary Kasparov battled an IBM computer named Deep Blue. Kasparov won the first match (4–2), but lost the second (3.5–2.5). Kasparov's defeat was subsequently rationalized as the result of his bad play or the limited intellectual challenge of chess itself, but the fact remains that Deep Blue for the first time beat a human champion in a difficult contest.[16]

The ancient game of Go, while deceptively simple to learn, is considerably more complex than is chess. The game is played on a board filled with squares, nineteen across and nineteen down. Players, one with black stones and one with white, place their stones at the intersections on the board, the main object being to surround, capture, and remove a stone of the opponent and thereby control more territory. As there are 361 intersections on the board, one player starts with 181 stones and the other begins with 180, and the number of possible board positions is incredibly large. In fact, that number was only determined in 2016 and approximates 10^{171} positions. The number of those positions is, incredibly, greater than the number of atoms in the universe![17]

In March of 2016, artificial intelligence known as AlphaGo beat world-champion Go player Lee Sedol (4–1).[18] Created by DeepMind and acquired by Google, AlphaGo was programmed to learn from mistakes and teach itself how to succeed. This victory was extraordinarily significant. It demonstrated more than mere computing strength, and displayed the state of "deep learning," a form of artificial intelligence architecture which, in a rudimentary way, mimics human neural networks and allows a machine to learn by observation, data collection, and analysis—in other words, like humans do.

At the same time, while AlphaGo was surprisingly successful, artificial neural networks are still "about a million times smaller than the

15. See Fattal, "Torah-Writing Robot."

16. See "Deep Blue."

17. See Moyer, "Google's AlphaGo," para. 7.

18. See Hassabis, "What We Learned in Seoul," paras. 5–6.

brain," with its one quadrillion neuron connections, or synapses.[19] Consequently, although it may be hyperbolic to say that the future of artificial intelligence is unlimited, the prospects envisioned even now are truly stunning.

ROBOTS, JEWS, AND JEWISHNESS

Advances in robotics, in mobility, dexterity, and intelligence, are raising the question about what it means to be human and Jewish. To our more parochial concern, if a robot can play the violin, compete at a high level at Go, and write a Torah scroll, why can't a robot be Jewish? Assume that the android looks human, is at least as mobile as ASIMO, and is thoroughly familiar not only with Jewish foundational texts like the Torah, the rest of the *Tanakh*, and the Talmud, but also all that followed, up to and including today's Jewish literature and philosophies. Assume further that it has a complete grasp of Jewish history from its roots in the second millennia BCE through the demographics of various resulting Jewish communities today. Why can't that droid be Jewish? Why can't it have a Jewish name, like Enosh ben Yehuda v'Rut, be called to read as a *bot mitzvah*,[20] be counted in a *minyan* (required prayer quorum), and sit on a *shul* (synagogue) board?

Objections to a Jewdroid are numerous, ranging from the silly to the sacred. They include arguments based on appearance, on ritual, on descent, on speciesism, and on foundational philosophy. How will the Jewish community react when an artificial entity is created that not only looks human, but is thoroughly versed in all things Jewish? Will the Jewdroid's presence be too much to bear, or is Judaism's tent big enough to hold him too? Shall we reject the Jewdroid whose existence is unprecedented, or shall we welcome the stranger? What assumptions and values shall inform us? Let's look at some objections to a proposed Jewdroid.

19. See Lee, "Meaning of AlphaGo," para. 16.

20. Traditionally, when a Jewish boy reaches the age of thirteen, he is considered to be a *bar mitzvah*, literally, a son of the commandment, entitled to the rights and subject to the obligations of an adult. He is eligible to mark the milestone by reading from the Torah publicly. A bot mitzvah would be an analogous ceremony for a qualified robot.

Not looking Jewish

The first, and most trivial argument, is that based on appearance: the droid does not "look Jewish." A similar objection was raised against the Bulbas at William Tenn's imagined interstellar Neo-Zionist convention. Whether coming from Jews or non-Jews, that line assumes that there is such a thing as a Jewish "look." Whether there ever was a "look" is doubtful, but today any argument based on a presumed Jewish look involving a distinctive set of physical traits shared by all Jews is not only obnoxious, it is contrary to the evidence of the varieties of contemporary Jewry.[21] In the world in which we live, Jews come in many shapes and sizes, each with a wide range of physical features. There are, as we have seen, even ginger Jews, and our droid, like other Jews, could come in any hue and be a Jew. Looks alone cannot compel a conclusion that our Jewdroid either can or cannot be Jewish.

Not circumcised

Then there is the argument based on ritual: the droid cannot be circumcised. This objection is premised on the recognition, early in the Torah text, that male circumcision is a sign of membership in a covenanted community (see Gen 17:9–14). Putting aside the obvious counter that the droid could be a fembot—that is, a humanoid robot gendered feminine—this argument cannot withstand scrutiny.

The contention seems to assume that our droid would not be formed as an outwardly anatomically correct human male, but there is no engineering impediment to doing so. In the movie *A.I.: Artificial Intelligence*, Gigolo Joe, one of the humanoids, was designed and apparently functioned completely and very well as a male lover. There is no obvious reason why our droid could not be similarly formed, and even provided initially with a section of synthetic skin which could be removed.

Even if he were formed somewhat less than anatomically correct from a human viewpoint, as Rabbi Mark Goldfeder, a senior lecturer at Emory Law School, suggests, he could still be accepted under the *halakhik* principle of *nolad mahul*, because he could be considered to have been formed "pre-circumcised."[22] Indeed, there is precedent for the belief

21. See, e.g., Mocha Juden's website at http://mochajuden.com/?p=4250.

22. See Soclof, "Should Robots Count," para. 17.

in some traditional circles that a number of major biblical figures were born pre-circumcised, starting with the first fully formed man, Adam (Michelangelo's depiction in the Sistine Chapel, notwithstanding), and including Moses and Jacob.[23] They further view the pre-circumcised condition to reflect perfection. Consequently, the argument from ritual should not bar our humanoid from being considered Jewish.

No Jewish mother

A third objection is the argument based on descent: our droid would not have a Jewish mother. It is true, of course, that the Jewdroid would not be born of a Jewish mother in a conventional, biological sense, and it is also true that, since the Mishnaic Period, in the first through the early third centuries of the Common Era, Jewish identity has, for the most part, been transmitted to a child by its mother. Conversely, according to Shaye J. D. Cohen, a professor of Hebrew literature and philosophy at Harvard, before the *Mishna*, one's Jewish identity was determined not by the mother but by the father.[24] The reasons for the transition are not clear, but the argument for matrilineal descent is one that is grounded on a custom developed in response to unique circumstances, ones which may no longer be particularly pressing, or even relevant, today. In fact, different social challenges in the United States over recent decades have led the Reconstructionist and Reform movements to accept conditional patrilineality, meaning recognition of a child as Jewish where the child's father, but not mother, is Jewish and where, in the phrasing of Reform rabbis in a 1983 resolution, there are accompanying "public and formal acts of identification with the Jewish faith and people."[25] Today, in some Jewish communities, though not everywhere, not having a Jewish biological mother does not preclude one from being Jewish.

Perhaps more importantly, if our Jewdroid is an adult, well-established and diverse procedures for conversion of individuals provide models for acceptance without biological Jewish ancestry.[26] These approaches involve issues both of personal identity and communal status. Once the conversion is complete, though, at least within the denomination in

23. See "Was Moses Born Circumcised?"

24. See generally Cohen, "Matrilineal Principle."

25. See Kaplan, "Who is a Jew," para. 8.

26. See Romanoff, "Denominational Differences on Conversion."

which it takes place, the convert is to be treated as no less a Jew than one born into the community. S/he is to be welcomed and embraced. If our Jewdroid commits to living a Jewish life, through study and action, s/he would seem to qualify for Jewish status by conversion, if not otherwise.

Not human

A fourth argument, similar to another raised against the Bulbas, acknowledges that while Jews may have different physical appearances and come to their Judaism other than through biological descent, at least Jews must be human. This is the argument based on species. In relying on a biological classification, however, the objectors display a cramped understanding both of the reality of the evolution of modern humans and the possibility of other life forms in the cosmos.

The authors of the origin stories in Genesis surely understood humankind to be the pinnacle of God's creation. What they did not know, and could not have known, but what we know quite well today, is that the first humans were neither created fully-formed nor fashioned from the dust of the Earth—nor, in the female's case fashioned from a rib of the male (compare Gen 1:27; 2:7, 22). Rather, as we noted in chapter 2, our genus evolved, slowly, over time, finally emerging just two to three million years ago.[27] Our species, *Homo sapiens*, appeared about two to three hundred thousand years ago.[28] While this seems like a distant period, if we imagine the history of our planet as occurring over the course of twenty-four hours, humans did not arrive until just over one minute before midnight.[29]

Looking inward, there is no reason to place special value on our species simply because of its relatively recent arrival in a multibillion year evolutionary chain. To the contrary, other species have been around much longer and could make a reasonable claim to supremacy through proven adaptability and survival. Humans have surely developed and refined skills such as communication and tool usage to a greater extent than have other creatures, but we have not yet demonstrated our ability to survive for long, nor even to take care of our home planet.

27. See also Lewis,"Where Did We Come From," para. 8.

28. See also Smithsonian Museum of Natural History, "What Does it Mean," para. 1.

29. See "History of Earth," https://flowingdata.com/2012/10/09/history-of-earth-in-24-hour-clock/.

Looking outward, we have previously referenced Rabbi Norman Lamm's consideration of humankind's place in a universe with the potential for extraterrestrial life. Recall that, while he agreed that Judaism has seen humanity as the purpose of creation and that humans are made in the image of God, Lamm also believed that "there is nothing in . . . the Biblical doctrine per se . . . that insists upon man's singularity."[30] Rather than restricting devotion to one species, Lamm understood that Judaism could accept other intelligent and spiritual residents in a universe created by the God in which he believed.

What we learn from a broader perspective about our place in the cosmos is that we and indeed everyone and everything on this planet owe our existence to the same source. Astrophysicists tell us that we and all else are made of stardust, the product of explosions of supernovae billions of years ago.[31] And theologians who accept science agree. Rabbi Bradley Shavit Artson has summarized the situation as follows: "All of the cosmos is our mother/father; we are the descendants and the cousins of galaxies . . . Everything everywhere is an expression of oneness. This is both scientific fact—inescapable and inspiring—and theological value—to be is to belong is to be community."[32] Therefore, whether we are carbon-based and naturally born or silicon-based and manufactured, we are cosmological relatives, distanced by history for sure, but bound nevertheless to the ultimate origin of all. The implication is profound. As author Arthur C. Clarke has written, "[w]hether we are based on carbon or on silicon makes no fundamental difference; we should each be treated with appropriate respect."[33]

Not made in the image of God

Despite this scientific fact and this theological value, though, the most fundamental objection to the Jewishness of our droid is indeed theological: the assertion that the droid is not made in the image of God. The biblical story of origins is quite emphatic that humans are God's crowning achievement. They are the last listed in the litany of creatures which were

30. See Lamm, *Faith and Doubt*, 99, 128.

31. See "Are We Really All Made of Stardust?"

32. Artson, *Renewing the Process*, 48.

33. See https://www.goodreads.com/quotes/787731-whether-we-are-based-on-carbon-or-on-silicon-makes.

made to live in our world, and they are charged to have dominion over all other animals, on land, in the air, and in the sea (see Gen 1:26). And if that is not explicit enough, they are told to fill the planet with offspring, to subdue it, to take all seed-bearing plants, all fruit-bearing trees, indeed, all green plants for food (see Gen 1:28–30).

Even greater than the commands to reproduce and to use other living creatures for their benefit is the essential nature of the human being. According to Genesis, God said, "Let us make humankind, in our image, according to our likeness!" (Gen 1:26). To emphasize this particular decision, the text shifts from prose to poetry, as it describes God's ultimate work in the first week of creation:

> God created humankind in his image,
> in the image of God did he create it,
> male and female did he create them. (Gen 1:27)

The poem is essentially repeated several chapters later at the beginning of the genealogy of the line of Adam and Eve's third son, Seth (see Gen 5:1–2).

The text could not be more clear in elevating the stature of humans to that of the creator God, yet the meaning of the phrase "image of God" is not so obvious. Part of the resolution of the puzzle depends on the time period we are considering. According to Rabbi Cassuto, "[t]here is no doubt that the original signification of this expression in the Canaanite tongue was . . . corporeal, in accordance with the anthropomorphic conception of the godhead among the peoples of the ancient East."[34] The Reform movement's commentary agrees that there is a relationship to another Ancient Near East language. It states that the Hebrew word for image, *tzelem*, is related to the Akkadian *tzalmu*, which "applied specifically to divine statues in human guise."[35] Similarly, Professor Leon Kass holds that the root of *tzelem* means to cut off or chisel, as from a statue, indicating a physical resemblance.[36]

Moreover, as just noted, when Seth is born, the narrative echoes the previous text: "Adam . . . begot one in his likeness, according to his image" (Gen 5:3). This is completely understandable. What would we expect, other than our children would look like us? So the concept that humans

34. Cassuto, *Commentary on Genesis Part One*, 56.

35. Plaut and Stein, *Torah*, 35n2. The commentary adds that the "biblical use is, of course, different."

36. See Kass, *Beginning of Wisdom*, 37.

are shaped like God, and, conversely, that God is shaped like humans, seems a quite plausible projection within an anthropocentric framework of the origin of the world and the place of humans in it.

Yet, if familiar external physical features are crucial, then any objection to a Jewdroid based on an anthropomorphic image of God surely fails. If Adam looked like God, and Seth looked like Adam, and our Jewdroid looks like the descendants of Seth, then it must pass the image test.

Belief in the corporeality of God, that God has a figure and shape, changed over time, but it persisted into the Middle Ages. It was taken seriously enough, and apparently was prevalent enough, that Maimonides opened his major work, *The Guide for the Perplexed*, with a refutation of the idea. He argued that in Hebrew there was a word for form other than *tzelem*, and that *tzelem* really meant the essential and distinctive quality of a human, his intellectual perception.[37] Maimonides also recognized that the capacity to learn and reason differs from one person to another.[38] He posited that there were four finds of wisdom, one involving cunning, another with the acquisition of moral principles, a third with workmanship, and, most importantly, a fourth—the "knowledge of those truths which lead to the knowledge of God."[39] Man will attain true perfection, he said, when the knowledge of God's ways and attributes leads man to commit "always to seek loving-kindness, judgment, and righteousness, and thus to imitate the ways of God."[40]

Like Maimonides, Professor Kass argues that "image" involves more than mere physical resemblance. Because he wants to understand how man could be more godlike, aside from appearance, Kass looks at God's activities and powers as described in Genesis. Among other attributes, he finds that God "exercises speech and reason, freedom in doing and making, and the powers of contemplation, judgment and care."[41]

We could argue at length and depth about whether the academic line from Maimonides to Kass concerning God's essential attributes is either complete or valid. Everyone from biological anthropologists to philosophers has an opinion. Rabbi Jonathan Sacks, referencing the movie *A.I.*, argues that the uniqueness of humanity lies *not* in intelligence, but

37. See Maimonides, *Guide for the Perplexed*, 13–14.

38. Maimonides, *Guide for the Perplexed*, 288–89.

39. Maimonides, *Guide for the Perplexed*, 393.

40. Maimonides, *Guide for the Perplexed*, 397.

41. Kass, *Beginning of Wisdom*, 38.

"in loving and being loved."[42] We will defer diving into that pool. What is important for now is that if the identified attributes are descriptors of being made in the "image of God," then our postulated droid, who would be created intelligent and learned and thoughtful and compassionate, would seem to qualify. How then, could our Jewishly knowledgeable and committed droid not be considered Jewish? And how then, if he is, could he not receive a Jewish name, celebrate his *bot mitzvah*, be counted in a *minyan*, and serve on the *shul* board?

TODAY AND TOMORROW

In his reported responses during an interview with the Jewish Telegraphic Agency, Rabbi Goldfeder appears to take a contextual approach to the sufficiency of evidence for the hypothetical robot who walks into his office and wants to be counted in a *minyan*. The rabbi is quoted as saying that such an event would "[n]ot necessarily" provide enough evidence for him, but adds "[w]hen something looks human and acts human, to the point that I think it might be human, then halachah might consider the threshold to have been crossed."[43] Moreover, he believes this tentative conclusion is supported by a Jewish ethical perspective.

Rabbi Goldfeder also stressed that he was engaging in a "theoretical outlaying of views." That qualification did not impress Rabbi Moshe Taub, an Orthodox rabbi, who literally summoned seven pages filled with chapters and verses to attack Rabbi Goldfeder's views as "bizarre and misguided."[44] This, in turn, prompted a reply from Rabbi Goldfeder, which included the statement that "[o]bviously I am not of the opinion that a robot can actually count in a minyan," but no exposition as to why the threshold he previously referenced could not be crossed.[45]

For all his citations and quotations, Rabbi Taub's position is quite simple: the robot may not be part of a *minyan* because it is neither human, nor male, nor Jewish. Of course, this approach would also bar women from being counted in a *minyan*, too. So it isn't just robots whose participation Rabbi Taub would prohibit. Women need not apply either. Presumably, he would preclude the robot from reading Torah at a *bot*

42. Sacks, "Human's Uniqueness Is Not in Intelligence," para. 6.

43. See Soclof, "Should Robots Count," paras. 13, 15.

44. See Taub, "*Cogito Ergo Sum*," para. 9.

45. See Taub, "*Cogito Ergo Sum*," para. 23.

mitzvah, just as he would ban a female from reading Torah at her *bat mitzvah*, here meaning a ceremony in a synagogue in which a twelve- or thirteen-year-old female reads a portion of the Torah in front of her community.[46] It is a defensible position, in the sense that it is based on thousands of years of tradition, but it is also a position that, in the United States, at least, has been rejected by the overwhelming majority of Jewish Americans.

The failure of the argument, then, is not that it lacks historic grounding, but that it is uniquely devoid both of imagination and pragmatism. The first failure is ironic given the fertile musings and considerable ingenuity of the sages upon whose stories and views Rabbi Taub relies. The second failure is that this approach is neither sensitive to, nor sensible for, most Jews today. Yes, this approach may work, even work well, in an insular community. In the world most Jews occupy today, it is stifling. Rather than empowering all Jews, or even simply allowing them to enhance their Jewish experience, it limits what some can achieve. Whether knowingly or not, the vast majority of American Jews follow a principle that Rabbi Mordecai Kaplan is said to have taught: the Jewish tradition has a voice, but not a veto. A corollary is that Jewish rules and rituals should function primarily to benefit the Jewish people, not the converse. When a rule or ritual fails to do so, it must be reevaluated and even discarded.

We are not yet at the point when a learned and committed Jewdroid is going to walk into a rabbi's office and asked to be counted. But we are, as Yuval Noah Harari has written, already in "a world in which culture is releasing itself from the shackles of biology."[47] Surely future developments will be unsettling and challenging, and perhaps disruptive, but they will also be amazing. So the day of the Jewdroid is coming, maybe within a generation, and we need to be prepared. We have all listened as a soon-to-drop-out-of-Hebrew-school teenager stumbles through a reading of the weekly *parashah* (Torah portion) and gives a less than

46. The practice of allowing young women to read from the Torah in their congregations can be traced to a day almost a century ago, specifically March 18, 1922, when Rabbi Mordecai Kaplan, in his capacity as rabbi of the Society for the Advance of Judaism in New York City permitted his eldest daughter Judith to read a portion of the Torah designated for recital that Shabbat morning. She did not stand on the *bima* (stage) or read from a Torah scroll. Rather, she stood below the *bima* and read from her own Bible. See "Jews in America." Nevertheless, she stood and read. Today, it is a common practice in all nonorthodox congregations. See Pelaia, "Bat Mitzvah Ceremony."

47. Harari, *Sapiens*, 409.

enthusiastic or challenging *d'var* (commentary on the portion). We have all seen the leader of a *minyan* scramble to find the tenth person to fill the required quorum. We have all encountered synagogues with inattentive, unproductive, and bored board members and search committees futilely looking for "someone else" to carry the load.[48] Don't we need to think more than reflexively about whether our Jewdroid can become a contributing part of our community?

A passage in the Talmud, attributed to Rabbi Eliezer the Great, asserts that there are literally dozens of decrees in Torah calling on us with negative phrasing to avoid oppressing the stranger, and also with affirmative phrasing urging us to treat the stranger with respect.[49] (See, e.g., Lev 19:34; Deut 10:18–19.) The Jewdroid of the future can be both capable and qualified to function constructively within the Jewish community. How shall we greet, how shall we treat him or her when s/he walks in the door? Whatever we do, the most important thing, as Rabbi Nachman of Breslov (1772–1810) teaches, is "not to be afraid."[50]

48. See, e.g., Temple Sholom Cincinnatti, "Be Someone Else."

49. See b. B. Meṣ. 59b, http://www.come-and-hear.com/babamezia/babamezia_59.html#PARTb.

50. See Zwolinski, "Reb Nachman Explains It All," para. 11.

Chapter 25

In the Beginning and in the End

Jewish tradition literally begins with the Beginning. The Judahites and Israelites and their ancestors who first began to contemplate the Beginning and their beginnings had no inkling either about the reality of the origin of the cosmos, or of the nature and duration of its development. But then, how could they have known what actually happened? Science as we understand it did not exist when the authors of the Torah stories put quill to scroll.

And, equally important, the purpose of those authors was not to observe, describe, and test natural phenomena. Rather, they were interested in the preservation and development of a particular people in a particular place during a particular period of time. As we saw in chapter 12 and elsewhere, their collected story was not so much about fact, as it concerned faith and future.

Consequently, they chose to begin their national saga both by significantly demythologizing the then current creation stories extant in the Ancient Near East and asserting, though not consistently or in philosophical terms, the idea that there was and ought to be order in the universe, an order established by a single, powerful god. They wrote of a deity who not only operated in history, but who initiated history through a series of acts of creation, differentiation, separation, and identification (see Gen 1:1–31). With key elements of the universe both established and ordered, their tale of the development of humanity generally and the destiny of a particular family, nation, and people could now unfold.

Still, while the creation story in Genesis is mythic (i.e., a traditional story that explains) and not scientific (i.e., analytic and predictive), it is a powerful myth. It is one of the two interventions in history, along with the story of an exodus from Egypt, that are invoked frequently as evidence of God's powers and achievement. Biblical prophets like Isaiah (Isa 40:12; 42:5) and psalmists (Pss 8:2–4; 19:2; 102:26; 121:1–2) maintained and embellished the theme of the creator God. That theme later became incorporated in the *siddur*, the Jewish prayer book, as part of prayers like the Sabbath *Kiddush* recited over a cup of wine, the evening prayer *Maariv Aravim*, the Sabbath morning *Psukei D'zimrah* (verses of praise), and the traditional closing prayers, *Aleinu* and *Kaddish Yatom* (the Mourner's *Kaddish*).

Discoveries in science over the last few centuries, and especially in the last few generations, have challenged the literal Genesis text. Yet, as we have seen, for some, the biblical story nevertheless remains the accurate report of events which took place within the last six thousand years, and they have rejected modern science. For others, the story retains such import that they have strained to reconcile the ancient words with contemporary knowledge. For still others, though, people who embrace modern science without reservation, the biblical myth serves as the basis for what some call creation or process theology,[1] or evolution theology,[2] or emergence theology.[3]

For instance, Rabbi Arthur Green begins one of his books with what he calls a "theological assertion." He believes (meaning he holds as a religious belief) "that the evolution of species is the greatest sacred drama of all time."[4] That is, Green not only completely accepts the "scientific consensus" on origins and natural history, he considers that unfolding process as both meaningful and a way to understand God.[5] By God, Green does not affirm a personal God or "a Being or a Mind that exists separate from the universe and acts upon it intelligently and willfully."[6] Consequently (and consistent with Stephen Jay Gould), he rejects the claim of the "intelligent design" community that more complex

1. See, e.g., Artson, *Renewing the Process.*
2. See, e.g., Green, *Radical Judaism.*
3. See, e.g., Nelson, *Emergence of God.*
4. Green, *Radical Judaism*, 16.
5. Green, *Radical Judaism*, 20.
6. Green, *Radical Judaism*, 17.

or higher forms of life, specifically human beings, emerged as a result of conscious planning.[7] But (and contrary to Gould) he does hold to "the presence of divinity within nature," to the "inner force of existence itself," who "underlies all being," who "*is* and dwells within . . . the evolutionary process."[8] Acknowledging the difficulty of language, Green elects to call this phenomenon "God."[9]

These and other related approaches seek to integrate newly revealed science with certain philosophical or theological perspectives. They incorporate the facts that we are all stardust, all related members of the same evolutionary tree of life. And they assert that we are bound by our connection to the Earth to preserve and protect it. Such contemporary Jewish thought demonstrates that the biblical creation story still resonates.

For more than two and a half millennia, then, Jewish thinkers have engaged energetically and enthusiastically with the beginning of time. During that same period, however, they have been less concerned about confronting the end of time. To understand just how limited Jewish thought is regarding the end of time, it will be helpful to understand what modern science teaches about our future, and that of our planet, our solar system, our galaxy, and the universe itself.

Putting aside the possibility of worldwide death by disease and the perhaps greater possibility of global demise due to thermonuclear war (the Doomsday Clock now sitting as of this writing at only two minutes to midnight),[10] history evidences no fewer than five mass extinctions of life in the last four hundred and fifty million years.[11] The last of these extinctions occurred, as noted previously in chapters 4 and 5, about sixty-five million years ago when an extraterrestrial object hit Earth.[12] The force of the blow, on the Yucatan Peninsula in Mexico, was more than a billion times in excess of the atomic bomb that leveled Hiroshima in World War II.[13] The impact crater it created, known as the Chicxulub Crater, is about 110 miles in diameter. While our shrewlike ancestors survived, and we

7. Green, *Radical Judaism*, 21.

8. Green, *Radical Judaism*, 17, 19 (emphasis original).

9. Green, *Radical Judaism*, 19 and n8.

10. See "Doomsday Clock," https://thebulletin.org/overview.

11. See "Mass Extinctions."

12. See Howell, "What Was the Impact."

13. See Palmer, "We Finally Know How Much," para. 15.

ultimately evolved, there can be no guarantee that we would survive the next such event. After all, Chicxulub may not even be the biggest impact crater on Earth.[14]

In any event, even assuming that we avoid disease, war, and asteroids, human life on Earth (indeed, Earth itself) is still doomed. Our Sun is about halfway through its own lifecycle. By most estimates, it has only another five billion years, give or take, to live.[15] During that time, it will consume all of its hydrogen, collapse, begin to fuse helium, and then enlarge into an immense red giant star. As the solar death process unfolds, the Sun will expand first to reach Mercury's orbit, then that of Venus, and, finally, Earth's. The atmosphere will dissipate, the seas will evaporate, and Earth may well spiral into the enlarged Sun and vaporize.[16] Ultimately, the Sun itself will collapse into a white dwarf with a carbon core.

Perhaps humankind will have left Earth by then and established bases on more distant planets or their moons, or even in other star systems in the Milky Way. Yet that may not be enough to save our species. Our spiral galaxy is on a collision course with our nearest and much larger galactic neighbor, Andromeda. Andromeda is presently about 2.5 million light years away from us. That equates to over seventeen million trillion miles, a goodly distance, but we are approaching each other at about a quarter of a million miles per hour. NASA tell us that in only four to five billion years the two star systems will collide.[17] Of course, by then, as we have seen, our Sun will already be dying and taking our birth planet with it. Will the new worlds on which we have landed survive the twists and turns of intergalactic gravitational forces?

But wait, there's less. That is, if an intergalactic collision is much too much to contemplate, there may be less, much less, in our longer term future anyway. In recent years, science has come to understand what the late, great New York Yankee catcher Yogi Berra, with no earned academic degrees, knew all along (though he may not have been the first to say so). "The future," Yogi once said, "ain't what it used to be."[18]

14. See Line, "Asteroid Impacts."

15. See O'Neill, "Sun as a White Dwarf Star," para. 3.

16. See "Red Giant Stars," para. 9.

17. See National Aeronautics and Space Administration, "Hubble Space Telescope."

18. See https://www.brainyquote.com/quotes/yogi_berra_102747.

Edwin Hubble's discovery in 1929 that galaxies in general were moving apart from each other was astonishing enough.[19] It indicated that our universe, space itself, was expanding. Less than twenty years ago, in 1998, astrophysicists determined that the rate of expansion is not steady, as once supposed, but is in fact accelerating due to a repulsive force of gravity called dark energy.[20] Because dark energy is not well understood, there is no consensus as to what the future holds for our home universe, but of the many possible scenarios, three (each with variations) are most discussed. They are often called the Big Crunch, the Big Rip, and the Big Freeze.

In the Big Crunch, at some point trillions of years in the future, perhaps because dark energy is less pervasive than thought or ceases to be as powerful as thought, gravity overcomes the expansion of the universe, a process of universal collapse and consolidation ensues, and all that is rushes back ultimately to a hot, dense state similar to that from which our universe emerged.[21]

In the Big Rip, the expansion of space proceeds to the point where first galaxies, then stars and planets, and ultimately the atoms of elements themselves cannot hold together. They simply rip apart, leaving a vast expanding universe of drifting subatomic particles.[22] Models of a future Big Rip vary considerably, but generally estimate that such an event, if it occurred, would not take place for tens of billions of years from now.

In the Big Freeze, the universe continues to expand, but not sufficiently fast to cause a cosmic tearing apart. Instead, as galaxies become more and more separated from each other, and stars within galaxies do as well, our descendants, if any, will lose the ability to see them, and the evening sky will become darker and darker. As the universe gets larger, its temperature will approach absolute zero. Existing stars ultimately will die and no more will be born. The universe will suffer from heat death and become an infinitely large area of dark husks and waste.[23]

Of course, much of this is speculative. And there are variations on these themes as well as other scenarios. For instance, some believe that the Big Crunch could lead to a Big Bounce and renewed life for the universe.

19. See "1929: Edwin Hubble."

20. See "Hubble Finds Evidence."

21. See Dickerson, "3 Ways the Universe Might End," paras. 10–14.

22. Dickerson, "3 Ways the Universe Might End," paras. 7–9.

23. See Trosper, "Four Ways," paras. 11–16.

Others talk about a Big Change[24] or a Big Slurp[25] where a bubble in our universe or from another universe suddenly appears and annihilates our universe. That said, and with all possible caveats, the Big Freeze scenario appears, for now, to be the most probable of all futures.[26]

Obviously, the authors of the Torah knew nothing about an expanding universe, or dark energy, or Big Crunches, Rips, or Freezes. If they ever conceived of the evolution and ultimate fate of their universe at all, they said nothing. For them, as it is most of the time for most of us, all concerns, like all politics, were local.

Similarly, biblical prophets and poets, and later Talmudic and medieval sages, were no more informed about modern astrophysics as it applies to the future and ultimate death of the universe than they were about the physics that applied to the origin of the universe or the biochemistry involved in the evolution of life. But they did speak on occasion about *acharit hayamim*, meaning the end of days or the days to come, an unspecified time in the distant future. Even as they did, though, they created no complete narrative analogous to the creation stories, and nothing that described the end of life on our home planet, much less the death of the universe as we know it.

When the days to come were envisioned in the book of Isaiah, they were seen as an idyllic time when the house of the God of Israel would be established on the highest mountain, God's word would go forth from Jerusalem, and none but the God of Israel would be worshipped (see Isa 2:2, 3, 17). At that time, the many nations of the world would beat their swords into pruning hooks and forego war, and the wolf would dwell with the lamb, just as the leopard would lie down with the kid (see Isa 2:4; 11:6). Similarly, on that day, the dispersed people of Israel would return from Assyria, Egypt, and other lands, and Judah and Israel would be reunited (see Isa 11:11–13). The reunited community would be rewarded with everlasting joy and gladness, with the smallest becoming a mighty clan, and the least, a mighty nation (see Isa 51:11; 60:22).

Invoking the metaphors of dried bones and sticks, Ezekiel, too, foresaw the reunification of Judah and Ephraim, never again to be divided (see Ezek 36:24; 37:1–23). Given a new heart and new spirit, the people would be cleansed and their land would once again become like the

24. See Becker, "How Will the Universe End," para. 39.

25. See Boule, "Will Our Universe End."

26. See Dickerson, "3 Ways the Universe Might End," para. 17. For a personal and poetic view, see the poem titled "Fire and Ice" in Frost, *Poetry*, 220.

garden of Eden, with abundant fields, but now populated and fortified (see Ezek 36:24–27, 30–35). In Zechariah's words, when the dispersed returned home, the squares of Jerusalem would be filled with old men and women with their staffs in hand, and crowded with young boys and girls playing (see Zech 8:4–5).

While the exile of Judahites to Babylon did end, and a Second Temple was constructed, ultimately that period ended in destruction and dispersal as well.[27] Not only had the prophesied time of peace and prosperity not come, much less been sustained, but the rabbis in the Talmudic period were faced with communal concerns quite different than those confronting the ancient prophets. They continued to discuss and elaborate on a hypothetical end of days, which they extrapolated into a Messianic Age, one which would be presided over by the anointed one, the *Mashiach* or Messiah. The new era would come, by some interpretations, no later than six thousand years after creation,[28] a view maintained by some today.[29] By a traditional count, the birthday of the world is dated to *Rosh HaShanah* in 3761 BCE, meaning that the Messianic Age would arrive no later than *Rosh HaShanah* in the year 2239 CE.[30] The rabbis were giving themselves and the Messiah plenty of slack.

But as Maimonides observed, whenever the Messianic Age might commence, even while there would be greater peace and more wisdom, there would also still be rich and poor, strong and weak. The essence of the Messianic Age for Rambam was that Jews would return to the land of Israel and regain their independence.[31] Life would go on. Even as some imagined horrific battles or other conditions that would presage the Messianic Age, there was no narrative about mass destruction during the Messianic Age itself. The whole point was the reconstitution of a united Jewish people in their ancestral homeland, free to build their society in peace.

27. According to the *Tanakh*, upon the return of exiled Judahites to Jerusalem, work began on the construction of a new sanctuary to replace that destroyed by the Babylonians early in the sixth century BCE. That structure, for which there is limited archaeological evidence, as later enhanced (especially by Herod) is known as the Second Temple. The Second Temple period extends from approximately 516 BCE to 70 CE.

28. See b. Roš Haš. 31a, https://juchre.org/talmud/rosh/rosh2.htm#31a, and b. Sanh. 97a, https://juchre.org/talmud/sanhedrin/sanhedrin5.htm#97a.

29. See, e.g., Winston, "Moshiach and the World Today."

30. See Solomin, "History of the World."

31. See "Maimonides' Thirteen (13) Principles," para. 65.

How could it be otherwise? After all, if it were foreseen that the entirety of what is, including life itself, would someday cease to be, and cease irrespective of the scope of adherence to a particular set of commandments, laws, and instructions, how would rabbis of that time have explained the purpose of the original creation? What would such an anticipated end say of the Creator of the Beginning? Or of God's later promise to Noah, his descendants, and all living creatures—a promise symbolized by the rainbow—to never again destroy the world? (See Gen 9:8–17.)

The same questions can be asked of rabbis today, of course, and they have less of an excuse of ignorance. At the same time, perhaps the Messianic Age is too important a topic to be left to theologians. Perhaps it is for a poet to provide the compelling lesson, as Danny Siegel does here:[32]

If you always assume
the person sitting next to you
is the Messiah
waiting for some simple human kindness –

You will soon come to weigh your words
and watch your hands.

And if the person chooses
not to be revealed
in your time –

It will not matter.

32. See "A Rebbe's Proverb (From the Yiddish)" in Siegel, *God Braided Eve's Hair*, 3.

Part Five

ACCEPTING THE CHALLENGE, EMBRACING THE OPPORTUNITY

Chapter 26

Einstein, Kaplan, and Heschel Walk into a Bar, and Other Hypotheticals

JUDAISM IS A CONTINUING dialogue by and among Jews. It is a conversation that no doubt began with stories told orally, but which well over two thousand years ago were beginning to be inscribed on papyrus and clay and stone. Texts begat other texts that conserved, reformed, and even reconstructed those earlier conversations. Today those texts are printed on paper and also transmitted electronically.

As the medium varied over the years, so did the context of the dialogue. At any moment, the conversation was always colored by the circumstances in which Jews found themselves, and tempered by time and the historical memory of the Jewish people. In these conversations, Jews were wrestling with matters of fact and fiction and faith. It is what Jews do and have always done.

This book reflects part of that grand dialogue. We have looked at many texts and traditions through analytic lenses of various shapes and sizes, and with the help of loads of data. And, by meeting a wide variety of Jews along the way, we have uncovered some of the many and different ways in which Judaism responds to science. It's been quite a journey.

We have met Jews who reject what science teaches. A notion as basic, and established, as the age of our universe being more than six thousand years old is apparently considered a threat to their sense of self, their understanding of where they are from and where they belong.

We have met Jews who attempt to accommodate science to text, and text to science, invoking mysticism and metaphor in an effort to provide some balance as they try to navigate between the tug of the past and the pull of modernity.

We have met Jews who accept text and science when it suits them, when a source supports their understanding of the way they think the world ought to be, but then disregard science when it runs counter to their preconceived narrative.

We have met Jews who take science as gospel, and in their guise as sophisticated "cultural Jews" reject core Jewish religious or ethical texts or precepts, or at least their idea of what those concepts entail.

And we have met Jews who are as comfortable with science as they are with breathing, and fully engaged in one or more aspects of the Jewish tradition as well.

In a sense, these differences should not be surprising. Judaism is not, and really never has been, monolithic in theory or practice. Rather, from its foundational texts forward, Judaism is replete with expressions of competing understandings of root issues theological, political, and social. You can see it in the biblical metamorphosis of the God of Israel, early on a creator god and a storm god, later a warrior god superior to the other gods, then a royal god who enters into covenants like Ancient Near Eastern kings, and ultimately the One, the Eternal, the master of the universe. You can see it in inconsistent statements about slavery in codes set forth in Exodus and in Deuteronomy, in the disparate treatment of priests of different tribes discussed in Leviticus and Deuteronomy, in the express and implicit calls to welcome strangers in Torah and the book of Ruth and the strident directives to eject them from the community in the books of Ezra and Nehemiah. In the Talmud, the disputations between the House of Hillel and that of Shammai regarding matters of Jewish law were so serious and persisted for so long that a truce had to be fashioned under the banner *Elu v' Elu*, asserting that these and those were the words of God.[1]

With such origins, and their evolution over millennia in numerous countries around the globe, what else should one have expected other than that a movement in the United States, which sought to reform theory and practice and to distance itself from the customs and thoughts of its European ancestors, would, in its excesses, give birth

1. See b. 'Erub. 13b, https://www.sefaria.org/Eruvin.13b?lang=bi.

to a more conservative approach, which in turn would generate a sort of synthesis among those who would reconstruct the entire apparatus. Moreover, because evolution is not a static process, over time, with self-reflection, each of these movements would recalibrate its approach, and other movements would arise.

And Jews are nothing if not human, burdened as humans tend to be by a wide variety of cognitive biases. Like others, we tend to live in limited and limiting communities and to adopt the beliefs and practices of the majority of our cohort. We often focus on outcomes rather than analysis, and we are guided by motivated reasoning in which the desired conclusion moderates our thought process. So, at times we treat information selectively. We are inclined to ignore information that may adversely affect our particular worldview and, conversely, seek and are comfortable only with information that confirms preexisting conceptions and the order with which we are familiar. Groupthink, the ostrich approach, outcome bias, familiarity bias, confirmation bias, and similar tendencies are all very powerful forces.

In such circumstances, for some, science can seem quite threatening, while for others it is liberating. And science, of course, is not static either. New discoveries and new technologies may seem to create a constant barrage of challenges to those who like things as they were, or, at least, as they are, that is, just the way they believe God intended the world to be. At the other extreme, such scientific revelations are simply more proof that God is, or ought to be, dead.

But does it have to be this way? Let's consider a thought experiment. Imagine for the moment if Judaism and science could speak to each other, at least if a couple of wise rabbis and a prominent scientist could get together. What might have happened if Albert Einstein, the preeminent scientist of the twentieth century, ever shared drinks with Mordecai Kaplan and Abraham Joshua Heschel, two of the finest Jewish scholars of that period? In this thought experiment, return with us to days of yesteryear, and imagine . . .

* * *

It was in 1953, or so. The exact date is lost to memory. The pub was somewhere just north of Columbia University. Albert Einstein, perhaps the greatest physicist of the century, picked the place in part because he was visiting an old friend at Columbia, as he was traveling from Princeton to his summer home on Long Island. Not coincidentally (for he did not believe in coincidences), it was also not too far from the Jewish Theological Seminary.

Einstein wanted to meet JTS luminaries Mordecai Kaplan and Abraham Joshua Heschel, and had heard that there was a booth in the back of the bar that was conducive to conversation. He was interested in Kaplan because he had heard of Kaplan's attempts to create a Jewish theology without supernaturalism. The idea of a naturalist philosophy (or transnaturalism, as Kaplan sometimes called it) appealed to Einstein. Anyone whose prayer book was radical enough to get burned, by Jews no less, was a bonus for the seventy-plus-year-old, but ever rebellious, Einstein.

Heschel was a different matter. A dozen or so years earlier, Heschel had been severely critical of Einstein because he thought that Einstein had dismissed the God from heaven. Einstein was aware of the criticism, but had also heard that Heschel was described by some as a pantheist and by others as a panentheist. Einstein's theology, such as it was, fell somewhere in there too, usually. It made no real difference to Einstein. All agreed that Heschel had a mystical bent. That approach made no sense at all to Einstein, but to some degree, that was one of the points of the whole pub exercise. He was there for a variation on one of his thought experiments. Except this time he was interested not so much in experimenting (though the idea of having a German Jew, a Polish Jew, and a Litvak at the same table was intriguing). He was just enjoying thinking about thinking.

Einstein had requested the meeting. Kaplan and Heschel knew each other, indeed were colleagues, even friends. And each of them knew who Einstein was, of course, but neither had met him despite the fact that each was prominent in the American Jewish community. Both accepted Einstein's invitation to join him without hesitation. After all, who could turn down the man who just turned down Ben Gurion's request to become president of the reconstituted State of Israel?

Einstein preordered drinks for everyone—atomics, they were called. Imp that he was, Einstein thought the selection was funny, but he really just wanted to save time.

Kaplan and Heschel came in together. Neither cared much for the other's philosophy, but they cared for each other. Apparently, you could do that in those days.

Heschel saw Einstein first. Even in the darkness of the pub, the light seemed to shine off the waves and curls that appeared to sprout randomly from Einstein's head. After some initial fumbling with Doctor, Herr Professor, Rabbi, they settled quickly into Al, Mordi, and Abe.

"Nice space, yes?" Einstein said matter-of-factly.

"Space," said Heschel, "is full of wonder."

"You're right, Abe," Einstein replied, "but I was talking about the space immediately around us—this tavern, not the whole universe."

"Of course," said Heschel, "but still . . ."

"And thanks for coming on time," added Einstein.

"Man transcends space, and time transcends man," Heschel interjected.

"Does he always talk like this, Mordi?"

"Always and always," replied Kaplan, with a shrug.

"I ordered atomics. I hope that is all right with you," Einstein said.

"I don't know much about your science, Al," said Heschel. "I do know that under the running sea of our theories and scientific explanations lies the aboriginal abyss of radical amazement."

With a wink to Kaplan, Einstein said "You have a gift, Abe."

"My gift," replied Heschel, "is my ability to be surprised."

"I am not much interested in surprises," Einstein offered. "I prefer mathematics and science. They disclose the order of the universe. I now realize that God himself could not have arranged these connections any other way than that which does exist."

Kaplan rose, glass in hand, and announced: "The so-called laws of nature represent the manner of God's immanent functioning. The element of creativity, which is not accounted for by the so-called laws of nature, and which points to the organic character of the universe or its life as a whole, gives us a clue to God's transcendent functioning. God is not an identifiable being who stands outside the universe. God is the life of the universe, immanent insofar as each part acts upon every other, and transcendent insofar as the whole acts upon each part." Kaplan then sat down.

"So, Abe, does he always talks like this?"

"Always and always," Heschel replied, wearily. Then, without missing a beat, Heschel added, "Mystery remains. The root of religion is the question of what to do with the feeling for the mystery of living, what to do with awe, wonder, or fear."

To which Kaplan responded: "Religion is as much a progressive unlearning of false ideas concerning God as it is the learning of the true ideas concerning God."

"Fellas," Einstein interjected, "Can we not agree on this: Try and penetrate with our limited means the secrets of nature and you will find, that, behind all the discernible laws and connections, there remains something subtle, intangible and inexplicable. It is a force worthy of veneration, yes?"

"Yes," said Kaplan, "This is the force, the process, I call God."

"And Abe, you agree, too, do you not, that the most beautiful emotion we can experience is the mysterious? To sense that behind anything that can be experienced there is something that our minds cannot grasp, whose beauty and sublimity reaches us only indirectly: this is religiousness."

"Beyond the grandeur is God." Heschel replied, adding, "God is a mystery, the mystery is not God."

"Mordi, now I think that I am getting into the swing. Science without religion is lame, and religion without science is blind. How's that?"

"That's it, Al. Now you sound like Heschel."

"You could learn from him, too, you know. After all, it does not mean a thing if it does not have that swing."

"I know, Al. If only I had been able to write like Abe, I could have been a contender."

"Say again?" asked a man in an Actors Studio T-shirt.

"I could have been a contender," boomed Kaplan.

"Thanks," mumbled the young fellow, as he seemed to scribble some note.

"You have done all right, Mordi," said Einstein, "Clear and to the point. And your ideas will help our people evolve."

"It's all relative, isn't that right, Al ?" Kaplan asked with a twinkle.

"Yes," Einstein agreed, chuckling, "It's all relative."[2]

* * *

Now, we cannot all be Einstein. He was, after all, an Einstein! Nor can most of us aspire to be a Kaplan or a Heschel. So, what can we learn from this hypothetical meeting? At least three points should be clear. First, though we have different backgrounds and skill sets, we often experience the same needs, desires, and emotions, but may have some difficulty

2. Statements attributed to Einstein, Kaplan, and Heschel, except for filler, are essentially their own words as quoted in one or more of the following: Isaacson, *Einstein*; Jammer, *Einstein and Religion*; Scult, *Radical American Judaism*; Kaplan, *Spiritual Radical.*

communicating with someone from different circumstances. Second, though each of us by temperament and training approaches matters of Judaism and science with different sensibilities, we have much to learn from each other. Third, more than listening in order to reply, it is important to listen in order to understand. And we need to listen actively, because the issues which confront us at the intersection of Judaism and science are serious.

What are those issues? Here is a short list of some of the matters being considered today:

- To what extent, if any, should we edit the genomes of the unborn? To cure malformations, prevent disease, or enhance appearance or mental or physical skills?
- To what extent, if any, should we alter human behavior by chemical or mechanical means? Does it matter if the purpose is to avoid pain or promote pleasure? Should any such alteration be voluntary, or could it be mandatory?
- Under what circumstances, if any, should we clone humans?
- Under what circumstances, if any, should be recreate extinct species?
- If climate change is a real, and existential, threat to our planet and ourselves, to what extent, if any, should humans, through biological, chemical, or geo-engineering, or other means, seek to alter such change? Who should decide and who should pay?
- To what extent, if any, do humans have a right to alter or inhabit another world? What, if any, rights do life forms on other worlds have?

Because we have argued that facts really matter, our approach to these and other questions might be altered considerably if a container holding two stone tablets carbon-dated to 3,300 years ago and inscribed with the Ten Commandments were found in an underground cavern near Jerusalem, or if the bones of a man from that period were found on Mount Nebo, with his skull exuding an unusual, but certain, radiance and his DNA consistent with the Cohen Modal Haplotype. And, needless to say, should God appear one day and announce God's self, we will all have to reevaluate our positions on these and a host of matters.

These potential discoveries and revelations, however, are currently just figments of our imagination. In the real world we inhabit, we still face educational and communication deficits. This leads to another thought experiment. What if we could ramp up our learning process? What if we could appreciate the necessity of learning more science and making sure that what we say to each other and to our children in all Jewish contexts is credible?

Toward that end:

- What if we heeded Yitz Greenberg's call and committed to rising up rather than dumbing down?
- What if we spoke and acted as if facts matter (because they do)?
- What if we opened our hearts and minds to new perspectives and new possibilities?
- What if Jewish seminaries were to partner with local medical schools and teaching hospitals to establish ongoing programs at the seminaries on bioethics, and require all future rabbis to attend?
- What if Jewish seminaries were to partner with local universities to establish ongoing programs on cosmogony and cosmology, on evolution and archaeology, and on the history, languages, and literature of the Ancient Near East, and require all future rabbis to attend?
- What if these seminaries, by themselves or with the congregational arms of their movements (if they are so affiliated), then produced curricula which incorporated modern science into lessons for all of their congregational schools and camps?
- What if Jewish educational institutions, from seminaries to synagogues to day schools to summer camps, were not just to incorporate science in their curricula, but to explore the relationship of the science taught to Jewish texts and values?
- What if, because words matter too, these institutions began to revise their Torah commentaries and *siddurim* to reflect what they have learned, to distinguish truth statements and value statements more clearly, and to clarify when we are quoting from our past and when we are affirming for our present because we value the honesty and the connection?
- What if Jewish adult education and lifelong learning programs were to invite scientists, and other critical thinkers, to discuss their respective fields and how, if at all, they and Judaism might inform each another?

- What if we did these things not as members of Jewish denominations, but as members of a broader Jewish community, assembled together?
- And because some individuals and institutions may be doing some of this already, what if those who are would share what they have learned with those who are not?
- And what if Jewish philanthropies invested seriously in one or more of the programs identified here, as if the life of the community hinged on the credibility of its words and the integrity of its deeds?

In a different context, Moses is said to have placed before the Israelites blessings and curses and asked them to make a choice (see Deut 30:11–20). Today, we, too, are faced with a choice, as science, in the form of critical thinking with empirical evidence, continues its rapid advance. It poses a challenge to some of Judaism's most cherished concepts, but it also offers an opportunity to those who would seize it.

This book has argued that, in many aspects of Jewish life, a reality-based Judaism is not only possible, but also a firmer foundation for the development of a vigorous and enduring Judaism than is myth-based Judaism. Because facts matter, because words matter, a reality-based Judaism can provide core structural support on which we can stand with intellectual integrity because it depends on credibility. But it is more than that—much more.

As Carl Sagan recognized a generation ago, with science, one can have "profound experiences of awe and reverence before the universe." So science, according to Sagan, "is not only compatible with spirituality; it is a profound source of spirituality." Summarizing his approach to the sacred, Dartmouth religion professor Nancy Frankenberry has written that Sagan was "utterly imbued with . . . a marvelous sense of belonging to a planet, a galaxy, a cosmos that inspires devotion as much as discovery."[3] And somewhat more mundanely, by helping us understand our roots and the origins of our traditions by providing the data which can inform our ethical decision-making, science can help us in the vital task of determining what in our texts we can comfortably quote for the sake of continuity and what we can enthusiastically affirm for the sake of credibility.

Reality-based Judaism is part of that grand Jewish tradition that teaches us not to adore idols, but to shatter them. Consequently, under

3. See Frankenberry, *Faith of Scientists*, 222–24.

its banner, we worship neither science not scroll, but seek to gain wisdom from both. The journey will follow empirical evidence as far as possible, and when the trail ends, it allows for faith or skepticism or both in tension with each other. But first, we must walk the walk.

Just as early humans stood up in Africa so that they could see farther across the veldt, just as our ancestors left Africa so that they could explore our world, so now we are beginning to leave our birth home, perhaps to find our place on distant planets, even among the stars. We have traveled many different roads, and we have many different customs, but our future is not to be found in a rebuilt temple in Jerusalem nor in a communal imitation of some isolated eighteenth-century European *shtetl* (small town), however idealized we imagine it. Neither will it be sustained by an untethered individualism, cloaked in feel-goodism. The reason is simple: a healthy and vital Judaism of the future can neither refuse to recognize reality nor lose touch with its origins, its communal norms, or its shared expectations and responsibilities. We need a different path.

If there is to be a vibrant and vigorous Judaism for the adults of Israel, one that their minds can engage, one worthy of being chosen by them as the educated members of the community that they are, then Jews cannot shrink from science. Rather, they must embrace it just as they embrace their heritage because science, like that heritage, can truly be a blessing and not a curse. Reality-based Judaism is, then, a path Jews can travel to fulfill their destiny.

Who will make the decision to abandon pediatric Judaism and select the more fulfilling route? And how will that choice be made? Two stories concerning the wise and compassionate leader of the people in the first century BCE point the way.

The first story tells of a fellow who one day decided to try to embarrass the Jews in town by picking on their leader, Rabbi Hillel. The man went to Hillel with his hands behind his back and told Hillel that he was holding a bird in his hands. He asked Hillel if the bird were dead or alive. Hillel knew what the man was up to and recognized the challenge. He also knew the answer to the question. He knew that if he said the bird was dead that the man would open his hands and let the bird fly away, but, if he said the bird was alive, then the man would snap the bird's neck and show Hillel a dead bird. So Hillel spoke wisely and truthfully. The answer is in your hands, he said, the answer is in your hands.

So, too, here.

The second story tells of a person who asked Hillel to explain to him the meaning of the entire Torah while he, the person asking, stood on one foot. Hillel responded with a negative formulation of the Golden Rule: "What is hateful to you, do not do to others." He then added that all the rest is commentary, and advised his inquirer to go study.[4]

So, too, here.

May your studies be sweet and satisfying.

4. See b. Šabb. 31a, https://www.sefaria.org/Shabbat.31a?lang=bi.

Bibliography

"1929: Edwin Hubble Discovers the Universe is Expanding." https://cosmology.carnegiescience.edu/timeline/1929.

"Abortion Rates by Race and Ethnicity." https://www.guttmacher.org/infographic/2017/abortion-rates-race-and-ethnicity.

Adams, Fred. *Origins of Existence: How Life Emerged in the Universe*. New York: Free Press, 2002.

"Admonitions of Ipuwer." http://www.reshafim.org.il/ad/egypt/texts/ipuwer.htm.

AgriMarketing. "Estimated U.S. Seed Market Shares Released." https://www.agrimarketing.com/ss.php?id=97703.

Albaugh, Janel M. "Golden Rice: Effectiveness and Safety, A Literature Review." https://ideaexchange.uakron.edu/cgi/viewcontent.cgi?article=1398&context=honors_research_projects.

Alter, Robert. *The Bible as Poetry: The Art of Biblical Poetry*. New York: Basic, 1985.

American Academy of Allergy, Asthma, and Immunology. "Food Allergy Overview." http://www.aaaai.org/conditions-and-treatments/allergies/food-allergies.

American Medical Association. "Reports Of The Council On Science And Public Health." https://www.ama-assn.org/sites/default/files/media-browser/public/hod/a12-csaph-reports_0.pdf.

Amin, Osama S. M. "Siege of Lachish Reliefs at the British Museum." http://etc.ancient.eu/photos/siege-lachish-reliefs-british-museum/.

"Ancient Mesopotamian Geography and Location." http://ancientmesopotamians.com/ancient-mesopotamia-geography.html.

Angel, Marc D. *Maimonides, Spinoza and Us: Toward an Intellectually Vibrant Judaism*. Woodstock: Jewish Lights, 2009.

"Annual Energy Outlook 2018 with Projections to 2050." https://www.eia.gov/outlooks/aeo/pdf/AEO2018.pdf.

Arad, Roy. "Finally, a Red Alert This Summer That (Most) Israelis Welcomed." *Haaretz*, January 11, 2019. https://www.haaretz.com/.premium-finally-a-red-alert-that-israelis-welcomed-1.5262135.

"Are We Really All Made of Stardust?" http://www.physics.org/article-questions.asp?id=52.

Artson, Bradley. "Judaism and the Human Body." https://www.myjewishlearning.com/article/lets-get-physical/.

———. *Renewing the Process of Creation: A Jewish Integration of Science and Spirit*. Woodstock: Jewish Lights, 2016.

Aumann, Robert J. "A Personal Perspective on the Work of the 'Gans' Committee." http://www.torah-code.org/experiments/dp_365_1.pdf.

Aviezer, Nathan. "The Anthropic Principle: What Is It and Why Is It Meaningful to the Believing Jews?" http://ou.org.s3.amazonaws.com/publications/ja/5759spring/anthropic.pdf.

Bar-Hillel, Maya, et al. "Solving the Bible Code Puzzle." https://projecteuclid.org/euclid.ss/1009212243.

Bar-Natan, Dror, and Brendan McKay. "Equidistant Letter Sequences in Tolstoy's 'War and Peace.'" http://users.cecs.anu.edu.au/~bdm/dilugim/WNP/.

Baron, Joseph L., ed. *A Treasury of Jewish Quotations*. New York: Crown, 1956.

Barskey, Albert E., et al. "Mumps Outbreak in Orthodox Jewish Communities in the United States." *New England Journal of Medicine* 367 (2012) 1704–13. https://www.nejm.org/doi/full/10.1056/NEJMoa1202865.

Becker, Adam. "How Will the Universe End, and Could Anything Survive?" *BBC Earth*, June 2, 2015. http://www.bbc.com/earth/story/20150602-how-will-the-universe-end.

Becker, Gary. "Fracking and Self-Sufficiency in Gas and Oil." *The Beckner-Posner Blog*, April 1, 2012. http://www.becker-posner-blog.com/2012/04/fracking-and-self-sufficiency-in-gas-and-oil-becker.html.

Beckett, Lois. "Meet America's Gun Super-Owners—with an Average of 17 Firearms Each." *The Guardian*, September 20, 2016. https://www.theguardian.com/us-news/2016/sep/20/gun-ownership-america-firearms-super-owners.

Berlin, Adele, and Marc Zvi Brettler, eds. *The Jewish Study Bible*. New York: Oxford University Press, 1985.

Berrin, Danielle. "Passover Proof Lies in Egyptian Hieroglyphs." *Jewish Journal*, March 24, 2010. http://jewishjournal.com/culture/religion/passover/77833/.

Biblical Archaeology Society Staff. "The Oldest Hebrew Script and Language." https://www.biblicalarchaeology.org/daily/biblical-artifacts/inscriptions/the-oldest-hebrew-script-and-language/.

Blue Cross Blue Shield. "Childhood Allergies in America." https://www.bcbs.com/the-health-of-america/reports/childhood-allergies-america.

Boule, Alan. "Will Our Universe End in a 'Big Slurp'? Higgs-Like Particle Suggests It Might." *NBC News*, November 2, 2015. https://www.nbcnews.com/science/science-news/will-our-universe-end-big-slurp-higgs-particle-suggests-it-f1C8415998.

Bratman, Raphael. "Water Fluoridation and Vaccinations are Contrary to Torah Principles." http://jewcology.org/2011/09/water-fluoridation-and-vaccinations-are-contrary-to-torah-principles/.

———. "Why Genetically Modified Foods Should Not Be Considered Kosher." http://jewcology.org/2010/12/why-genetically-modified-foods-should-not-be-considered-kosher/.

Brettler, Marc Zvi. *How to Read the Jewish Bible*. New York: Oxford University Press, 2005.

Britt, Robert Roy. "Milky Way's Age Narrowed Down." https://www.space.com/263-milky-age-narrowed.html.

Brodie, Yehuda. "Why the Fluenz Vaccine Is Not Treif." *Jewish Chronicle*, October 10, 2013. https://www.thejc.com/comment/analysis/why-the-fluenz-vaccine-is-not-treif-1.49559.

Bronner, Leila. "Is Abortion Murder? Jews and Christians Will Answer Differently." http://www.bibleandjewishstudies.net/articles/abortion.htm.

Brown, Jeremy. "Can Judaism and Science Co-Exist? Yes." https://www.onfaith.co/onfaith/2013/09/16/can-judaism-and-science-co-exist-yes/11030.

Brown, Nino. "Plant Domestication." http://plantbreeding.coe.uga.edu/index.php?title=2._History_of_Plant_Breeding.

Bucklin, Stephanie. "A Conductor of Evolution's Subtle Symphony." *Quanta Magazine*, November 3, 2016. https://www.quantamagazine.org/long-term-evolution-the-richard-lenski-interview-20161103/.

Bulliet, Richard W. *The Camel and the Wheel*. New York: Columbia University Press, 1990.

Byrd, Deborah. "How Long to Travel to Alpha Centauri?" http://earthsky.org/space/alpha-centauri-travel-time.

Cantor, Geoffrey, and Marc Swetlitz. *Jewish Tradition and the Challenge of Darwinism*. Chicago: The University of Chicago Press, 2006.

Caouette, Justin. "The 'Free Will Problem.'" *A Philosopher's Take* (blog), August 13, 2012. https://aphilosopherstake.com/2012/08/13/the-free-will-problem/.

———. "Neuroscience and Free Will: New Study Debunks Libet's Interpretation." *A Philosopher's Take* (blog), August 10, 2012. https://aphilosopherstake.com/2012/08/10/neuroscience-and-free-will-new-study-debunks-libets-interpretation/.

Carroll, Lewis. *The Annotated Alice: The Definitive Edition*. New York: Norton, 2000.

Carroll, Sean. *From Eternity to Here: The Quest for the Ultimate Theory of Time*. New York: Dutton, 2010.

———. "Welcome to the Multiverse." *Discover*, October 18, 2011. http://blogs.discovermagazine.com/cosmicvariance/2011/10/18/column-welcome-to-the-multiverse/#.XDkIJ89KjOQ.

Cassuto, Umberto. *A Commentary on the Book of Genesis Part One: From Adam to Noah*. Translated by Israel Abrahams. Skokie, IL: Varda, 2005.

———. *A Commentary on the Book of Genesis Part Two: From Noah to Moses*. Translated by Israel Abrahams. Skokie, IL: Varda, 2005.

Centers for Disease Control and Prevention. "Abortion Surveillance—United States, 1998." https://www.cdc.gov/MMWR/PREVIEW/MMWRHTML/ss5103a1.htm.

———. "Achievements in Public Health, 1900–1999 Impact of Vaccines Universally Recommended for Children—United States, 1990–1998." https://www.cdc.gov/mmwr/preview/mmwrhtml/00056803.htm.

———. "Data and Statistics." https://www.cdc.gov/reproductivehealth/data_stats/index.htm.

———. "Firearm Mortality by State." https://www.cdc.gov/nchs/pressroom/sosmap/firearm_mortality/firearm.htm.

———. "Measles Cases and Outbreaks." https://www.cdc.gov/measles/cases-outbreaks.html.

———. "*QuickStats*: Average Number of Deaths from Motor Vehicle Injuries, Suicide, and Homicide, by Day of the Week." https://www.cdc.gov/mmwr/volumes/66/wr/mm6622a5.htm.

———. "Suicide and Self-Inflicted Injury." https://www.cdc.gov/nchs/fastats/suicide.htm.

———. "What Would Happen If We Stopped Vaccinations?" https://www.cdc.gov/vaccines/vac-gen/whatifstop.htm.

Central Conference of American Rabbis. "CCAR Responsa 5774.5." https://www.ccarnet.org/ccar-responsa/.

———. "CCAR Responsa: When Is Abortion Permitted?" https://www.ccarnet.org/ccar-responsa/carr-23-27/.

"The Chaldean Empire." http://www.thelatinlibrary.com/imperialism/notes/chaldean.html.

Chivers, Tom. "The Vatican Joins the Search for Alien Life." *Telegraph*, November 10, 2009. https://www.telegraph.co.uk/news/science/space/6536400/The-Vatican-joins-the-search-for-alien-life.html.

Choi, Charles Q. "Earth's Sun: Facts About the Sun's Age, Size, and History." https://www.space.com/58-the-sun-formation-facts-and-characteristics.html.

———. "Our Expanding Universe: Age, History & Other Facts." https://www.space.com/52-the-expanding-universe-from-the-big-bang-to-today.html.

Chow, Denise. "The Universe: Big Bang to Now in 10 Easy Steps." https://www.space.com/13320-big-bang-universe-10-steps-explainer.html.

CNN Library. "Abortion Fast Facts." https://www.cnn.com/2013/09/18/health/abortion-fast-facts/index.html.

Cohen, Shaye J. D. "The Matrilineal Principle in Historical Perspective." http://archive.jewishrecon.org/resource-files/files/Shaye%20Cohen%20-%20the%20Matrilineal%20Principle%20in%20Historical%20Perspective.pdf.

Collins, Lorence G. "Twenty-One Reasons Noah's Worldwide Flood Never Happened." *Skeptical Inquirer* 42.2 (March/April 2018). https://www.csicop.org/si/show/twenty-one_reasons_noahs_worldwide_flood_never_happened.

———. "Yes, Noah's Flood May Have Happened, but Not Over the Whole Earth." https://ncse.com/library-resource/yes-noahs-flood-may-have-happened-not-over-whole-earth.

Cook, Troy, et al. "Hydraulically Fractured Horizontal Wells Account for Most New Oil and Natural Gas Wells." https://www.eia.gov/todayinenergy/detail.php?id=34732.

Cornell, Dewey, and Nancy G. Guerra. "Gun Violence: Prediction, Prevention, and Policy." http://www.apa.org/pubs/info/reports/gun-violence-prevention.aspx.

Cosgrove, Elliot J., ed. *Jewish Theology in Our Time: A New Generation Explores the Foundations and Future of Jewish Belief.* Woodstock: Jewish Lights, 2010.

"Cosmos: The Big Bang." http://www.scienceforthepublic.org/things-to-know/cosmos/cosmos-the-big-bang.

Coyne, Jerry A. "No Faith in Science." *Slate*, November 14, 2013. http://www.slate.com/articles/health_and_science/science/2013/11/faith_in_science_and_religion_truth_authority_and_the_orderliness_of_nature.html.

———. "Rabbi Sacks Is an Ignorant Fool." https://whyevolutionistrue.wordpress.com/2013/06/13/rabbi-sacks-is-an-ignorant-fool/.

———. *Why Evolution Is True.* New York: Penguin, 2010.

———. "You Don't Have Free Will." *The Chronicle of Higher Education*, March 18, 2012. https://www.chronicle.com/article/jerry-a-coyne-you-dont-have/131165.

Crimesider Staff. "Chicago Saw More 2016 Murders than NYC, LA Combined." *CBS News*, January 2, 2017. https://www.cbsnews.com/news/chicago-murders-shootings-2016-more-than-new-york-city-los-angeles-combined/.

Dann, Jack, ed. *Wandering Stars: An Anthology of Jewish Fantasy & Science Fiction.* Woodstock: Jewish Lights, 1998.

"Dates and Sources: When Shakespeare Wrote Hamlet and the Stories that Inspired Him." https://www.rsc.org.uk/hamlet/about-the-play/dates-and-sources.

Davies, Paul. "Taking Science on Faith." *New York Times*, November 24, 2007. https://www.nytimes.com/2007/11/24/opinion/24davies.html.

Dawkins, Richard. *The God Delusion*. New York: First Mariner, 2008.

———. *The Greatest Show on Earth: The Evidence for Evolution*. New York: Free Press, 2009.

"Deep Blue." http://www-03.ibm.com/ibm/history/ibm100/us/en/icons/deepblue/.

Demsky, Aaron. "Historical *Hakhel* Ceremonies and the Origin of Public Torah Reading." https://thetorah.com/historical-hakhel-ceremonies-and-the-origin-of-public-torah-reading/.

Dever, William. *The Lives of Ordinary People in Ancient Israel: Where Archeology and the Bible Intersect*. Grand Rapids: Eerdmans, 2012.

Dickerson, Kelly. "3 Ways the Universe Might End." *Business Insider*, February 3, 2015. http://www.businessinsider.com/how-will-the-universe-end-2015-2.

di Cristopher, Tom. "American Oil Drillers' Output Could Top Saudi Arabia and Rival Russia by 2019, US Forecast Shows." *CNBC*, January 9, 2018. https://www.cnbc.com/2018/01/09/us-oil-drillers-could-beat-saudi-arabia-and-rival-russia-by-2019.html.

D'Onofrio, Jessica, et al. "102 Shot, 15 Fatally, in Chicago over July 4 Weekend in 2017." *ABC7 Eyewitness News*, July 5, 2017. https://abc7chicago.com/news/102-shot-15-fatally-in-chicago-over-july-4-holiday-weekend/2184156/.

"Doomsday Clock." https://thebulletin.org/overview.

Drosnin, Michael. *The Bible Code*. New York: Simon & Schuster, 1997.

———. *The Bible Code II: The Countdown*. New York: Viking, 2002.

———. *The Bible Code III: Saving the World*. New York: Worldmedia, 2010.

Duwe, Grant. "Mass Shootings Are Getting Deadlier, Not More Frequent." *Politico*, October 4, 2017. https://www.politico.com/magazine/story/2017/10/04/mass-shootings-more-deadly-frequent-research-215678.

Ecklund, Elaine Howard. *Science vs. Religion: What Scientists Really Think*. New York: Oxford University Press, 2010.

Economides, Michael. "Don't Be Swayed By Faucets On Fire and Other Anti-Fracking Propaganda." https://www.forbes.com/sites/greatspeculations/2011/03/07/dont-be-swayed-by-faucets-on-fire-and-other-anti-fracking-propaganda/#8262556165fa.

Edelstein, Avraham. "Judaism and Science—Harmony or Conflict." http://nleresources.com/2013/11/judaism-science-harmony-conflict/#.WsJGMojwbcs.

Efron, Noah J. *A Chosen Calling*. Baltimore: Johns Hopkins University Press, 2014.

———. *Judaism and Science: A Historical Introduction*. Westport: Greenwood, 2007.

Eisenberg, Daniel. "Abortion in Jewish Law." http://www.aish.com/ci/sam/48954946.html.

Ellenberg, Jordan. *How Not to Be Wrong: The Power of Mathematical Thinking*. New York: Penguin, 2014.

"The Emergence of Complex Life." https://globalchange.umich.edu/globalchange1/current/lectures/complex_life/complex_life.html.

Entine, Jon. *Abraham's Children: Race, Identity and the DNA of the Chosen People*. New York: Grand Central, 2007.

"The Epic of Atraḥasis." https://www.livius.org/sources/content/anet/104-106-the-epic-of-atrahasis/?#Mankind_Punished.

"The Exodus." https://www.chabad.org/library/article_cdo/aid/1663/jewish/The-Exodus.htm.

Falk, Gerhard. "American Jews." http://jbuff.com/c052302.htm.

Fattal, Isabel. "Torah-Writing Robot Speeds Past Human Scribes." *Tablet*, July 11, 2014. http://www.tabletmag.com/scroll/178935/torah-writing-robot-speeds-past-human-scribes.

Federal Bureau of Investigation. "Expanded Homicide." https://ucr.fbi.gov/crime-in-the-u.s/2015/crime-in-the-u.s.-2015/offenses-known-to-law-enforcement/expanded-homicide.

———. "Expanded Homicide Table 1." https://ucr.fbi.gov/crime-in-the-u.s/2016/crime-in-the-u.s.-2016/tables/expanded-homicide-data-table-1.xls.

———. "Expanded Homicide Data Table 2." https://ucr.fbi.gov/crime-in-the-u.s/2016/crime-in-the-u.s.-2016/tables/expanded-homicide-data-table-2.xls.

———. "Expanded Homicide Data Table 10." https://ucr.fbi.gov/crime-in-the-u.s/2015/crime-in-the-u.s.-2015/tables/expanded_homicide_data_table_10_murder_circumstances_by_relationship_2015.xls.

———. "Looking for Life in All the Right Places." https://exoplanets.nasa.gov/the-search-for-life/life-signs/.

———. "Murder Victims by Weapon, 2012–2016." https://ucr.fbi.gov/crime-in-the-u.s/2016/crime-in-the-u.s.-2016/tables/expanded-homicide-data-table-4.xls.

———. "Victims." https://ucr.fbi.gov/hate-crime/2017/topic-pages/victims.

"Federal Rules of Evidence." https://www.uscourts.gov/sites/default/files/Rules%20of%20Evidence.

Ferris, Timothy. *The Whole Shebang: A State-of-the-Universe(s) Report*. New York: Simon & Schuster, 1997.

Finkelstein, Israel. "Digging for the Truth: Archaeology and the Bible." In *The Quest for the Historical Israel: Debating Archaeology and the History of Early Israel*, edited by Brian B. Schmidt, 9–20. Archaeology and Biblical Studies 17. Atlanta: Society of Biblical Literature, 2007.

Finkelstein, Israel, and Neil Asher Silberman. *The Bible Unearthed: Archaeology's New Vision of Ancient Israel and the Origin of its Sacred Texts*. New York: Free Press, 2001.

Fishberg, Maurice. "Physical Anthropology of the Jews II—Pigmentation." *The American Anthropologist* 5.1 (1903) 89–106. https://anthrosource.onlinelibrary.wiley.com/doi/epdf/10.1525/aa.1903.5.1.02a00110.

Fleischacker, Samuel. "Making Sense of the Revelation at Sinai—Revisiting Maimonides' Eighth Principle of Faith." http://thetorah.com/making-sense-of-the-revelation-at-sinai/.

Follman, Mark, et al. "The True Cost of Gun Violence in America." *Mother Jones*, April 15, 2015. https://www.motherjones.com/politics/2015/04/true-cost-of-gun-violence-in-america/.

Food Allergy Research and Education. "Facts and Statistics." https://www.foodallergy.org/life-with-food-allergies/food-allergy-101/facts-and-statistics.

"The Four Elements in Kabbalah." http://www.tzfat-kabbalah.org/whatis.asp?p=1096.

Fowler, Katherine A., et al. "Childhood Firearm Injuries in the United States." *Pediatrics* 140.1 (July 2017). http://pediatrics.aappublications.org/content/early/2017/06/15/peds.2016-3486.

———. "Firearm Injuries in the United States." *Preventative Medicine* 79 (2015) 5–14. https://www.ncbi.nlm.nih.gov/pmc/articles/PMC4700838/.

Fox, Everett. *The First Five Books of Moses: Genesis, Exodus, Leviticus, Numbers, Deuteronomy*. New York: Schocken, 1995.

Frankenberry, Nancy K., ed. *The Faith of Scientists: In Their Own Words*. Princeton: Princeton University Press, 2008.

Freedman, David Noel, et al., eds. *The Anchor Bible Dictionary, Volume 1: A–C*. New York: Doubleday, 1992.

Freeman, Tzvi. "Are Genetically Modified Foods Kosher?" http://www.askmoses.com/en/article/277,345/Are-genetically-modified-foods-kosher.html.

Fried, Lisbeth. "Sukkot in Ezra-Nehemiah and the Date of the Torah." https://thetorah.com/sukkot-in-ezra-nehemiah-and-the-date-of-the-torah/.

Friedman, Richard Elliott. *The Bible with Sources Revealed*. New York: HarperOne, 2005.

———. *Commentary on the Torah*. New York: HarperCollins, 2001.

———. *Who Wrote the Bible?* New York: HarperCollins, 1997.

Frost, Robert. *The Poetry of Robert Frost: The Collected Poems*. Edited by Edward Connery Lathem. New York: Holt, 1969.

Frydman-Kohl, Baruch. "Priests and Levites in the Bible and in Jewish Life." In *Etz Hayim: Study Companion*, edited by Jacob Blumenthal and Janet L. Liss, 290–300. New York: The Rabbinical Assembly, 2005.

Gani, Faiz. "The Price Of Gun Violence." *Health Affairs* (blog), November 2, 2017. https://www.healthaffairs.org/do/10.1377/hblog20171031.874550/full/.

Gardner, Martin. *The Night is Large: Collected Essays, 1938–1995*. New York: St. Martin's, 1996.

Ginsburg, Joseph, and Herman Branover. "Appendix 4: Evolution: Myth and Facts." https://www.chabad.org/library/article_cdo/aid/113105/jewish/Appendix-4-Evolution-Myths-and-Facts.htm.

Glatt, Charles A. "Patriarchal Life Span Exponential Decay by Base e." www.ldolphin.org/lifespans.pdf.

Goddard, Harrold C. *The Meaning of Shakespeare: Volume 1*. Chicago: Phoenix, 1960.

Gold, Russell. *The Boom: How Fracking Ignited the American Energy Revolution and Changed the World*. New York: Simon & Schuster, 2014.

———. "The Texas Well That Started the Fracking Revolution." *Wall Street Journal*, June 29, 2018. https://www.wsj.com/articles/the-texas-well-that-started-a-revolution-1530270010?emailToken=1bf7093faa8ea3cc47254f2222a15228XfbBKOv424V1ZLkXaGC0QoXhNpokQlUIwo2Bdcg.

Golden Rice Project. "The Golden Rice Project Wins the Patents for Humanity Award 2015." http://www.goldenrice.org/.

Goldman, Gretchen, et al. "Toward an Evidence-Based Fracking Debate: Science, Democracy, and Community Right tt Know in Unconventional Oil and Gas Development." https://www.ucsusa.org/sites/default/files/legacy/assets/documents/center-for-science-and-democracy/fracking-report-full.pdf.

Goldsmith, Mirele. "Fracking Not the Answer to Energy Security." *Greenstrides Consulting* (blog), April 28, 2013. http://www.greenstridesconsulting.com/greenstrides-copenhagen-blog/2013/4/28/fracking-not-the-answer-to-energy-security.html.

———. "Why Jews Should Be Against Hydrofracking." https://jewsagainsthydrofracking.org/jewish-perspectives/why-jews-should-be-against-hydrofracking.

Goldstein, David B. *Jacob's Legacy: A Genetic View of Jewish History.* New Haven: Yale University Press, 2008.

Goldstein, Rebecca Newberger. *36 Arguments for the Existence of God: A Work of Fiction.* New York: Pantheon, 2010.

———. *Betraying Spinoza: The Renegade Jew Who Gave Us Modernity.* New York: Schocken, 2006.

Goodman, Steven, and Sander Greenland. "Assessing the Unreliability of the Medical Literature: A Response to 'Why Most Published Research Findings are False.'" https://biostats.bepress.com/cgi/viewcontent.cgi?article=1135&context=jhubiostat.

Gordis, Daniel. "Revelation: Biblical and Rabbinic Perspectives." In *Etz Hayim: Torah and Commentary*, edited by David L. Lieber et al., 1394–98. New York: Rabbinical Assembly and United Synagogue of Conservative Judaism, 2001.

Gordon, Dan. "Brain Is 10 Times More Active than Previously Measured, UCLA Researchers Find." *UCLA Newsroom*, March 9, 2017. http://newsroom.ucla.edu/releases/ucla-research-upend-long-held-belief-about-how-neurons-communicate.

Gould, Stephen Jay. *Rock of Ages: Science and Religion in the Fullness of Life.* New York: Ballantine, 1999.

"Great Flood: Sumerian Version." http://www.livius.org/articles/misc/great-flood/flood2/.

Green, Arthur. *Radical Judaism: Rethinking God and Tradition.* New Haven: Yale University Press, 2010.

———. *Seek My Face, Speak My Name: A Contemporary Jewish Theology.* Northvale, NJ: Aronson, 1992.

Greenberg, Irving. "Meeting the Challenge of Critical Scholarship with Leviticus." https://thetorah.com/meeting-the-challenge-of-critical-scholarship-with-leviticus/.

———. *The Third Great Cycle of Jewish History: Voluntary Covenant: The Third Era of Jewish History: Power and Politics.* New York, National Jewish Resource Center, 1981.

Greenblatt, Jonathan. "Why the Holocaust Has No Place in the Gun Debate." *ADL* (blog), October 12, 2015. https://www.adl.org/blog/why-the-holocaust-has-no-place-in-the-gun-debate.

Greene, Brian. *The Fabric of the Cosmos: Space, Time, and the Fabric of Reality.* New York: Knopf, 2004.

Gupta, Ruchi S., et al. "Geographic Variability of Childhood Food Allergy in the United States." *Clinical Pediatrics* 51.9 (2012) 856–61. http://journals.sagepub.com/doi/pdf/10.1177/0009922812448526.

Haidt, Jonathan. *The Righteous Mind: Why Good People Are Divided by Politics and Religion.* New York: Vintage, 2013.

Harari, Yuval Noah. *Sapiens: A Brief History of Humankind.* New York: HarperCollins, 2015.

———. "Yuval Noah Harari on Big Data, Google and the End of Free Will." *Financial Times*, August 26, 2016. https://www.ft.com/content/50bb4830-6a4c-11e6-ae5b-a7cc5dd5a28c.

Harman, Oren. *Evolutions: Fifteen Myths that Explain Our World.* New York: Farrar, Strauss and Giroux, 2018.

Harris, Ben. "Jewish Groups Take On Fracking Debate." *Times of Israel*, January 31, 2012. https://www.timesofisrael.com/jewish-groups-take-on-fracking-debate/.

Harris, Sam. *The End of Faith: Religion, Terror, and the Future of Reason*. New York: Norton, 2004.

———. *Free Will*. New York: Free Press, 2012.

Hart, Ari. "The Weapon's Shame: A Case for Gun Control in Jewish Law." *Huffington Post*, December 18, 2012. https://www.huffingtonpost.com/ari-hart/the-weapons-shame-a-case-for-gun-control-in-jewish-law_b_2313505.html.

Hassabis, Demis. "What We Learned in Seoul with AlphaGo." *Google* (blog), March 16, 2016. https://blog.google/technology/ai/what-we-learned-in-seoul-with-alphago/.

Hayes, Christine. *Introduction to the Bible*. New Haven: Yale University Press, 2012.

Hazony, Yoram. *The Philosophy of Hebrew Scripture*. New York: Cambridge University Press, 2012.

Herculano-Houzel, Suzana. "The Human Brain in Numbers: a Linearly Scaled-Up Primate Brain." *Frontiers in Human Neuroscience* 3 (2009) 31. https://www.frontiersin.org/articles/10.3389/neuro.09.031.2009/full.

Hewings-Martin, Yella. "How Many Cells Are in the Human Body?" *Medical News Today*, July 12, 2017. https://www.medicalnewstoday.com/articles/318342.php.

"Hezekiah's Defeat: The Annals of Sennacherib on the Taylor, Jerusalem, and Oriental Institute Prisms, 700 BCE." http://cojs.org/hezekiah-s_defeat-_the_annals_of_sennacherib_on_the_taylor-_jerusalem-_and_oriental_institute_prisms-_700_bce/.

Hilbring, Veronica. "The Number Of Black Women Applying For Concealed-Carry Gun Permits Surges In Chicago." *Essence*, September 13, 2017. https://www.essence.com/news/black-women-chicago-concealed-carry-gun-permits/.

Hiltzik, Michael. "Measles Is Spreading, and the Anti-Vaccine Movement Is the Cause." *Los Angeles Times*, March 17, 2014. http://www.latimes.com/business/hiltzik/la-fi-mh-measles-20140317-story.html.

Hirsch, Emil G., et al., "Hair." http://www.jewishencyclopedia.com/articles/7061-hair.

"History of ASIMO." http://asimo.honda.com/asimo-history/.

"History of Earth in 24-Hour Clock." https://flowingdata.com/2012/10/09/history-of-earth-in-24-hour-clock/.

"History of Plant Breeding." http://cls.casa.colostate.edu/transgeniccrops/history.html.

Hoffman, Joel M. *And God Said: How Translations Conceal the Bible's Original Meaning*. New York: Dunne, 2010.

Holtz, Shalom E. "The Flood Story in Its Ancient Near Eastern Context." https://thetorah.com/flood-story-in-its-ancient-near-eastern-context/.

Hong, S. W., et al. "Safety Investigation of Noah's Ark in a Seaway." https://creation.com/safety-investigation-of-noahs-ark-in-a-seaway.

Horton, Alex. "The Las Vegas Shooter Modified a Dozen Rifles to Shoot Like Automatic Weapons." *Washington Post*, October 3, 2017. https://www.washingtonpost.com/news/checkpoint/wp/2017/10/02/video-from-las-vegas-suggests-automatic-gunfire-heres-what-makes-machine-guns-different/.

"How Much Carbon Dioxide Is Produced when Different Fuels Are Burned?" https://www.eia.gov/tools/faqs/faq.php?id=73&t=11.

Howell, Elizabeth. "What Was the Impact That Killed the Dinosaurs?" *Universe Today*, February 3, 2015. https://www.universetoday.com/36697/the-asteroid-that-killed-the-dinosaurs/.

"Hubble Finds Evidence for Dark Energy in the Young Universe." http://hubblesite.org/news_release/news/2006-52.

Igielnik, Ruth, and Anna Brown. "Key Takeaways on Americans' Views of Guns and Gun Ownership." http://www.pewresearch.org/fact-tank/2017/06/22/key-takeaways-on-americans-views-of-guns-and-gun-ownership/.

Ingall, Marjorie. "Going Nuts." *Tablet*, March 15, 2010. http://www.tabletmag.com/jewish-life-and-religion/28135/going-nuts.

Ingraham, Christopher. "There Are More Guns Than People in the United States, According to a New Study of Global Firearm Ownership." *Wonkblog*, June 19, 2018. https://www.washingtonpost.com/news/wonk/wp/2018/06/19/there-are-more-guns-than-people-in-the-united-states-according-to-a-new-study-of-global-firearm-ownership/?utm_term=.f33d55991ba3.

International Service for the Acquisition of Agri-biotech Applications. "Global Status of Commercialized Biotech/GM Crops in 2017." http://www.isaaa.org/resources/publications/briefs/53/download/isaaa-brief-53-2017.pdf.

Isaacs, Jacob. "The Exodus from Egypt." https://www.chabad.org/library/article_cdo/aid/1663/jewish/The-Exodus.htm.

Isaacson, Walter. *Einstein: His Life and Universe*. New York: Simon & Schuster, 2007.

"Is There a Link between Fracking and Contaminated Groundwater?" *Libertarian Jew* (blog), December 19, 2016. http://libertarianjew.blogspot.com/2016/12/is-there-link-between-fracking-and.html.

Jacobovici, Simcha. "Proof for the Biblical Exodus!" http://www.simchajtv.com/proof-for-the-biblical-exodus/.

Jameson, Kenneth, Jr. "The Sandy Hook Atrocity (Part 2)." https://americancultureshock.wordpress.com/the-sandy-hook-atrocity-jfk-and-the-154-magic-bullets-part-1-dec-1/the-sandy-hook-atrocity-part-2/.

Jammer, Max. *Einstein and Religion: Physics and Theology*. Princeton: Princeton University Press, 1999.

Jansen, Bart. "Florida Shooting Suspect Bought Gun Legally, Authorities Say." *USA Today*, February 15, 2018. https://www.usatoday.com/story/news/2018/02/15/florida-shooting-suspect-bought-gun-legally-authorities-say/340606002/.

Jerman, Jenna, et al. "Characteristics of U.S. Abortion Patients in 2014 and Changes Since 2008." https://www.guttmacher.org/report/characteristics-us-abortion-patients-2014.

"Jews in America: The First Bat Mitzvah in the United States." http://www.jewishvirtuallibrary.org/the-first-bat-mitzvah-in-the-united-states.

"The Jewish Calendar Year." https://www.chabad.org/library/article_cdo/aid/526875/jewish/The-Jewish-Year.htm.

Jewish Council for Public Affairs. "Resolution on Hydrofracking." http://engage.jewishpublicaffairs.org/blog/comments.jsp?blog_entry_KEY=6341&t=.

"Jewish Genetics: Abstracts and Summaries." http://www.khazaria.com/genetics/abstracts-jews.html.

Jewish Virtual Library. "Hebrew: History of the Aleph-Bet." http://www.jewishvirtuallibrary.org/history-of-the-hebrew-aleph-bet.

Johnson, Edgar. "Dickens, Fagin and Mr. Riah: The Intention of the Novelist." *Commentary*, January 1950. https://www.commentarymagazine.com/articles/dickens-fagin-and-mr-riahthe-intention-of-the-novelist/.

Jonah, Edmund. "A Shake-Up of Shakespeare's Shylock." http://www.jewishmag.com/134mag/shakespeare_shylock/shakespeare_shylock.htm.

Jones, David S., et al. "The Burden of Disease and the Changing Task of Medicine." *New England Journal of Medicine* 366 (2012) 2333–38. https://www.nejm.org/doi/full/10.1056/NEJMp1113569.

Jones, Robert P., and Daniel Cox. "Chosen for What? Jewish Values in 2012." https://www.prri.org/wp-content/uploads/2012/04/Jewish-Values-Report.pdf.

Joosten, Jan. "How Hebrew Became a Holy Language." Biblical Archaeology Review 43.1 (January/February 2017) 44–49, 62.

JPS Hebrew–English Tanakh. Philadelphia: The Jewish Publication Society, 1999.

Kahneman, David, "Of 2 Minds: How Fast and Slow Thinking Shape Perception and Choice [Excerpt]." *Scientific American*, June 15, 2012. https://www.scientificamerican.com/article/kahneman-excerpt-thinking-fast-and-slow/.

Kalmanofsky, Jeremy. "Cosmic Theology and Earthly Religion." In *Jewish Theology in Our Time: A New Generation Explores the Foundation and Future of Jewish Belief*, edited by Elliot Cosgrove, 23–30. Woodstock: Jewish Lights, 2010.

Kaplan, Aryeh. *The Aryeh Kaplan Reader: The Gift He Left Behind*. Brooklyn: Mesorah, 1983.

———. "Understanding God." www.aish.com/jl/p/g/48942416.html.

Kaplan, Dana Evan. "Who is a Jew: Patrilineal Descent." https://www.myjewishlearning.com/article/patrilineal-descent/.

Kaplan, Edward K. *Spiritual Radical: Abraham Joshua Heschel in America, 1940–1972*. New Haven: Yale University Press, 2007.

Kaplan, Mordecai M. *Judaism as a Civilization: Toward a Reconstruction of American-Jewish Life*. Philadelphia: The Jewish Publication Society, 2010.

Karlamangla, Soumya. "Measles Outbreak Grows in L.A.'s Orthodox Jewish Community Despite California's Strict New Vaccination Law." *Los Angeles Times*, January 21, 2017. http://www.latimes.com/local/california/la-me-ln-measles-20170120-story.html.

Kass, Leon R. *The Beginning of Wisdom*. New York: Free Press, 2003.

Kelly, Jason. "Only Human." *UChicago Magazine*, 2014. https://mag.uchicago.edu/science-medicine/only-human.

King, L. W., trans. "Hammurabi's Code of Laws." http://www.g2rp.com/pdfs/Hammurabi.pdf.

King, Zachary, et al. "Transgenic Breeding." http://plantbreeding.coe.uga.edu/index.php?title=17._Transgenic_Breeding.

Knott, Elizabeth. "The Amarna Letters." https://www.metmuseum.org/toah/hd/amlet/hd_amlet.htm.

"Kohanim Forever." http://www.cohen-levi.org/the_tribe/kohanim_forever.htm.

Kohler, Kaufmann, and Isaac Broydé. "Predestination." http://www.jewishencyclopedia.com/articles/12338-predestination.

Kovacs, Maureen Gallery, trans. "The Epic of Gilgamesh." http://www.ancienttexts.org/library/mesopotamian/gilgamesh/tab11.htm.

Kvasnica, Andrew P. "The Ages of the Antediluvian Patriarchs in Genesis 5." Paper presented at the 2005 Student Academic Conference, Dallas Seminary, Dallax, TX, April 28, 2005. https://bible.org/article/ages-antediluvian-patriarchs-genesis-5.

Lamm, Norman. *Faith and Doubt: Studies in Traditional Jewish Thought*. New York: KTAV, 1971.

Lane, Nick. *Life Ascending: The Ten Great Inventions of Evolution*. New York: Norton, 2009.

Lavazza, Andrea. "Free Will and Neuroscience: From Explaining Freedom Away to New Ways of Operationalizing and Measuring It." *Frontiers in Human Neuroscience* 10 (2016) 262. https://www.frontiersin.org/articles/10.3389/fnhum.2016.00262/full.

Lee, Adrian. "The Meaning of AlphaGo, the AI Program That Beat a Go Champion." https://www.macleans.ca/society/science/the-meaning-of-alphago-the-ai-program-that-beat-a-go-champ/.

Lee, Seung. "The Race to Map the Brain So We Can Upload It Into a Computer." *Motherboard*, June 11, 2015. https://motherboard.vice.com/en_us/article/ezvwzn/the-race-to-map-the-brain-so-we-can-upload-it-into-a-computer.

Lewis, Dyani. "Where Did We Come From? A Primer on Early Human Evolution." *Cosmos*, June 9, 2016. https://cosmosmagazine.com/palaeontology/where-did-we-come-from-a-primer-on-early-human-evolution.

Libet, Benjamin, et al. "Time of Conscious Intention to Act in Relation to Onset of Cerebral Activities (Readiness-Potential): the Unconscious Initiation of a Freely Voluntary Act." *Brain* 106 (1983) 623–42. http://www.trans-techresearch.net/wp-content/uploads/2015/05/Brain-1983-LIBET.pdf.

Libresco, Leah. "I Used to Think Gun Control Was the Answer. My Research Told Me Otherwise." *Washington Post*, October 3, 2017. https://www.washingtonpost.com/opinions/i-used-to-think-gun-control-was-the-answer-my-research-told-me-otherwise/2017/10/03/d33edca6-a851-11e7-92d1-58c702d2d975_story.html?utm_term=.4da5d96107af.

Lieber, Chavie. "The Other Torah." http://www.tabletmag.com/jewish-life-and-religion/132004/the-other-torah.

Lieber, David L., et al., eds. *Etz Hayim: Torah and Commentary*. New York: Rabbinical Assembly and United Synagogue of Conservative Judaism, 2001.

Line, Brett. "Asteroid Impacts: 10 Biggest Known Hits." *National Geographic*, February 15, 2013. https://news.nationalgeographic.com/news/2013/13/130214-biggest-asteroid-impacts-meteorites-space-2012da14/.

Lopez, German. "The Research Is Clear: Gun Control Saves Lives." *Vox*, October 4, 2017. https://www.vox.com/policy-and-politics/2017/10/4/16418754/gun-control-washington-post.

Lucido, Gary. "Chicago's Safest And Most Dangerous Neighborhoods: Homicide Rates." *Chicago Now* (blog), July 28, 2016. http://www.chicagonow.com/getting-real/2016/07/chicagos-safest-and-most-dangerous-neighborhoods-homicide-rates/.

Luckenbill, David Daniel, trans. "The Annals of Sennacherib." https://oi.uchicago.edu/sites/oi.uchicago.edu/files/uploads/shared/docs/oip2.pdf.

"Lunar Flood, Solar Year." https://www.chabad.org/parshah/article_cdo/aid/2596/jewish/Lunar-Flood-Solar-Year.htm.

Maimonides, Moses. *The Guide for the Perplexed*. Translated by Michael Freidländer. New York: Cosimo, 2007.

"Maimonides' Thirteen (13) Principles of Faith." http://www.bestjewishstudies.com/13-Principles-of-Faith.

Maltz, Judy. "World Jewish Population on Eve of New Year—14.7 Million." Haaretz, September 9, 2018. https://www.haaretz.com/jewish/.premium-world-jewish-population-on-eve-of-new-year-14-7-million-1.6464812.

Marx, Ryan. "Data: Chicago Homicide Data Since 1957." *Chicago Tribune*, March 2, 2016. http://www.chicagotribune.com/news/local/breaking/ct-chicago-homicides-data-since-1957-20160302-htmlstory.html.

Marzano, Gilberto. "The Turing Test and Android Science." *Journal of Robotics and Automation* 2.1 (2018) 64–68. http://scholarlypages.org/Articles/robotics/jra-2-008.php?jid=robotics.

Masci, David. "American Religious Groups Vary Widely in Their Views of Abortion." http://www.pewresearch.org/fact-tank/2018/01/22/american-religious-groups-vary-widely-in-their-views-of-abortion/.

"Mass Extinctions." https://www.nationalgeographic.com/science/prehistoric-world/mass-extinction/.

"Mathematicians' Statement on the Bible Codes." https://web.archive.org/web/20150104034034/http:/www.math.caltech.edu/code/petition.html

Matt, Daniel Chanan. *God and the Big Bang: Discovering Harmony between Science and Spirituality*. Woodstock: Jewish Lights, 1996.

Maugh, Thomas H., II. "Benjamin Libet, 91; Physiologist Probed Consciousness." *Los Angeles Times*, August 27, 2007. http://articles.latimes.com/2007/aug/27/local/me-libet27.

Mayo Clinic. "Food Allergy." https://www.mayoclinic.org/diseases-conditions/food-allergy/symptoms-causes/syc-20355095/.

McAleer, Phelim. "The Gasland Movie: A Fracking Shame—Director Pulls Video to Hide Inconvenient Truths." https://wattsupwiththat.com/2011/06/04/the-gasland-movie-a-fracking-shame-director-pulls-video-to-hide-inconvenient-truths/.

McCoy, Krisha. "Peanut Allergies on the Rise—What Every Parent Should Know." https://www.everydayhealth.com/allergies/peanut-allergies.aspx.

McCullough, David. *John Adams*. New York: Simon & Schuster, 2001.

McDonald, John H. "Red Hair Color: The Myth." http://udel.edu/~mcdonald/myth redhair.html.

McGraw, Seamus. "Is Fracking Safe? The 10 Most Controversial Claims about Natural Gas Drilling." https://www.popularmechanics.com/science/energy/g161/top-10-myths-about-natural-gas-drilling-6386593/.

McKay, Brendan. "Assassinations Foretold in Moby Dick!" http://users.cecs.anu.edu.au/~bdm/dilugim/moby.html.

McKay, Brendan, et al., "Scientific Refutation of the Bible Codes." http://users.cecs.anu.edu.au/~bdm/dilugim/torah.html.

Meko, Tim, and Laris Karklis. "The United States of Oil and Gas." *Washington Post*, February 14, 2017. https://www.washingtonpost.com/graphics/national/united-states-of-oil/.

Melina, Remy. "Can Saltwater Fish Live in Fresh Water?" *Live Science*, September 28, 2012. https://www.livescience.com/32167-can-saltwater-fish-live-in-fresh-water.html.

Mergler, Wayne. "Dickens' Great Dark Villain: Fagin the Fierce." http://weeklyhubris.com/dickens%E2%80%99-great-dark-villain-fagin-the-fierce/.

"Merneptah Stele." https://www.allaboutarchaeology.org/merneptah-stele-faq.htm.

Mindel, Nissan. "Rabbi Judah Loew—'The Maharal of Prague.'" https://www.chabad.org/library/article_cdo/aid/111877/jewish/Rabbi-Judah-Loew-The-Maharal-of-Prague.htm.

Minow, Newton N. "If Your Mother Says She Loves You . . ." *Chicago Tribune*, November 19, 2000. http://articles.chicagotribune.com/2000-11-19/news/0011190214_1_exit–polls-illinois-delegate-answer.

Mirabile, Francesca. "Chicago Still Isn't the Murder Capital of America." *Trace*, January 18, 2017. https://www.thetrace.org/2017/01/chicago-not-most-dangerous-city-america/.

Mobilia, Michael. "The United States Exported More Natural Gas than It Imported in 2017." https://www.eia.gov/todayinenergy/detail.php?id=35392&src=email.

Mohl, Reuven. "Tsimtsum in the Writings of Rabbi Eliezer Berkovits." https://library.yctorah.org/files/2016/09/Tsimtsum-in-the-Writings-of-Rabbi-Eliezer-Berkovits.pdf.

Moore, Jason J., et al. "Dynamics of Cortical Dendritic Membrane Potential and Spikes in Freely Behaving Rats." *Science* 355.6331 (2017) 1–15. http://science.sciencemag.org/content/early/2017/03/08/science.aaj1497?rss=1.

Moore, Nicole Casal. "Fracktopia." https://news.engin.umich.edu/features/fracktopia/.

Moyer, Christopher. "How Google's AlphaGo Beat a Go World Champion." *The Atlantic*, March 28, 2016. https://www.theatlantic.com/technology/archive/2016/03/the-invisible-opponent/475611/.

Moyer, Melinda Wenner. "More Guns Do Not Stop More Crimes, Evidence Shows." *Scientific American*, October 1, 2017. https://www.scientificamerican.com/article/more-guns-do-not-stop-more-crimes-evidence-shows/.

Murphy, Sherry L., et al. "Deaths: Final Data for 2015." *National Vital Statistics Reports* 66.6 (2017) 1–73. https://stacks.cdc.gov/view/cdc/50011.

Nadler, Steven M. *Spinoza: A Life*. Cambridge: Cambridge University Press, 1999.

National Institute of Allergy and Infectious Diseases. "How Do Vaccines Work?" https://www.niaid.nih.gov/research/how-vaccines-work.

———. "Vaccine Benefits." https://www.niaid.nih.gov/research/vaccine-benefits.

National Aeronautics and Space Administration. "ESO Discovers Earth-Size Planet in Habitable Zone of Nearest Star." https://www.nasa.gov/feature/jpl/eso-discovers-earth-size-planet-in-habitable-zone-of-nearest-star.

———. "Exoplanet Exploration." https://exoplanets.nasa.gov/.

———. "Exoplanets 101." https://exoplanets.nasa.gov/the-search-for-life/exoplanets-101/.

———. "First Earth-Sized Planet in the Habitable Zone." https://exoplanets.nasa.gov/alien-worlds/historic-timeline/#first-earth-sized-planet-in-the-habitable-zone.

———. "First Exoplanets Discovered." https://exoplanets.nasa.gov/alien-worlds/historic-timeline/#first-exoplanets-discovered.

———. "First Planet Found within the 'Habitable Zone.'" https://exoplanets.nasa.gov/alien-worlds/historic-timeline/#first-planet-found-within-the-habitable-zone.

———. "Hubble Space Telescope." https://www.nasa.gov/mission_pages/hubble/science/milky-way-collide.html.

———. "Kepler Planet-Finding Mission Launches." https://exoplanets.nasa.gov/alien-worlds/historic-timeline/#kepler-planet-finding-mission-launches.

———. "Kepler's First Rocky Exoplanet Discovered." https://exoplanets.nasa.gov/alien-worlds/historic-timeline/#keplers-first-rocky-exoplanet-discovered.

———. "NASA Discovers First Earth-Size Planets." https://www.nasa.gov/mission_pages/kepler/news/kepler-20-system.html.

———. "Neutron Stars, Pulsars, Millisecond Pulsars, and Gravitational Radiation: A Primer." https://www.nasa.gov/centers/goddard/news/gsfc/spacesci/pictures/2003/0702pulsarspeed/0702ssu_primer.html.

———. "Precocious Earth." http://science.nasa.gov/science-news/science-at-nasa/2001/ast17jan_1/.

———. "R2 Robonaut." https://robonaut.jsc.nasa.gov/R2/.

———. "Tests of Bing Bang: The CMB." https://wmap.gsfc.nasa.gov/universe/bb_tests_cmb.html.

NASA Jet Propulsion Laboratory. "Dying Stars." http://www.spitzer.caltech.edu/mission/219-Dying-Stars.

Natanson, Hannah. "Louisiana Judge Says Jews Are a Race and Protected by Anti-Racial-Discrimination Laws." *Washington Post*, July 20, 2018. https://www.washingtonpost.com/news/acts-of-faith/wp/2018/07/20/louisiana-judge-says-jews-are-a-race-are-protected-by-anti-racial-discrimination-laws/?noredirect=on&utm_term=.f2271b5958d2.

Nathan-Kazis, Josh. "Fracking Comes to Jewish Summer Camp." *Forward*, July 13, 2011. https://forward.com/news/139831/fracking-comes-to-jewish-summer-camp/.

Nelson, David W. *The Emergence of God: A Rationalist Jewish Exploration of Divine Consciousness*. Lanham: University Press of America, 2015.

———. *Judaism, Physics and God: Searching for Sacred Metaphors in a Post–Einstein World*. Woodstock: Jewish Lights, 2005.

Nevins, Daniel S. "Halakhic Perspectives on Genetically Modified Organisms." http://www.rabbinicalassembly.org/sites/default/files/public/halakhah/teshuvot/2011-2020/nevins-gmos.pdf.

"New Study Sheds Light on How and When Vision Evolved." http://www.bristol.ac.uk/biology/news/2012/240.html.

Newcomb, Alyssa. "New IBM Robot Holds Its Own in a Debate with a Human." *NBC News*, June 18, 2018. https://www.nbcnews.com/mach/tech/new-ibm-robot-holds-its-own-debate-human-ncna884536.

Ngo, Robin. "Precursor to Paleo-Hebrew Script Discovered in Jerusalem." *Bible History Daily*, October 8, 2018. https://www.biblicalarchaeology.org/daily/biblical-artifacts/inscriptions/precursor-to-the-paleo-hebrew-script-discovered-in-jerusalem/.

Nicolia, Alessandro, et al. "An Overview of the Last 10 Years of Genetically Engineered Crop Safety Research." Critical Reviews in Biotechnology 34.1 (2014) 77–88. https://www.tandfonline.com/doi/abs/10.3109/07388551.2013.823595.

"On Science and Its Truths." https://www.chabad.org/therebbe/letters/default_cdo/aid/664329/jewish/On-Science-and-Its-Truths.htm.

O'Neill, Ian. "The Sun as a White Dwarf Star." *Universe Today*, March 19, 2009. https://www.universetoday.com/25669/the-sun-as-a-white-dwarf-star/.

Ost, Laura. "Einstein was Right (again): Experiments Confirm that $E=mc^2$." NIST, December 21, 2005. https://www.nist.gov/news-events/news/2005/12/einstein-was-right-again-experiments-confirm-e-mc2.

Oz, Amos, and Fania Oz-Salzberger. *Jews and Words*. New Haven: Yale University Press, 2012.

Palmer, Jane. "We Finally Know How Much the Dino-Killing Asteroid Reshaped Earth." *Smithsonian*, February 25, 2016. https://www.smithsonianmag.com/science-nature/we-finally-know-how-much-dino-killing-asteroid-reshaped-earth-180958222/.

"Pamphlet 9—The Letters of the Torah." http://www.aishdas.org/toratemet/en_pamphlet9.html.

Parker, Kim, et al. "America's Complex Relationship With Guns." http://www.pewsocialtrends.org/2017/06/22/americas-complex-relationship-with-guns/.

Parks, Jake. "The First Stars Formed When the Universe Was Less Than 2% Its Current Age." *Astronomy*, May 16, 2018. http://www.astronomy.com/news/2018/05/first-stars-from-oxygen.

"Pearl." https://www.roboticstoday.com/robots/pearl-description.

Pelaia, Ariela. "The Bat Mitzvah Ceremony and Celebration." *ThoughtCo.*, January 11, 2019. https://www.thoughtco.com/what-is-a-bat-mitzvah-2076848.

Penkower, Jordan S. "The Development of the Masoretic Text." In *The Jewish Study Bible*, edited by Adele Berlin and Marc Zvi Brettler, 2077–84. New York: Oxford University Press, 2004.

Penrose, Roger. *Cycles of Time: An Extraordinary New View of the Universe*. New York: Knopf, 2011.

———. *The Emperor's New Mind: Concerning Computers, Minds and The Laws of Physics*. Oxford: Oxford University Press, 1989.

Perakh, Mark. "Not a Very Big Bang about Genesis." www.talkreason.org/PrinterFriendly.cfm?article=/articles/schroeder.cfm.

———. "The Rise and Fall of the Bible Code." http://www.talkreason.org/articles/Codpaper1.cfm.

Perrin, Jack, and Troy Cook. "Hydraulically Fractured Wells Provide Two-Thirds of U.S. Natural Gas Production." https://www.eia.gov/todayinenergy/detail.php?id=26112.

Pew Research Center. "Chapter 4: Religious Beliefs and Practices." http://www.pewforum.org/2013/10/01/chapter-4-religious-beliefs-and-practices/.

———. "Future of World Religions: Population Growth Projections, 2010–2050—Jews." http://www.pewforum.org/2015/04/02/jews/.

———. "When Americans Say They Believe in God, What Do They Mean?" http://www.pewforum.org/2018/04/25/when-americans-say-they-believe-in-god-what-do-they-mean/.

Philologos, "Redheaded Warrior Jews." *Forward*, August 12, 2009. https://forward.com/culture/111973/redheaded-warrior-jews/.

Pinholster, Ginger. "AAAS Board of Directors: Legally Mandating GM Food Labels Could 'Mislead and Falsely Alarm Consumers.'" https://www.aaas.org/news/aaas-board-directors-legally-mandating-gm-food-labels-could-mislead-and-falsely-alarm.

Planetary Habitability Laboratory. "Habitable Expolanets Catalog." http://phl.upr.edu/projects/habitable-exoplanets-catalog.

Plaut, W. Gunther, and David E. Stein, eds. *The Torah: A Modern Commentary*. Rev. ed. New York: Union for Reform Judaism, 2005.

Pomeroy, Ross. "Massive Review Reveals Consensus on GMO Safety." *Real Clear Science* (blog), September 30, 2013. https://www.realclearscience.com/blog/2013/10/massive-review-reveals-consensus-on-gmo-safety.html.

Popik, Barry. "Guns Don't Kill People—People Kill People." https://www.barrypopik.com/index.php/new_york_city/entry/guns_dont_kill_people_people_kill_people.

"Post-*Heller* Litigation Summary." http://lawcenter.giffords.org/wp-content/uploads/2017/04/Post-Heller-Litigation-Summary-2017-April.pdf.

Prager, Dennis. "Jews and Abortion." http://jewishjournal.com/opinion/110132/.

"Premature Birth Statistics." http://www.preemiesurvival.org/info/index.html.

Prescott, Alexander, et al. "Readiness Potentials Driven by Non-Motoric Processes." *Consciousness and Cognition* 39 (2016) 38–47. https://www.dartmouth.edu/~peter/pdf/69.pdf.

Price, Roger. "Gould in the Fullness of Life." https://www.judaismandscience.com/gould-in-the-fullness-of-life/.

———. "Jewish Atheism and Jewish Theism: The Data and the Dilemma." https://www.judaismandscience.com/jewish-atheism-and-jewish-theism-the-data-and-the-dilemma/.

———. "Science and Judaism: WWMD? What Would Maimonides Do?" https://www.judaismandscience.com/science-and-judaism-wwmd-what-would-maimonides-do/.

———. "What If Cyrus Had Not Freed the Jews?" https://www.judaismandscience.com/what-if-cyrus-had-not-freed-the-jews/.

Prouser, Joseph H. "Compulsory Immunization in Jewish Day Schools." https://www.rabbinicalassembly.org/sites/default/files/public/halakhah/teshuvot/20052010/prouser_immunization.pdf.

Pundit Planet. "Pope Francis Blesses Golden Rice." *Pundit from Another Planet*, February 21, 2015. https://punditfromanotherplanet.com/2015/02/21/pope-francis-blesses-golden-rice/.

"Queen Elizabeth I: Biography, Facts, Portraits & Information." https://englishhistory.net/tudor/monarchs/queen-elizabeth-i/.

"Questions and Answers about EPA's Hydraulic Fracturing Drinking Water Assessment: Hydraulic Fracturing Drinking Water Assessment Report." https://www.epa.gov/hfstudy/questions-and-answers-about-epas-hydraulic-fracturing-drinking-water-assessment.

Rabbinical Assembly. "Resolution on Reproductive Freedom in the United States." https://www.rabbinicalassembly.org/story/resolution-reproductive-freedom-united-states?tp=378.

"The Racial Theory." http://international.aish.com/seminars/whythejews/yjd05600.htm.

"Red Giant Stars: Facts, Definition & the Future of the Sun." https://www.space.com/22471-red-giant-stars.html.

"Red Hair Gene." http://www.myredhairgene.com/.

Redd, Nola Taylor. "How Old Is the Universe?" https://www.space.com/24054-how-old-is-the-universe.html.

Redden, Elizabeth. "Why More Colleges Want Jewish Students." *Inside Highered*, October 29, 2008. https://www.insidehighered.com/news/2008/10/29/why-more-colleges-want-jewish-students.

Rees, Martin. *Just Six Numbers: The Deep Forces that Shape the Universe*. New York: Basic, 2000.

Reform Jewish Voice. "Hydrofracturing." https://web.archive.org/web/20121120035428/http://rac.org:80/advocacy/rjv/issues/fracking/index.cfm.

"Religion and Science." https://www.chabad.org/library/article_cdo/aid/271670/jewish/Religion-and-Science.htm.

Religious Action Center of Reform Judaism. "Gun Violence Prevention: Jewish Values." https://rac.org/gun-violence-prevention-jewish-values.

———. "Reform Jewish Leader Disappointed by Supreme Court Ruling in DC Gun Ban Case." https://rac.org/reform-jewish-leader-disappointed-supreme-court-ruling-dc-gun-ban-case.

Remy, Vanessa, et al. "Vaccination: The Cornerstone of an Efficient Healthcare System." *Journal of Market Access & Health Policy* 3.1 (2015). https://www.ncbi.nlm.nih.gov/pmc/articles/PMC4802703/.

Rendsburg, Gary. "The Book of Genesis." https://archive.org/stream/BookOfGenesis/Book%20of%20Genesis%20Guidebook_djvu.txt.

———. "Reading the Plagues in their Ancient Egyptian Context." https://thetorah.com/plagues-in-their-ancient-egyptian-context/.

———. "YHWH's War against the Egyptian Sun-God Ra." https://thetorah.com/yhwhs-war-against-the-egyptian-sun-god-ra/.

Renstrom, Joelle, "The Humanoid Robot Space Explorer." *Daily Beast*, September 11, 2016. https://www.thedailybeast.com/the-humanoid-robot-space-explorer.

———. "Robot Nurses will Make Shortages Obsolete." *Daily Beast*, September 24, 2016. https://www.thedailybeast.com/robot-nurses-will-make-shortages-obsolete.

"Review of 10 Years of GMO Research—They're Safe." *Skeptical Raptor* (blog), February 28, 2015. http://www.skepticalraptor.com/skepticalraptorblog.php/review-10-years-gmo-research-no-significant-dangers/.

Reynolds, Alan. "Are Mass Shootings Becoming More Frequent?" https://www.cato.org/blog/are-mass-shootings-becoming-more-frequent.

Ridlington, Elizabeth, and John Rumpler. "Fracking by the Numbers: Key Impacts of Dirty Drilling at the State and National Level." https://environmentamerica.org/sites/environment/files/reports/EA_FrackingNumbers_scrn.pdf.

"Robotic Knight." http://www.da-vinci-inventions.com/robotic-knight.aspx.

Romanoff, Lena. "Denominational Differences on Conversion." https://www.myjewishlearning.com/article/cross-denominational-differences regarding-conversion/.

Römer, Thomas. *The Invention of God*. Translated by Raymond Geuss. Cambridge: Harvard University Press, 2015.

Rosenberg-Douglas, Katherine, and Tony Briscoe. "2016 Ends with 762 Homicides; 2017 Opens with Fatal Uptown Gunfight." *Chicago Tribune*, January 2, 2017. http://www.chicagotribune.com/news/local/breaking/ct-two-shot-to-death-in-uptown-marks-first-homicide-of-2017-20170101-story.html.

Rosenblum, Jonathan. "Charlie Darwin's Angels." *Aish*, January 21, 2006. http://www.aish.com/ci/sam/48952141.html.

Ross, Hugh. "Design and the Anthropic Principle." http://www.s8int.com/anthropic.html.

———. "Fine-Tuning For Life In The Universe." http://wordblessings.com/pdf/Reasons-To-Believe%20_%20Fine-Tuning-For-Life-In-The-Universe-AUG-2006.pdf.

Rotkop, Noa. "Toyota's Violin Playing Robot." http://thefutureofthings.com/5779-toyotas-violin-playing-robot/.

Rubenstein, WS. "Hereditary Breast Cancer in Jews." *Fam Cancer* 3.3–4 (2004) 249–57. https://www.ncbi.nlm.nih.gov/pubmed/15516849.

Rubright, Sam. "34 States Have Active Oil & Gas Activity in the U.S. Based on 2016 Analysis." https://www.fractracker.org/2017/03/34-states-active-drilling-2016/.

Ruse, Michael, and Joseph Travis, eds. *Evolution: The First Four Billion Years*. Cambridge: Belknap, 2009.

Ruttenberg, Danya. "Why Are Jews So Pro-Choice?" *Forward*, January 30, 2018. https://forward.com/opinion/393168/why-are-jews-so-pro-choice/.

Saber, Ashraf S. "The Camel in Ancient Egypt." https://www.researchgate.net/publication/265796991_The_Camel_in_Ancient_Egypt.

Sacks, Jonathan. "Freewill (Vaera 5775)." http://rabbisacks.org/freewill-vaera-5775/.

———. *The Great Partnership: Science, Religion, and the Search for Meaning*. New York: Schocken, 2011.

———. "Human's Uniqueness Is Not in Intelligence Which Can Be Artificial, But in Loving and Being Loved." http://rabbisacks.org/humans-uniqueness-is-not-in-intelligence-which-can-be-artificial-but-in-loving-and-being-loved/.

Sagan, Carl. *Cosmos*. New York: Random House, 1980.
Samson, David. "Are Vaccines Kosher?" https://www.yeshiva.co/ask/?id=10.
Samuelson, Norbert M. *Jewish Faith and Modern Science: On the Death and Rebirth of Jewish Philosophy*. Plymouth: Rowman & Littlefield, 2009.
———. *Judaism and the Doctrine of Creation*. Cambridge: Cambridge University Press, 1994.
"Santorini Volcano Eruption Date." http://www.santorini-volcano.net/santorini-volcano-eruption-date.html.
Sarna, Nahum M. *Exploring Exodus: The Heritage of Biblical Israel*. New York: Schocken, 1986.
———. *Understanding Genesis: The Heritage of Biblical Israel*. New York: Schocken, 1972.
Sassoon, Hacham Isaac. "Source Criticism Enhances our Acceptance of the Torah." https://thetorah.com/source-criticism-enhances-our-acceptance-of-the-torah/.
"Saul Smilansky." http://www.informationphilosopher.com/solutions/philosophers/smilansky/.
Schapiro, Jeff, "Professor Asks: 'Did Jesus Die for Klingons Too?'" Christian Post, October 5, 2011. https://www.christianpost.com/news/professor-asks-did-jesus-die-for-klingons-too-57285/.
Scherman, Nosson, and Meir Zlotowitz, eds. *The Chumash (The Stone Edition)*. Brooklyn: Mesorah, 1995.
Schmidt, Brian B., ed. *The Quest for the Historical Israel: Debating Archaeology and the History of Early Israel*. Archaeology and Biblical Studies 17. Atlanta: Society of Biblical Literature, 2007.
Schniedewind, William M. *How the Bible Became a Book: The Textualization of Ancient Israel*. New York: Cambridge University Press, 2005.
Schroeder, Gerald L. "The Age of the Universe." http://www.geraldschroeder.com/AgeUniverse.aspx.
———. *Genesis and the Big Bang: The Discovery of Harmony Between Modern Science and the Bible*. New York: Bantam, 1990.
———. *The Science of God: The Convergence of Scientific and Biblical Wisdom*. New York: Free Press, 1997.
Schulweis, Harold. "Keruv, Conversion and the Unchurched." https://www.vbs.org/worship/meet-our-clergy/rabbi-harold-schulweis/sermons/keruv-conversion-and-unchurched.
Schwab, Ivan R. *Evolution's Witness: How Eyes Evolved*. New York: Oxford University Press, 2011.
Schwartz, Amy E. "In What Ways, If Any, Do Science and Judaism Conflict?" Moment, December 31, 2013. https://www.momentmag.com/ask-rabbis-religion-science/.
Scult, Mel. *The Radical American Judaism of Mordecai M. Kaplan*. Bloomington: Indiana University Press, 2014.
Segal, Eliezar. "Kabbalah: The Ten Sefirot, of the Kabbalah." https://www.jewishvirtuallibrary.org/the-ten-sefirot-of-the-kabbalah.
Seidenberg, David. "How Fracking Conflicts with Kabbalah." *Forward*, July 16, 2013. https://forward.com/opinion/180507/how-fracking-conflicts-with-kabbalah/?p=all&p=all.
Seitz-Wald, Alex. "The Hitler Gun Control Lie." *Salon*, January 11, 2013. https://www.salon.com/2013/01/11/stop_talking_about_hitler/.

Shakespeare, William. "Hamlet." In *Shakespeare: The Complete Works*, edited by G. P. Harrison, 885–934. New York: Harcourt Brace, 1958.

Shan, Li. "Robots Are Becoming Security Guards." *Los Angeles Times*, September 2, 2016. http://www.latimes.com/business/la-fi-robots-retail-20160823-snap-story.html.

Shapiro, Mark Dov. "The God Survey." *Reform Judaism*, Summer 2012. https://issuu.com/reformjudaism/docs/rj_sum12.

Sharp, Tim. "Alpha Centauri: Nearest Star System to the Sun." https://www.space.com/18090-alpha-centauri-nearest-star-system.html.

Sheinman, Anna. "Fluenz Flu Vaccine Containing Pork Product Is Kosher." *The JC*, October 10, 2013. https://www.thejc.com/news/uk-news/fluenz-flu-vaccine-containing-pork-product-is-kosher-1.49557.

Sheskin, Ira, and Arnold Dashefsky. "Jewish Population in the United States, 2011." https://www.jewishdatabank.org/content/upload/bjdb/612/Jewish_Population_in_the_United_States_20111.pdf.

Shim, Eileen. "3 Diseases That Have Suddenly Made a Comeback Thanks to Anti–Vaccine Truthers." *Mic*, March 18, 2014. https://mic.com/articles/85525/3-diseases-that-have-suddenly-made-a-comeback-thanks-to-anti-vaccine-truthers#.37nqAQas7.

Shubin, Neil. *Your Inner Fish: A Journey into the 3.5 Billion Year History of the Human Body*. New York: Pantheon, 2008.

Shurger, Aaron, et al. "An Accumulator Model for Spontaneous Neural Activity prior to Self-Initiated Movement." *Proceedings of the National Academy of Sciences of the United States of America* 109.42 (2012) E2904–13. http://www.pnas.org/content/109/42/E2904.full.

Shurkin, Joel N. "Prenatal Whole Genome Sequencing Technology Raises Jewish Ethical Questions." *Jewish Telegraph Agency*, November 21, 2012. https://www.jta.org/2012/11/21/life-religion/prenatal-whole-genome-sequencing-technology-raises-jewish-ethical-questions.

Seeskin, Kenneth. *Maimonides on the Origin of the World*. New York: Cambridge University Press, 2007.

"Sefer Torah." http://www.esofer.com/infoSeferTorah.asp.

Siegel, Danny. *And God Braided Eve's Hair*. Spring Valley, NY: Town House, 1976.

Siegel, Ethan. "There Was No Big Bang Singularity." *Forbes*, July 27, 2018. https://www.forbes.com/sites/startswithabang/2018/07/27/there-was-no-big-bang-singularity/#4c570ab57d81.

———. "When Did the First Stars Appear in the Universe?" *Forbes*, December 21, 2017. https://www.forbes.com/sites/startswithabang/2017/12/21/when-did-the-first-stars-appear-in-the-universe/#36d205a13bcf.

Siegel, Jonathan P. "The Evolution of Two Hebrew Scripts." *Biblical Archaeological Review* 5.3 (1979) 28–33.

Simon, Barry. "The Case against the Codes." http://web.archive.org/web/20140112092910/http:/www.khunwoody.com/biblecodes/TheCase.htm.

Ska, Jean-Louis. *Introduction to Reading the Pentateuch*. Translated by Pascale Dominique. Winona Lake: Eisenbrauns, 2006.

Slifkin, Natan. *The Challenge of Creation: Judaism's Encounter with Science, Cosmology, and Evolution*. 2nd ed. Brooklyn: Zoo Torah, 2008.

———. "Strange Reactions." http://www.rationalistjudaism.com/2012/11/strange-reactions.html?m=1.

Smith, Tom W., and Jaesok Son. "Trends in Gun Ownership in the United States, 1972–2014." http://www.norc.org/PDFs/GSS%20Reports/GSS_Trends%20in%20Gun%20Ownership_US_1972-2014.pdf.

Smithsonian National Museum of Natural History. "Genetic Evidence." http://humanorigins.si.edu/evidence/genetics.

———. "What Does it Mean to be Human?" http://humanorigins.si.edu/evidence/human-fossils/species/homo-sapiens.

Soclof, Adam. "Should Robots Count in a Minyan? Rabbi Talks Turing Test." *Jewish Telegraphic Agency*, June 12, 2014. https://www.jta.org/2014/06/12/life-religion/should-robots-count-in-a-minyan-rabbi-talks-turing-test.

SoftBank Robotics. "Pepper." https://www.softbankrobotics.com/emea/en/pepper.

Solomin, Rachel M. "History of the World." https://www.myjewishlearning.com/article/history-of-the-world/.

Solomon, Norman. "The Torah's Version of the Flood Story." http://thetorah.com/noahs-flood-story/.

Soon, et al. "Unconscious Determinants of Free Decisions in the Human Brain." *Nature Neuroscience* 11 (2008) 543–45. https://www.nature.com/articles/nn.2112.

Sperling, S. David. *The Original Torah: The Political Intent of the Bible's Writers.* Reappraisals in Jewish Social and Intellectual History. New York: New York University Press, 1998.

Stenger, Victor J. *The Fallacy of Fine-Tuning: Why the Universe is Not Designed for Us.* Amherst: Prometheus, 2011.

———. "Flew's Flawed Science." *Free Inquiry* 25.2 (2005). http://www.freerepublic.com/focus/f-news/1333347/posts.

———. *God: The Failed Hypothesis: How Science Shows that God Does Not Exist.* Amherst: Prometheus, 2008.

Stephens, Tim. "Four Earth-Sized Planets Detected Orbiting the Nearest Sun-Like Star." *UC Santa Cruz Newscenter*, August 8, 2017. https://news.ucsc.edu/2017/08/tau-ceti-planets.html.

Stern, Philip. "Assyrian March against Judah." http://www.historynet.com/assyrian-march-against-judah.htm.

Student, Gil. "On the Text of the Torah." http://www.aishdas.org/toratemet/en_text.html.

Sugarmann, Josh. "For Women, Gun Violence Often Linked to Domestic Violence." *Huffington Post*, October 1, 2014. https://www.huffingtonpost.com/josh-sugarmann/for-women-gun-violence-of_b_5913752.html.

Sukel, Kayt. "The Synapse-A Primer." http://www.dana.org/News/Details.aspx?id=43512.

Swanson, Jeffery W., et al. "Mental Illness and Reduction of Gun Violence and Suicide: Bringing Epidemiologic Research to Policy." *Annnals of Epidemiology* 25.5 (2015) 366–76. https://www.ncbi.nlm.nih.gov/pmc/articles/PMC4211925/.

Tabor, James. "The Only Ancient Jewish Male Hair Ever Found, DNA, and the Tomb of the Shroud." https://docs.wixstatic.com/ugd/886cd7_78e528bd9d6b47dd9c61487344c30890.pdf.

"Talmud—Mas. Gittin 2a." https://www.halakhah.com/pdf/nashim/Gittin.pdf.

Taub, Moshe. "'*Cogito Ergo Sum*': Robots, Minyan and Halakhic Sensationalism: Can Robots Be Deemed People? Jewish?" https://www.bvkkosher.com/robots-in-halacha.

Tech Museum of Innovation. "Genes in Common." http://genetics.thetech.org/online-exhibits/genes-common.

Telushkin, Joseph. *The Book of Jewish Values: A Day-By-Day Guide to Ethical Living.* New York: Bell Tower, 2000.

Temple Sholom Cincinnatti. "Be Someone Else." https://vimeo.com/177462266.

Tenn, William. "On Venus, Have We Got a Rabbi." In *Wandering Stars: An Anthology of Jewish Fantasy & Science Fiction*, edited by Jack Dann, 7–40. Woodstock: Jewish Lights, 1998.

Teutsch, David A. *A Guide to Jewish Practice: Volume 1—Everyday Living.* Wyncote, PA: Reconstructionist Rabbinical College Press, 2011.

"A Textual Study of Noah's Flood." http://thetorah.com/textual-study-of-noahs-flood/.

Thompson, Derek. "Mass Shootings in America Are Spreading Like a Disease." *The Atlantic*, November 6, 2017. https://www.theatlantic.com/health/archive/2017/11/americas-mass-shooting-epidemic-contagious/545078/.

Tigay, Jeffrey H. "The Bible 'Codes': A Textual Perspective." https://www.sas.upenn.edu/~jtigay/codetext.html.

"Timeline of the Big Bang." https://www.physicsoftheuniverse.com/topics_bigbang_timeline.html.

Tobin, Jonathan S. "Loving Us to Death: How America's Embrace is Imperiling American Jewry." https://www.commentarymagazine.com/articles/loving-us-to-death/.

"Torah Codes Explained." http://www.aish.com/atr/Torah_Codes_Explained.html.

Trosper, Jaime. "Four Ways That Our Universe Might End, According to Science." https://futurism.com/four-ways-the-universe-could-end/.

Troster, Lawrence. "Redemption and Hydraulic Fracturing." https://jewsagainsthydrofracking.org/jewish-perspectives/the-return-of-the-healing-waters.

———. "The Return of the Healing Waters of Eden." http;//jewsagainsthydrofracking.org/jewish-perspectives/the-return-of-the-healing-waters/.

"The Turing Test, 1950." http://www.turing.org.uk/scrapbook/test.html.

Turk, Victoria. "Scientists Made a New Map of the Human Brain." https://motherboard.vice.com/en_us/article/qkjzvp/scientists-made-a-new-map-of-the-human-brain-cerebral-cortex.

Tyson, Neil DeGrasse. *Astrophysics for People in a Hurry.* New York: Norton, 2017.

Tyson, Neil DeGrasse, and Donald Goldsmith. *Origins: Fourteen Billion Years of Cosmic Evolution.* New York: Norton, 2005.

"Unconventional Oil and Natural Gas Development." https://www.epa.gov/uog.

United States Census Bureau. "U.S. and World Population Clock." https://www.census.gov/popclock/.

United States Department of Justice Bureau of Alcohol, Tobacco, Firearms and Explosives. "Firearms Commerce in the United States: Annual Statistical Update 2018." https://www.atf.gov/file/130436/download.

"U.S. Anti-Semitic Incidents Spike 86 Percent So Far in 2017 after Surging Last Year, ADL Finds." *Anti-Defamation League*, April 24, 2017. https://www.adl.org/news/press-releases/us-anti-semitic-incidents-spike-86-percent-so-far-in-2017.

USDA Economic Research Service. "Adoption of Genetically Engineered Crops in the United States, 1996-2018." https://www.ers.usda.gov/webdocs/charts/58020/biotechcrops_d.html?v=7149.

———. "Genetically Engineered Crops in the United States." https://www.ers.usda.gov/webdocs/publications/45179/43668_err162.pdf.

———. "Recent Trends in GE Adoption." https://www.ers.usda.gov/data-products/adoption-of-genetically-engineered-crops-in-the-us/recent-trends-in-ge-adoption/.

"Vaccine Excipient & Media Summary." https://www.cdc.gov/vaccines/pubs/pinkbook/downloads/appendices/B/excipient-table-2.pdf.

Verbruggen, Robert. "Reducing Gun Violence: Suggestions That Do Not Involve Gun Control." *National Review*, October 23, 2017. https://www.nationalreview.com/2017/10/reducing-gun-violence-focus-high-crime-areas-likely-criminals/.

"Viktor E. Frankl Quotes." https://www.goodreads.com/author/quotes/2782.Viktor_E_Frankl.

Vos, Kathleen D., and Roy F. Baumeister. "Addiction and Free Will." *Addict Res Theory* 17.3 (2009) 231–35. https://www.ncbi.nlm.nih.gov/pmc/articles/PMC2757759/.

Walsh, Jon. "Dickens' Greatest Villain: The Faces of Fagin." *The Independent*, October 7, 2005. http://web.archive.org/web/20081205101924/http:/www.independent.co.uk/arts-entertainment/films/features/dickens-greatest-villain-the-faces-of-fagin-509906.html.

"Was Moses Born Circumcised?" https://www.chabad.org/parshah/article_cdo/aid/1391191/jewish/Was-Moses-Born-Circumcised.htm.

Washofsky, Mark, ed. *Reform Responsa for the Twenty–First Century: Sh'eilot Ut'shuvot, Volume 1*. New York: CCAR, 2010.

Waskow, Arthur. "A Tu B'Shvat Seder to Heal a Wounded Earth." https://theshalomcenter.org/tu-bshvat-seder-heal-wounded-earth.

Watanabe, Teresa. "Doubting the Exodus Story." *Los Angeles Times*, April 13, 2001. http://articles.latimes.com/2001/apr/13/news/mn-50481.

Weintraub, David A. *How Old Is the Universe?* Princeton: Princeton University Press, 2011.

Weisberg, Josh. "The Hard Problem of Consciousness." http://www.iep.utm.edu/hard-con/.

Wesson, Kenneth. "A Primer on Neuroplasticity: Experience and Your Brain." *Brain World*, May 18, 2018. https://brainworldmagazine.com/a-primer-on-neuroplasticity-experience-and-your-brain/.

"What Comes After Mass Extinctions?" https://evolution.berkeley.edu/evolibrary/news/120901_afterextinction.

"What Is a Viable Pregnancy and a Nonviable Pregnancy?" https://www.babymed.com/what-is-a-viable-nonviable-periviability-viability-pregnancy.

"What Really Happened at the Sea." https://thetorah.com/what-really-happened-at-the-sea/.

Whitfield, Philip. *From So Simple a Beginning: The Encyclopedia of Evolution*. New York: Macmillan, 1993.

Wiener, Noah. "The Animals Went In Two by Two, According to the Babylonian Ark Tablet." https://www.biblicalarchaeology.org/daily/biblical-topics/hebrew-bible/the-animals-went-in-two-by-two-according-to-babylonian-ark-tablet/.

Wilson, Alison, and Jonathan Latham. "GMO Golden Rice Offers No Nutritional Benefits Says FDA." *Independent Science News*, June 3, 2018. https://www.independentsciencenews.org/news/gmo-golden-rice-offers-no-nutritional-benefits-says-fda/.

Wind, Rebecca. "U.S. Abortion Rate Continues to Decline, Hits Historic Low." *Guttmacher Institute*, January 17, 2017. https://www.guttmacher.org/news-release/2017/us-abortion-rate-continues-decline-hits-historic-low.

Winston, Pinchas. "Moshiach and the World Today." *Aish*, June 23, 2001. http://www.aish.com/jw/s/48883092.html.

Witze, Alexandra. "Einstein's 'Time Dilation' Prediction Verified." *Scientific American*, September 22, 2014. https://www.scientificamerican.com/article/einsteins-time-dilation-prediction-verified/.

Witztum, Doron, et al. "Equidistant Letter Sequences in the Book of Genesis." *Statistical Science* 9.3 (1994) 429–38. https://projecteuclid.org/download/pdf_1/euclid.ss/1177010393.

Wolpe, David. "Did the Exodus Really Happen?" http://www.beliefnet.com/faiths/judaism/2004/12/did-the-exodus-really-happen.aspx.

Wood, Robert. "Food Allergies More Widespread among Inner-City Children." https://www.hopkinsmedicine.org/news/media/releases/food_allergies_more_widespread_among_inner_city_children.

"World Flood Myths." https://arkencounter.com/flood/myths/.

World Health Organization. ""Frequently Asked Questions on Genetically Modified Foods." http://www.who.int/foodsafety/areas_work/food-technology/faq-genetically-modified-food/en/.

Yaffe, Shlomo. "What Does Judaism Say About Gun Control?" https://www.chabad.org/library/article_cdo/aid/507002/jewish/What-Does-Judaism-Say-About-Gun-Control.htm.

Yong, Ed. "How Brain Scientists Forgot That Brains Have Owners." *The Atlantic*, February 27, 2017. https://www.theatlantic.com/science/archive/2017/02/how-brain-scientists-forgot-that-brains-have-owners/517599/.

Yuter, Alan. "The Abortion Rhetoric within Orthodox Judaism: Consensus, Conviction, Covenant." https://www.jewishideas.org/article/abortion-rhetoric-within-orthodox-judaism-consensus-conviction-covenant.

Zaklikowski, Dovid. "Torah Scroll Facts." https://www.chabad.org/library/article_cdo/aid/351655/jewish/Torah-Scroll-Facts.htm.

Zarins, Juris. "Camel." In Freedman, *The Anchor Bible Dictionary* 1:824–26.

Zeligman, Naftali. "A List of Some Problematic Issues Concerning Orthodox Jewish Belief." http://www.talkreason.org/articles/list.cfm.

Zevit, Ziony. "First Kings: Introduction." In *The Jewish Study Bible*, edited by Adele Berlin and Marc Zvi Brettler, 668–71. New York: Oxford University Press, 1985.

Zimmer, Carl. "How Many Cells Are in Your Body?" *National Geographic*, October 23, 2013. https://www.nationalgeographic.com/science/phenomena/2013/10/23/how-many-cells-are-in-your-body/.

Zwolinski, Chaya Rivka. "Reb Nachman Explains It All." *Tablet*, May 25, 2012. http://www.tabletmag.com/jewish-arts-and-culture/books/100002/reb-nachman-explains-it-all.

Index

CPSIA information can be obtained
at www.ICGtesting.com
Printed in the USA
FFHW011314190819
54427581-60104FF